Microprocessors in Signal Processing, Measurement and Control

International Series on
MICROPROCESSOR-BASED SYSTEMS ENGINEERING

Microprocessors in Signal Processing, Measurement and Control

edited by

SPYROS G. TZAFESTAS

Electrical Engineering Department,
University of Patras, Patras, Greece

D. REIDEL PUBLISHING COMPANY

A MEMBER OF THE KLUWER ACADEMIC PUBLISHERS GROUP

DORDRECHT / BOSTON / LANCASTER

Library of Congress Cataloging in Publication Data

CIP

Main entry under title:

Microprocessors in signal processing, measurement and control.

(International series on microprocessor-based systems engineering)
Includes index.
1. Microprocessors. 2. Signal processing—Digital techniques.
3. Digital control systems. I. Tzafestas, S. G., 1939– . II. Series.
TK7895.M5M53 1983 001.64 83–11067
ISBN-13: 978-94-009-7009-0 e-ISBN-13: 978-94-009-7007-6
DOI: 10.1007/978-94-009-7007-6

Published by D. Reidel Publishing Company
P.O. Box 17, 3300 AA Dordrecht, Holland

Sold and distributed in the U.S.A. and Canada
by Kluwer Academic Publishers,
190 Old Derby Street, Hingham, MA 02043, U.S.A.

In all other countries, sold and distributed
by Kluwer Academic Publishers Group,
P.O. Box 322, 3300 AH Dordrecht, Holland

PREFACE

In recent years the LSI technology has witnessed a revolutionary development, and allowed substantial reductions in the size and cost of digital logic circuitry. Computer system building blocks have progressed from the level of discrete components to the level of complex ICs involving many logic circuits on a single "chip". The invention and wide applications of microprocessors have changed the philosophy of the signal processing, measurement and control engineering fields. The microprocessor-based digital signal processing systems and controllers have replaced the conventional ones based on standard analog and digital computing equipment. The first microprocessors and "on-chip" computers have appeared towards the end of 71 beginning 72. Their evolution since then and the number of applications, in which they have been utilized, have both been extremely spectacular. New system concepts and hardware/software tools are steadily under development to support the microprocessor in its multiple and complex tasks.

The goal of this book is to provide a cohesive and well-balanced set of contributions dealing with important aspects and applications of microprocessors to signal processing, measurement and system control. The majority of contributions include sufficient review material and present rather complete treatments of the respective topics. Although an exhaustive presentation of all existing microprocessor-based applications in the signal processing, measurement and control field is not possible in a single volume, the reader can find here the major aspects, tools and techniques employed in most applications. An outline of the material contained in the book, along with a general introduction to microprocessors and their applications is given in chapter 1.

This book is certainly not intended to cover introductory aspects of microprocessors and the application fields dealt with. The reader is assumed to have acquired some background know-

S. G. Tzafestas (ed.), Microprocessors in Signal Processing, Measurement and Control, v–vi.

ledge prior to reading the book. Of course each chapter is self-contained, and hence the reader can restrict his study to the chapters of his special interest without any particular difficulty. Since the book involves a blend of the experience of a large number of experts working in different environments, a multiplicity of approaches and angles of attack are offered which are of great value to those seeking for alternative designs and solutions.

I would like to thank my colleagues for their acceptance to offer some of their valuable experience for the book. Without their contribution this book would not have been possible.

Despite the variations in the organization and flavour of the particular chapters, I am convinced that students, scientists and engineers will find this volume a useful guide to current microprocessor-based modelling,signal processing, measurement, control system design, and automation techniques.

Patras 1983 Spyros G. Tzafestas

CONTRIBUTORS

M. BERTINO, System and Software Engineering, Soft Power, Torino, Italy.

F. M. CADY, Department of Electrical Engineering and Computer Science, Montana State University, Bozeman, USA.

C. S. CHEN, Electrical Engineering Department, The University of Akron, Akron, Ohio, USA.

P. CIVERA, Department of Electronics, Politecnico di Torino, Torino, Italy.

D. DELCORSO, Department of Electronics, Politecnico di Torino, Torino, Italy.

F. GREGORETTI, Department of Electronics, Politecnico di Torino, Italy.

R. GRONDIN, Systems Engineering and Electronics, Research Institute of Hydro-Quebec (IREQ), Varennes, Quebec, Canada.

F. HARASHIMA, Institute of Industrial Science, University of Tokyo, Tokyo, Japan.

R. M. HODGSON, University of Canterbury, Christchurch, New Zealand.

A. HOLMBERG, Systems Theory Laboratory, Helsinki University of Technology, Espoo, Finland.

R. D. JACKSON, Engineering Department, Cambridge University, Cambridge, England.

A. ST. JACQUES, Systems Engineering and Electronics, Research Institute of Hydro-Quebec (IREQ), Varennes, Quebec, Canada.

H. H. LE, Systems Engineering and Electronics, Research Institute of Hydro-Quebec (IREQ), Varennes, Quebec, Canada.

C. T. NGUYEN, Systems Engineering and Electronics Research, Institute of Hydro-Quebec (IREQ), Varennes, Quebec, Canada.

I. OKUMURA, Technical Research Institute, Japanese National Railways, Tokyo, Japan.

J. ORAVA, Systems Theory Laboratory, Helsinki University of Technology, Espoo, Finland.

J. B. PLANT, Graduate Studies and Research, Royal Military College of Canada, Kingston, Ontario, Canada.

E. J. PRENDERGAST, Engineering Department, Cambridge University, Cambridge, England.

N. K. SINHA, Department of Electrical and Computer Engineering, McMaster University, Hamilton, Canada.

S. SONE, Engineering Research Institute, Faculty of Engineering, University of Tokyo, Tokyo, Japan.

A. SOWIŃSKI, Industrial Institute of Electronics, Warsaw, Poland.

D. TABAK, Department of Electrical and Computer Engineering, Ben-Gurion University of the Negev, Beer Sheva, Israel.

G. A. TENDULKAR, Department of Control and Automation Systems, Siemens AG, Erlangen, FRG.

H. TOFFER, Physics and Engineering Analysis, Operations Division, UNC Nuclear Industries, Richland, Washington, USA.

S. K. TSO, Electrical Engineering Department, University of Hong Kong, Hong Kong.

S. G. TZAFESTAS, Control Systems Laboratory, Electrical Engineering Department, University of Patras, Patras, Greece.

P. URONEN, Department of Process Engineering, University of Oulu, Linnanmaa, Oulu, Finland.

CONTENTS

PART II SIGNAL PROCESSING

Chapter 4
DESIGN AND IMPLEMENTATION OF THE DIGITAL FILTER VIA THE SIGNAL PROCESSOR
C. S. Chen

Chapter 5
MICROCOMPUTER-BASED WAVEFORM ANALYZERS AND THEIR APPLICATIONS
R. D. Jackson and E.J. Prendergast

Chapter 6
MICROCOMPUTER-BASED INTERACTIVE IMAGE PROCESSING
F. M. Cady and R. M. Hodgson

Chapter 7
STATE OBSERVER IMPLEMENTATION VIA MICROCOMPUTERS WITH APPLICATIONS
N. K. Sinha

Chapter 8
STATE ESTIMATION USING MICROPROCESSORS FOR PROCESS SUPERVISION AND CONTROL
A. Holmberg and J. Orava

PART III MEASUREMENT

Chapter 9
MICROPROCESSOR-BASED MEASUREMENT SYSTEMS
A. Sowiński

Chapter 10
MICROCOMPUTERS IN NUCLEAR PLANT DATA MEASUREMENT
H. Toffer

Part I

GENERAL ASPECTS

CHAPTER 1

MICROPROCESSORS AND THEIR APPLICATIONS—OUTLINE OF THE BOOK

S. G. Tzafestas
Control Systems Laboratory
University of Patras
Patras, Greece

1. INTRODUCTION

Microprocessors have already been recognized to be important tools for the electronic and system designer [1-9]. Microprocessor-based computing systems have shown a great influence on signal processing and control, similar to that of large-scale digital computers in the decade of sixties. Their sucess is mainly due to the cost reduction and design flexibility increase they offer. After the introduction of the PDP-5 (1963) and PDP-8 (1965) minicomputers a steady reduction rate in the cost of signal-processing and computer-control has occured, (about 30 per cent per year) owing primarily to the improvements of semiconductor *IC* technology. The transition from *PDP-8* (which is programmable mainly in machine and assembly language) to *VAX-11* (a high-level language machine) needed 15 years, whereas analogous progress in microprocessors was made in less than half that time. The microcomputer offers a unique compactness and modularity in instrumentation, signal processing, and control, which makes now possible many new applications, that previously needed much more expensive dedicated computers. Actually, the application of microprocessors is limited more by the imagination of their users than by their technology. As more and more system designers are familiarized with the capabilities of microprocessors the number of new microprocessor-based applications increases very rapidly. It is remarkable that in only a few years we had a transition from the laboratory utilization of the microprocessor to a mass production and utilization of microprocessor based systems. Of course, in the early stages of the microprocessor revolution, the new technology has been used merely to replace TTL implemented systems, but presently an increasing number of products are surfaced which would be impractical without the microprocessors.

S. G. Tzafestas (ed.), Microprocessors in Signal Processing, Measurement and Control, 3–12.

2. MICROPROCESSOR CAPABILITIES

Microprocessors are the result of the semiconductor industry effort to accomodate an ever-increasing number of transistors in a single integrated circuit, and can implement a given function with a smaller number of chips than older devices. They have replaced hardwired logic with programmed logic that can be placed into ROMs offering even more capability per chip. Any basic microprocessor module can be enriched with a great logic power by simply adding of few more IC's. The most important characteristics of today's microprocessor usage is not so much in replacing hardwired logic or the low-end minicomputers, as it is in a substantial shift in the philosophy of electronic design and in using them in the new applications where use of big computers is not economically justified. The major microprocessor disadvantage is its low speed. Compared to minis, the micros have smaller istruction repertory, fewer features, and weaker software support.

Compared to the hardwired logic, the microprocessor offers greater flexibility, since wired connections and SSI/MSI devices are being replaced by bits in memories. Thus, modifications, improvements, and product expansions, instead of requiring new hardware designs and/or replacing logic gates and interconnections, simply require reprogramming and substituting bits in memory. It is noted that, although microprocessors have in general low speed, quite fast real-time programs can be written especially for slow industrial applications (such as temperature sensing in a heating plant, reading synchro signals, etc). When higher speed requirements exist, for which the conventional single chip μP is not fast enough, one can use multi-chip microprocessors with bit sliced architectures, most of which are intended for microprogrammed applications.

Microcomputers can replace minis for controlling industrial processes, reducing the cost by a factor greater than two. Originally the microcomputers were scale-down versions of existing minicomputers. This allowed the utilization of the mini control software with nearly no change. Today many microcomputer systems are being designed from the beginning with unique properties.

The major advantages of microprocessor-based systems over their random logic counterparts are the following:

(i) Lower manufacturing cost (60 to 20 percent of the TTL equivalents).
(ii) Lower time and cost of original design and development.
(iii) Faster production of new systems in closer corres-

spondence with the actual needs and demands.

(iv) Increased functional capability at lower costs (i.e. better products can be available at the same or lower prices).

(v) Smaller number of components and connections needed for a given system, resulting in increased overall system reliability.

(vi) Possibility of incorporating self-diagnostic procedures for failure detection, resulting in a substantial reduction in service and maintenance costs.

3. MICROPROCESSOR APPLICATIONS

Looking at the current repertory of microprocessor applications we see that the majority of them are in the fields of measurement and instrumentation, automatic testing, data acquisition systems, industrial control, automation, signal processing, communications, aerospace, and military systems. Microprocessors are also encountered in office equipment, business machines, calculators, data collection and communication equipment, and point-of-sale terminals. Since increasing the functional capabilities and enriching the job repertory of microprocessor-based systems can be done with small additional costs, there is presently a tendency to accomodate more functions on the same system than would have been realistic with older design techniques. This is particularly true in measurement and instrumentation systems, which are becoming increasingly more sophisticated and powerful by, say, the addition of remote control, programmability, peripheral interfaces, etc. Microprocessors are currently finding application in cases where previously the job was being performed by electromechanical, not electronic, means. Examples of such cases are building lifts, traffic lights, appliances, cocktail mixing, etc. It is anticipated that an ever increasing repertory of systems, presently working with electromechanical devices, will be replaced by microprocessor-based ones.

A representative list of current microprocessor applications is the following.

- Aerospace
- Automation/Robotics
- Communications
- Computing
- Consumer
- Domestic
- Education
- Industrial control
- Instrumentation
- Management
- Measurement
- Medical
- Military
- Numerical machines
- Power systems
- Office equipment
- Signal processing
- Testing systems
- Transportation
- Other

4. TEACHING CONTROL SYSTEM DESIGN USING MICROPROCESSORS

Among the applications of microprocessors we pick up here the one dealing with the teaching of computer-aided design of control systems, which is not discussed elsewhere in the book. The basic questions are: what to teach and how to teach it? The modern trend is to teach the design of control systems (after the required analysis background) [4]. This means that students must mostly be faced with "open design problems", i.e. problems with no unique solution. In an open problem one has to choose what type of compromise must be accepted among the alternative solutions, and to study the constraints trying to make them softer and reduce their number as much as possible. Of course the students have first to be familiarized with the solution of closed problems that have a unique and known solution.

In [2] an educational system is presented which is based on the Zilog Z-80 microprocessor developed into the 380Z system by Research Machines Ltd (UK). The software programs (in Basic) have been produced at the University of Reading. This system was named *"Cycloid"*, since it is a system for the *"Cybernetic Computerized Learning of Interactive Design"*.

The control system designer can obtain numerical and graphical information on a video unit, concerning the response of the system to various input commands and disturbances. An interesting feature of the system is that the program has a "command structure" and transmits to the designer the prompt: "What next?". The designer can then ask for help by simply keying the message HELP, and obtains a list of acceptable replies. In this sense no manual or operating instructions are needed for the designer to start using the system.

The program was written using classical frequency domain techniques, and allows the student to design SISO control systems using Nyquist and inverse Nyquist plots, and to test the system time responses to various inputs.

The actual closed-loop configuration incorporated into the program involves a transfer function under control of the type

$$G(s)=K_p e^{-T_d s}/s^q(1+\tau_1 s)(1+\tau_2 s)(1+\tau_3 s),$$

a series compensator of the type

$$G_c(s)=(K_1+K_2/s)(1+T_e s)/(1+T_a s),$$

and a feedback compensator around G(s) of the velocity type, i.e.

$$G_f(s)=K_f s.$$

The structure of the system under control is adjustable by setting to zero any time constant τ_1, τ_2 and τ_3. The system type index q must be chosen as 0,1, or 2. Details on the interface, the cycloid software, and on some applications, experiences, and extensions can be found in [2].

The command structure of *cycloic*, which permits the user to demand a response from the computer, is the main advantage over many other programs. This is in contrast to the "option" oriented programs where the designer must follow a specific path dictated by the computer to achieve a specific response. It is noted that,although microcomputer software is not directly transportable, the conversion of cycloid to operate on other microcomputers will not involve major restructuring of the program.

Another control educational system is described in [9]. This system is appropriate for teaching digital control design using both classical sampled data control theory and modern multivariable control methodology (see Chapter 13 of the present book). This is important due to increasing acceptance of digital processors by the industry to provide more reliable, economic and effective control of a variety of processes. Final year and Master degree students in systems and control engineering are becoming more and more interested on the subject of digital control.

In the past, the performance of digital control techniques was being studied in the laboratory using analog/hybrid setups, or via direct access to a general purpose minicomputer. Currently, attention is focused towards the utilization of microcomputers in cooperation with analog simulators for the process under control.

The teaching system of [9] is based on a Texas Instruments 990/4 single board microcomputer employing a 16-bit TMS 9900 microprocessor. It has 4K words of RAM and 512 words of PROM. All programs were written in Assembly language. A memory mapped interface accomodates six analog input channels (with 10-bit ADC), four analog outputs (with 8-bit DAC), and sixteen digital control inputs. The control programs are stored on magnetic tape, and are loaded in memory with the aid of a cassette data recorder and standard serial interface. The student can revise the program parameters using the video terminal in an interactive and conversational way.

The following program packages were developed.

(i) *Digital transfer function program:* This allows one to

study a variety of SISO control design methods based on the control algorithm

$$u_k=b_o e_k+b_1 e_{k-1}+\ldots+b_n e_{k-n}-(a_1 u_{k-1}+\ldots+a_n u_{k-n})$$

where e_k is the sampled error, and u_k the computed process control input at time t=kT, T being the sampling interval.

(ii) *Three-term (PID) direct digital control (DDC) program.*

(iii) *Optimal digital state feedback control program:* This allows one to study a variety of optimal controllers (quadratic, time-optimal, etc.).

This system has been used to complement and reinforce the material of control theory courses. It is really valuable for the student, since it helps him to obtain better insight into the various practical aspects (hardware and software) of digital control. The repertory and complexity of the systems under control can very much be increased if the designer has access to a control design package through a local computer terminal. A program package of this type is being currently developed by the writer, based on the theory of self-tuned digital controllers, for a DAI (Data Applications International S.A.) development microcomputer system.

5. OUTLINE OF THE PRESENT BOOK

The present book is divided in five parts. *PART I (General Aspects)* involves three chapters. The present chapter serves as a general introduction to the book, and as we have seen discusses the microprocessor capabilities, their applications, and their role in teaching control system design. This section actually provides a scan of the contents of the book.
CHAPTER 2, by *Civera, Del Corso* and *Gregorretti* provides an introduction to the problems arised when using microprocessor systems in real-time applications. Real-time systems are distinguished in two categories, namely "soft" and "hard" real-time systems. In the first, a computation is regarded to be correct when it is completed (in the average) within a given time period, whereas in the second, when the computer fails to finish the calculations within the desired time period, a general system failure occurs. The issues investigated are "hardware design" (data transfer, I/O techniques, hardware/software trade off, peripheral controllers), "real-time programming" (languages, concurrent run-time support, debugging support), and "distributed real-time systems".
CHAPTER 3, by *Tabak* discusses the basic principles of hardware realization and the corresponding software preparation for microprocessor-based control.

The presentation is based on a general block diagram of a microcontroller, and the problems of sampling period selection, timing organization and quantization effects are discussed.

PART II (Signal Processing) involves five chapters. *CHAPTER 4*, by *Chen*, gives a review of the basic "infinite-impulse response *(IIR)* and "finite-impulse response" *(FIR)* filters, emphasizing the various A/D conversion methods in the IIR filter design, and the windowing and frequency response methods for the FIR filter design. An application of the INTEL 2920 signal processor is included for demonstrating the implementation of the four basic digital filter operations namely I/O, add/subtract, multiply and delay. *CHAPTER 5*, by *Jackson* and *Prendergast*, is devoted to an investigation of the microprocessor capabilities for waveform analysis using digitized signals. The study is first concerned with "wide-band" signals requiring D/A converters with very fast conversion rates. Then, the various constraints imposed by the quantization operation of analog signals are examined. Two particular applications (semiconductor device testing and signal spectral analysis) are included. *CHAPTER 6*, by *Cady* and *Hodgson* provides an analysis of interactive image processing systems including their hardware and software issues. After a description of some typical applications, three new microprocessor-based image processing systems are described, namely a 4-channel multispectral remote sensing system, a high resolution image processing system (HIP) developed at Cantenbury University, and a similar system which is under construction at Montana State University. *CHAPTER 7*, by *Sinha*, discusses the microcomputer implementation of Luenberger state observers of dynamic systems. Two types of implementation are considered, which are based on a discrete-time model and on a block-function approximation of the system at hand. Extensions to nonlinear systems are provided and some simulation results are presented, including a brief description of the hardware and software issues. The final chapter of Part II is *CHAPTER 8* by *Holmberg* and *Orava*, where the microprocessor-based state estimation is treated in connection with process supervision and control. A brief presentation of the state-of-the-art is provided based on the current literature. The estimation algorithms considered are Luenberger observer, Kalman-Bucy filter, extended Kalman filter, and on-line nonlinear filters. Important applications are included concerning the state estimation of an activated sludge water treatment process, and a pH control process.

PART III (Measurement) Contains two chapters. In *CHAPTER 9*, by *Sowiński*, the basic general issues and techniques utilized in microprocessor-based measurement systems are presented. These concern the structure and organization of microprocessor-based systems, the tasks performed by a μP-based measurement system, the system testing and system development problems, and finally the hardware versus software tradeoff issues. *CHAPTER 10*, by *Toffer*, focuses on uses of microcomputers for acquisition and display of nuclear plant data. Specifically the uses of microprocessors at the Hanford N Reactor operated by UNC Nuclear Industries are examined, namely: the acquisition and analysis of flux data using intelligent terminal, the use of a network of microcomputers to monitor vibration levels on rotating machinery, and the use of a digital image processor to display nuclear reactor control parameters. These particular applications are sufficiently general to illustrate the use of microprocessors and minicomputers to other nuclear or industrial systems as well.

PART IV (Controllers) involves two chapters. *CHAPTER 11*, by *Tendulkar*, is devoted to the study of microprocessor-based industrial controllers in stand-alone and hierarchical configurations. After a general introduction to the application of industrial controllers and to the microprocessor architectures, the class of programmable controllers is considered, which offer considerable advantages over hard-wired systems. A "universal" programmable controller for both logic and fast closed-loop control is described in some detail. Finally the controller architecture trends, the influence of software and the architectural convergence are examined. *CHAPTER 12*, by *Tzafestas*, provides the design equations for a variety of digital controllers, involving the classical compensators and PID controllers, the minimum variance/self-tuned controllers, and a number of state controllers (optimal, eigenvalue assignment, model matching, dead-beat, minimum variance, and model-reference). The chapter includes a general description of microcontrollers, a short account of parallel-organized controllers, an overview of the latest microprocessors which are appropriate for controller implementation and finally a representative set of examples of microprocessor-implemented digital controllers (PID, servo-controller, self-tuning and self-selecting, optimal and multilevel controllers).

PART V (Application to Control and Automation) involves chapters 13 to 19. *CHAPTER 13*, by *Harashima*, presents the state-of the art on speed control systems of motor systems using microprocessors. After a brief review of the recent progress of μP-based motor control systems, the applications of modern control techniques are examined with reference to speed accuracy, robustness, and optimality. The chapter includes three important practical examples of microprocessor-based motor speed control systems. *CHAPTER 14*, by *Plant*, presents a new algorithm for the microprocessor control of three-phase SCR controlled DC motor systems. Two look-up tables are given for determining the open-loop nominal steady-state SCR firing angle and current in terms of the demanded speed. A feedback correction to the nominal SCR firing angles is then introduced, based on sampling at the instant of firing each SCR. A linearization procedure of the system equations is provided and employed, together with simulation of a three horsepower motor, to investigate the effectiveness of the control algorithm. In *CHAPTER 15*, by *Tso*, the principles of phase control as applied to converters and cycloconverters are reviewed and some insight into the microprocessor-based implementation is given. Development in this direction was greatly motivated by the wide range of converter and cycloconverter applications combined with the rapid acceptance of microprocessors as basic computing and control elements. *CHAPTER 16*, by *Bertino*, provides a general discussion on industrial robots used in automation processes, with regard to their microcomputer-based controls. The discussion starts with a presentation of the types of industrial robots classified according to mechanical structure, degree of freedom, hardware/software characteristics, and area of use. Then the path control for anthropomorphic manipulators is investigated, and the hardware structure of a control unit for such manipulators is studied. The chapter closes with a presentation of the robot applications and some possible future developments of them. In *CHAPTER 17*, by *Nguyen*, *Grondin*, *Jacques* and *Le*, the power system applications of microcomputers are described, with particular emphasis on power substation automation. The principal operations needed in power substation automation are outlined and two particular applications are described, namely a digital system for transmission line protection, and a voltage regulation system controlling tap changing power transformers. The discussion includes the basic issues for hardware design and the microcomputer implementation aspects. *CHAPTER 18* by *Uronen*, deals with some typical applications of microprocessors in the control and automation of the pulp and paper industry. It is concluded that classical process control will be increasingly integrated both horizontally to other operative functions (material handling, maintenance, marketing) and vertically to management and electronic data processing functions. Thus a hierarchical or corporate wide control scheme will be used, along with decision support systems which help top management in unstructured decision making situations. The chapter discusses several aspects of process automation such as distributed digital control, data highways, intelligent sensors, software implementation, identification/filtering, etc. The book closes with *CHAPTER 19* by *Sone*, which is devoted to the study of the application of microprocessors in railway systems control and automation. Firstly the important applications to monitoring on-board equipment, traffic control, passenger information systems, control at marchalling yards are briefly described. Then the state-of-the-art of the field is presented including energy saving operation of intercity trains, interference-free AC motor drive systems, and μComputerized realization of fail-safe functions of signalling systems.

REFERENCES

1. Alexandridis, N.A. and Tzafestas, S.G.:1978, "Microcompute Aspects for Control Applications", Proc. 20th Midwest Symp on Circuits and Systems, Texas, Aug. 15-17.
2. Atkinson, P. and Allen, A.J.:1980, "Low-Cost Microprocesso Based Systems for Teaching the C.A.D. of Control Systems", Proc.IEE, pt.A, pp.111-118.
3. Donaghev, L.F.:1976, "Microcomputer Systems for Chemical Process Control", Proc. IEEE, Vol. 64, pp. 975-987.
4. Franksen, O.:1972, "On the Future in Automatic Control Education", Automatica, Vol.8, 517-524.
5. Jones, M.R. and Pooch, U.W.:1976, "Evolution and Design of Microcomputers", Proc. Symp. Trends and Applications 76: Micro and Mini Systems", Gaithersburg, Maryland, May.
6. LeBoss, B.:1975, "Microprocessors Spark a Quiet Revolution in Instrumentation", Electronic News, August, p.1.
7. Neth, J.:1975, "Microprocessors Today:A Promising Past Turns into a Positive Future", EDN, pp. 82-88.
8. Nichols, A.J.:1976, "An Overview of Microprocessor Applications", Proc. IEEE, Vol. 64, pp. 951-953.
9. Walker, P.A.A.:1980, "A Microcomputer Facility for Teaching Digital Control Systems Design", IEEE Trans. on Education, Vol.E-23, pp.193-196.

CHAPTER 2

MICROCOMPUTER SYSTEMS IN REAL-TIME APPLICATIONS

P.Civera, D. Del Corso and F. Gregoretti
Dipartimento di Elettronica
Politecnico di Torino
Torino, Italy

1. REAL TIME PROCESSING

The term "REAL-TIME" was introduced some years ago to distinguish computations where there is a direct interaction of the processor with the real world from "batch" processing, where data and programs are supplied as batches to the computer and results are produced without reference to the external environment.

Interactive computations are in general composed of three operations: the acquisition of data or events from the surrounding environment, the processing of them and the output of results or actions back to the external world. Thus the term "real-time" derives from the fact that these computations are triggered by external events and that the results may depend, apart from on the input data and the computations, also on the value of the variable "time" at execution or on the duration of the execution itself. Such a definition is extremely general and actually may apply to a wide spectrum of computer applications ranging from computer networks for banking transactions to the control of industrial processes or from commercial video games to the control of complex weapon systems [MORR80,ARNO82].

Nevertheless real time systems may be divided into two main categories. For systems in the first a computation is considered to be correct when the mean execution time over a defined time interval is lower than a given maximum. In general the characteristics of these systems are specified in a statistical form such as :"97% operations must be executed within 3 seconds" or: "The average execution rate is 50 operations per second".

For this category of systems, since timing constraints are not absolute, the term "soft real-time" has been coined to describe them.
A typical example is that of a statistical multiplexer,as shown in figure 1., which concentrates the outputs of several low speed end terminals into a single high speed line connected to a computer. The number N of allowed terminals is in general greater than the ratio

S. G. Tzafestas (ed.), Microprocessors in Signal Processing, Measurement and Control, 13–32.

between the line speed A and the terminal speed B, because for most of the time only a few terminals will be transmitting or receiving simultaneusly. When many terminals are operating at the same time then some of them will be temporarily locked out and have to wait.

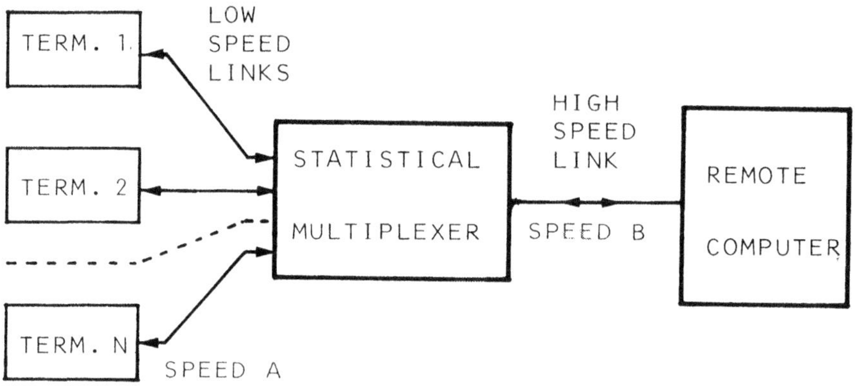

Fig.1. A statistical multiplexer

For the systems of the second category the applications are in general structured in such a way that the failure of the computer to complete a calculation within a specified time always causes a general system failure. To this class belong many of the so-called "event driven" systems used in the fields of machine automation, process control, image recognition, weapons and defence. The time limit may vary from a few milliseconds to many seconds. It apply continuously in the normal functioning of the system or become active only in particular cases (like power failures), but can never be exceedes without causing a severe malfunctioning of the system. Hence the term reserved for such systems, "hard real-time" .

In figure 2 some examples are given showing the extremely wide range of time limits for different real-time applications.

The obvious answer to real-time requirements must come from processing hardware with enough power and speed, in conjunction with proper programming techniques. We must however consider that to increase the power of the hardware and to improve the performance of a program are expensive processes. Therefore the actual goal is a tradeoff between complexity, program refinement, and system cost. A universal solution which can solve all the real-time problems obviously does not exist; the equipment (hardware and software) must be tailored for the specific kind of application. We can however describe the hardware design techniques that are widely used in real time machines to increase the processing speed, and the techniques for writing reliable real-time programs which are reliable, that is modular, documented and mantainable.

Image acquisition and analysis in an airborne defence radar system	100 microseconds
Control of the sparking time in a fuel engine	2 milliseconds
Control of the movement of mechanical parts in a machine tool	20 milliseconds
Interaction with a videogame terminal	100 milliseconds
Handling of a BSC type communication protocol	< 1 second
Interaction with a terminal for banking transactions	some seconds
Control of the temperature in a room	some minutes

Fig. 2. Some typical time limits

2 REAL-TIME HARDWARE DESIGN

A microcomputer system is composed of a Central Processing Unit (CPU), the memory, and I/O interfaces, connected by a bus as shown in figure 3.

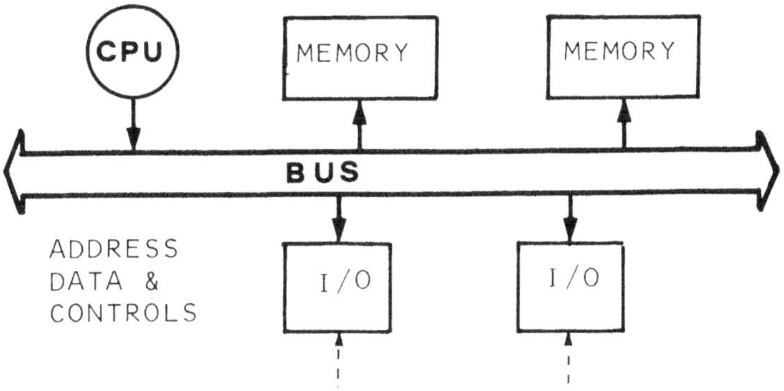

Fig 3 - General organization of a microcomputer system

The memory contains both the program and the data used or produced by the program itself. Any interaction with the external world is performed through a set of input/output interfaces. All the operations are ruled by the CPU which, while executing a program controls the transfer of information (data and instructions) through the bus. A module which can control the bus channel is called MASTER; the others are the SLAVEs. In figure 3. the only master is the CPU. Any slave unit (memory cell or I/O register) is identified by a unique ADDRESS.

The execution of a program consists of a coordinated sequence of data transfers from and to the memory, I/O operations, and logical or arithmetic computations. Therefore, to improve the execution speed, we must carry out faster data transfers, faster I/O handling, and faster computations. In the following sections we shall see how the speed can be increased by moving functions from the software to the hardware, that is by using "distributed processing" techniques at various levels.

2.1 Data transfer

The CPU is often used for repetitive tasks such as moving blocks of data between the memory and the I/O interfaces. In order to execute these operations the processor loops on a program segment where, besides transferring data, it fetches and decodes instructions, updates pointers and counters, checks for END conditions etc. A more efficient solution is to assign all these functions to a special unit that implements pointers and counters directly using the hardware. In this case the system is rid of the long process of program execution, because the data transfer is performed by the dedicated unit. This technique is called Direct Memory Access (DMA); the specialized unit which handles the transfer is the DMA controller.

A DMA controller manages all the activities required to exchange data blocks between the memory and I/O devices, or between internal and external memory, such as disks and tapes; therefore it is a master. This technique is useful expecially for transfering large amounts of information, because these operations require a long time if executed by the CPU under program control.
We must also consider that during the transfer of a packet of data under the control of the CPU, the execution of other instructions is blocked. With the DMA technique the CPU only has to initialize the parameters of the DMA controller (source and destination address, block size, etc), then wait for the end of DMA operations or, if the system bus supports interleaved transfer cycles, continue with other computations.

In most cases the system has only one bus, shared between the DMA and other CPU activities. This slows the execution of programs during DMA transfers, because the CPU can access the memory only when the bus is free. However, even with this limitation, the DMA technique must be used whenever high memory transfer rates are required. It must be pointed out that for each transfer the DMA controller has to select two slave units: the memory and the I/O interface. Usually the memory is addressed through the bus, while the I/O interface is handled via direct point-to-point lines, independent from the bus. For this reason the DMA circuitry is built on the I/O module itself, which must be designed from the beginning for this purpose. These connections are shown in figure 4.

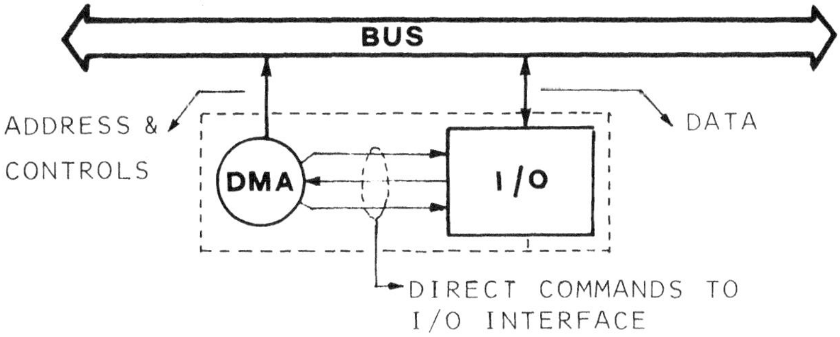

Fig 4 - Connections of a DMA controller.

2.2 I/O handling techniques

Let us consider as a first example the interfacing of the end-of-run switch of a moving arm. The switch state is mapped into one of the bits of an input register of the microcomputer. This register is read by the CPU to check the position of the arm. A simple solution is to perform this operation periodically in the main program loop. This technique is called "status polling" and has some disadvantages in a real time environment:

- the activation of the input is sensed only after a variable delay, which depends on the length of the loop;
- the check operations are performed even when the input does not change state, thus wasting precious time.

Both these features represent serious handicaps in a real-time system, because the result is to reduce the speed and the processing power. They can become unacceptable if there are "hard" real-time contraints: in our example, when the arm reaches the end-of-run, it must be stopped immediately, not after some time !

A better solution is to avoid the check operation by sensing the input variable using an asynchronous input to the processor. Most processors can do this by means of the INTERRUPT structure: when the interrupt input signal of the CPU is activated, the current program is suspended and the execution is transferred to another program section at a predetermined address , called the "interrupt routine", when the interrupt routine terminates, the control is transferred back to the program section which has been interrupted. In our example the end-of-run signal is tied to the interrupt input, and the interrupt routine performs the required operations. No software polling is required, and the response to flag activation is immediate.

As a further step let us consider a system with many inputs like the end-of-run referred to above. An interrupt servicing of these flags could be done with the structure of figure 5.

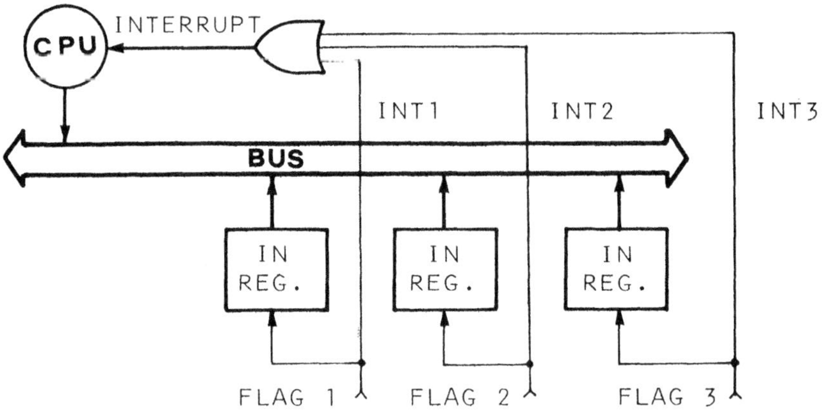

Fig 5 - Multiple interrupt structure.

All flags are connected to input registers and are ORed to the interrupt input. The interrupt routine will poll the status registers to find the input active at the moment.

A more advanced technique makes it possible to identify the interrupting device with a single read operation. The technique is called "vectored interrupt" and uses the structure shown in figure 6.

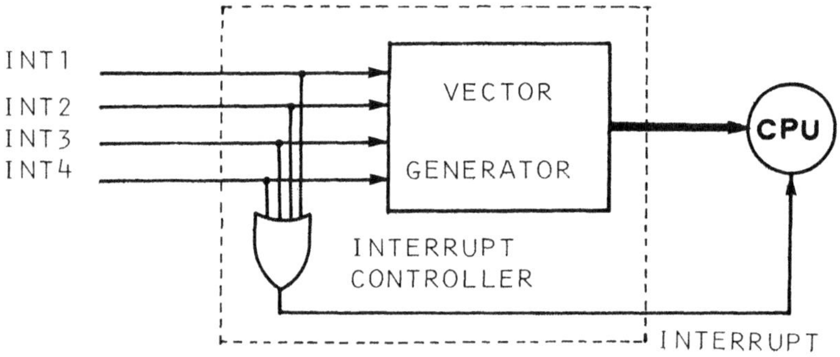

Fig 6 - Vectored interrupt structure

At any interrupt, a word is generated in the "interrupt vector register". This word identifies the highest priority active interrupt. The CPU reads this register and can use the vector to jump, directly or through an address table, to the different service routines of the various interrupts.

The logic diagram shown in figure 6 and this short description are oversimplifications of the actual structures. The interrupt controller can be integrated within the CPU itself, or it can be a special device, or it can even be fully distributed on the peripheral interfaces. In any case it must:

- select the highest priority active interrupt;
- generate a vector that identifies the requester.

A centralized controller limits the number of separate interrupts and requires a set of point-to-point lines for the independent interrupt requests. Control systems must often handle a large amount of I/O; the current trend is towards distributed vectored structures, which can be upgraded simply by adding new units.

2.3 The hardware/software tradeoff and peripheral controllers

Many functions in a processing system can be implemented with different tradeoff between the hardware and the software parts. Let us consider, for instance, an interface which must send 8-bit data on an asynchronous serial line in the format shown in figure 7.

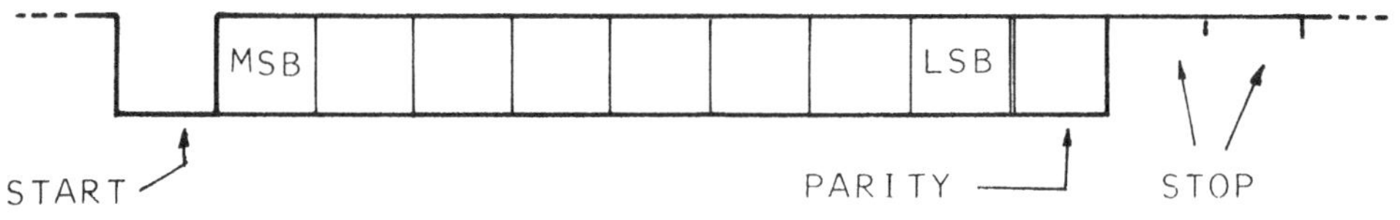

Fig 7 - Asynchronous serial 8-bit character

The complete process includes the following operations:

1) Take data from memory;
2) Format data according to the character protocol (start and stop bits, etc);
3) Serialize the character at the correct rate;
4) Convert electrical levels from TTL to EIA.

The first three operations can be performed either by a program which takes, formats, and serializes data, or by special hardware units: a DMA controller for operation 1); a special circuit called Universal Asynchronous Receiver/ Trasmitter (UART) for 2) and 3). When the functions are implemented with the use of software, the special hardware is no longer needed, but the CPU of the system is kept busy by clerical jobs, such as wait loops to set the bit rate in the serializer program. On the other hand the complete hardware implementation is more expensive because it requires two additional units: the DMA controller and the UART.

Since the hardware implementation of any function increases both the processing power of the system and the cost, because of the additional devices needed, mixed solutions are often the best ones. In the above example, we a very common solution is to design an interface that receives data from the CPU (i.e. using a software module), while it formats and serializes them with the hardware (that is using a UART).

Other more complex peripheral functions such as floppy disk controllers, video interfaces and link controllers, are now widely available. These I/O controllers are very often processors themselves, and execute a fixed internally stored program. They represent the first step towards a complete distribution of functions among the system units.

As a further step, besides I/O handling (interrupts),and data transfers (DMA), one can also move to functions that operate on data to dedicated units. For instance, the CPU can be relieved of heavy arithmetic computations by the use of special peripheral processors that are seen from the main processor as I/O ports. The CPU sends data and commands to them and reads the results back. The operation of a peripheral processor is faster because they are designed especially for this job.

Another technique is to extend the instruction set of the main processor, by means of "attached" or "extended" processors. They execute special instructions that are not implemented in the main CPU; when the extended processor does not exist, an interrupt-like structure calls a program segment which performs the same operations.

In systems that use these devices only the functions are distributed between many units; the control is still centralized into a single CPU which rules all the interactions and activates the peripheral processors. These are called "MASTER/SLAVE" MULTIPROCESSOR systems and are typical of the current level of industrial control systems.

Distributed systems with true decentralized control are still confined to laboratories or to special applications and will be described in section 4.

3. REAL TIME PROGRAMMING ENVIRONMENT

With the rapid diffusion of microprocessor systems the cost of the hardware components of complex systems has dramatically decreased, whereas the costs of the software, more tied to manpower cost have grown up to 70% of the whole project cost. This figure is even higher for real-time systems where computations are closely related to each specific environment with particular attention to timing constraints.

People have discovered at their own expense that in the real-time field the testing and maintenance costs of a program may be comparable or even greater than the costs of the design. Without a programming discipline "programmed logic" may become even less flexible than "hardwired logic". The use of intricate interrupt structures and of program tricks for optimizing the execution time or the memory requirements has often led to programs that after some time cannot be modified even by the designers themselves.

Hence, since the birth and the rapid growth of the Software Engineering discipline, there has been a considerable effort made by software designers to find standard procedures for the development, testing and maintenance of real-time software for microprocessors.

This effort followed three main guidelines: programming languages, concurrent run time supports and hardware and software debugging supports, all of which will be examined in the following sections.

3.1 Programming Languages

Programming languages may be classified in several ways. Taking into account the use for which they are best suited, people often distinguish between system-oriented and application-oriented languages (i.e. COBOL for business applications or FORTRAN for scientific computations). Nearly all real-time microcomputer applications seem to be in the area of system programming or control applications (i.e. they require direct access to the machine to perform I/O or control function or to achieve maximum efficiency) [CRES80].

In this area the lowest level is represented by ASSEMBLY LANGUAGES (or ASSEMBLERS). These have poor readability, are far from being self documenting and are not transportable across different machines. Moreover they permit the use of bad programming techniques, like the use of jump instructions, instruction modifications, etc.., which increase the cost of the maintenance of programs. In spite of these severe shortcomings assemblers are still widely used, partly for real and partly for claimed reasons of efficiency. It is still common in the area of real-time programming to find big programs written in assembly language in which the really critical parts are composed of a few routines that could have been called as procedures by a high level language.

The first step towards a more disciplined programming came from STRUCTURED ASSEMBLERS where control structures like IF... ENDIF, DO .... OD, REPEAT .... UNTIL, were added to the assembler statements.
In some cases these structures are defined as macroinstructions to be expanded at assembly time by a macroassembler; in other cases they are transformed into assembly statements by a small compiler program or preprocessor like in the case of the PLZ-ASM language [BASS78].

A further significant improvement came with the development of microprocessor oriented high level languages [BASS78, BROW78] which do not have many of the resource consuming features of traditional mainframe languages. They allow the definition of a rich set of data structure, perform type checking at compile time, and contain some machine-oriented operations (i.e. pointer operations in the PLZ-SYS) and tight control structures for memory allocation. One of the things that previously discouraged the use of high level languages was the size of the memory required. With the decreasing cost of memory chips these languages have now become widely used.

```
SCAN:   LD      HL,TABLE        ; Table address
        LD      B,0             ; Initializes the counter
SCANLO: LD      A,(HL)
        CP      REFER           ; Compares with the reference value
        JR      Z,SCANOU
        INC     B
        CP      CNTMAX          ; Checks if within the table bounds
        JR      NZ,SCANLO
        LD      B,ERROR         ; Error if out of the bounds
SCANOU: RET
```

Fig.8 Example of an assembler program

```
SCAN    PROCEDURE
ENTRY
LD      HL,TABLE
LD      B,0
DO                      ; Main loop
        LD      A,(HL)
        CP      REFER
        IF      EQ
                THEN
                EXIT    ; If equal to reference exits
        FI
        INC     B
        LD      A,B
        CP      CNTMAX
        IF      EQ
                THEN
                EXIT
                LD  B,ERROR ; Error if out of the bounds
        FI
OD
RET
END     SCAN
```

Fig.9 Example of a Structured Assembler

In order to underline the difference between the different programming languages three different examples are given in figures 8, 9 and 10. They represent the programming in three different languages (Assembler, Structured Assembler and PLZ-SYS) of a simple problem like the scanning of a table for a specified value, giving as result the position of the value in the table. The assembler used in the examples is that of the Z80 processor, while the high level language is the PLZ-SYS.

```
CONSTANT
REFER               := .... ! Reference Value !
COUNT_MAX           := .... ! Table limits    !

INTERNAL
! Data Declaration        !
TABLE   ARRAY [ CNTMAX BYTE]
TEMP    SHORT_INTEGER
SCAN    PROCEDURE
        RETURNS ( INDEX SHORT_INTEGER)
ENTRY
TEMP := 0
DO
        IF TABLE[TEMP] = REFER ! Test if eq. to reference !
           THEN
           INDEX := TEMP
           EXIT
        FI

        TEMP := TEMP + 1
        IF TEMP = COUNT_MAX     ! Check limits    !
           THEN
           INDEX := ERROR
        FI
OD
END     SCAN
```

Fig. 10 Example of microprocessor high level language

3.2 Concurrent Run-Time Support

By the term "Run-Time Support" we mean a set of software facilities that reside as a layered structure over the bare hardware machine and that transform it into a different "virtual" processor that supports at execution time the application software at a higher level of abstraction than the real machine. The supports may be algorithmic, like a set of floating point routines, or language-oriented, like interpreters that translate and execute high level language statements. The most widespread run-time support for real time programming are multitasking EXECUTIVES that provide the means for the "parallel" execution of several programs.

In fact most real time applications can be seen as a set of separate, loosely coupled activities (PROCESSES or TASKS), each one managing a function or a set of related functions. From the programmer point of view each process may be considered as an independent program which may communicate with the others to exchange information or to synchronize activities.

Let us consider, for instance, a system for the control of an industrial process which is required to activate in sequence the different phases of a working cycle, to monitor continuously the different parts of the system in order to detect faults and to record production data. These are three different and partially interacting activities which must proceed in parallel on the same machine.

In small systems a restricted number of concurrent activities may be handled by a single program, but normally it is impossible to have a centralized control of the system when there are more than three or four processes. Hence each process is written as a separate program, just as if it had at its disposal the whole processor, and communicates with the external world and with the other processes by means of requests to the executive. The executive controls the execution of the processes and its main activities are shown in figure 11.

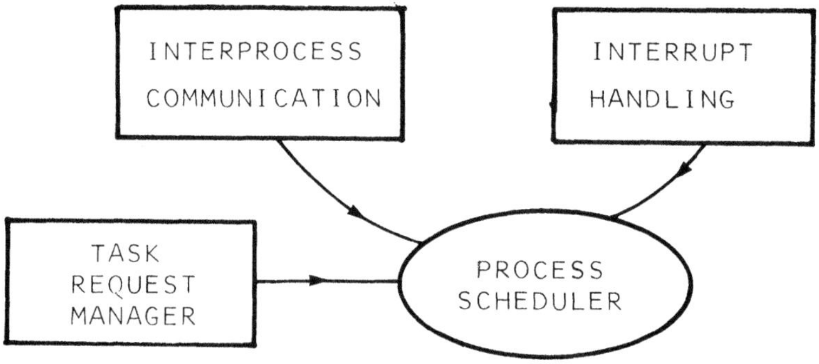

Fig.11 Main activities of an executive

The primary function is the SCHEDULING, that is the assignment of the CPU resource to the different processes on the basis of their relative importance (priority). Each process is provided with a protected environment (VIRTUAL MACHINE) consisting of a copy of the CPU registers, a Program Counter, a pointer to a Stack. This virtual machine is tranferred into CPU registers when the task is put into execution and back into memory when the task is suspended.This operation is called CONTEXT SWITCHING.

The second main function is interprocess communication handling between different tasks. One task may send or receive a message from another. In the latter case it will be suspended and wakened when the message is available from the sender. The executive is responsible for the transfer of the message from the sender to the receiver. Messages may be passed by VALUE, that is when the entire message is transferred from one memory area to another, or by REFERENCE when only a pointer to a memory area is passed from the sender to the receiver. Hardware interrupts are normally treated as messages from peripheral devices, and a task can wait for them as for any other message.

A task may have at any instant one of the states indicated in fig.12

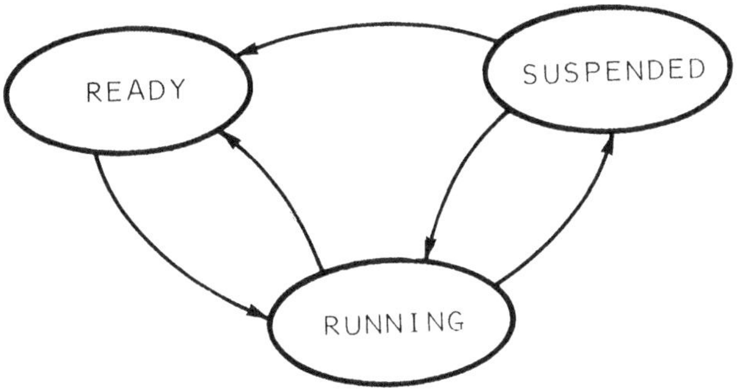

Fig.12 Possible states and transitions of a task

- READY when it has no outstanding executive requests. The CPU will be assigned to it when no higher priority task is ready. The executive maintains a list of ready tasks;

- WAITING when it requires the occurrence of an event like an interrupt from a peripheral device or a message from another task;

- RUNNING when it is the highest priority ready task.

A change of the running task (rescheduling) takes place on the occurence of one of the following events:

- the running task waits for a message or for an interrupt;
- a higher priority task is resumed by the arrival of an interrupt;
- the running task sends a message to a higher priority task which is thereby resumed.

Some of the comments made in the previous section on high level languages are also valid for real time executives. They are a very valuable tool for disciplined programming and provide a structured mainframe into which the programmers "plug" a set of application processes, without concerning themselves about scheduling and interprocess communication. The modularity introduced by the partitioning of the application into parallel tasks, increases the lifetime and the ease of maintenance of the product, by making it possible to update each module separately. The speed of program development is also increased because programs can be written as independent parts by different programmers.

Most problems of real time executives come from the overhead introduced by the context switching and by the delay in response to interrupts due to the fact that the executive acts as a filter between the peripheral devices and the interrupt managing tasks.

Many commercial real-time executives have been developed both by microprocessor manufacturers and by independent software suppliers [KAHN78,LIND80]. In some cases, when a high level language was available, system calls at high level were also provided. This was mostly achieved by supplying a package of routines in assembly language to the user. These routines can be called as procedures by an HLL and handle the interfaces with the real time executive.

3.3 Debugging Support

Although very often the costs for testing and maintenance may be a significant part of the total cost of the software, the debugging requirements are nearly always underestimated and therefore the inclusion of debugging aids is very seldom considered before or during the design of a system. The debugging aids are software tools which often require the addition of hardware modules in order to match real-time requirements [PLAT81]. At the lowest level we have the so-called "system monitors", which are simple programs, residing in general in ROM and which, by the use of a terminal console allow the execution of the following functions:

- Display and modification of memory and I/O registers;
- Software Breakpoint, that is the substitution of an instruction in the user program by a call to a routine giving back the control to the monitor and saving the context of the user program;
- Step-by-step execution of a program;
- Display of user program registers after a breakpoint or during step-by-step execution.

Some monitors have been developed in connection with run time executives and may be inserted as system processes in concurrent systems. Their functions have been extended to include process suspension or resumption, analysis of message queues, interrupt simulation, etc.

It is clear that all these monitors have many drawbacks when debugging real-time applications. First they do not allow breakpoints to be put in ROM and in some cases in ROM oriented systems, which are very common in real time applications, step-by-step execution is not made possible. Secondly, real time systems require visibility not only of the sequential execution of a program, but also of the time relations between the software and the external world. Very often it is necessary to "catch on the fly" an event and a sequence of instruction executions around it.

For these reasons many programmable hardware modules have been developed, with functions ranging from hardware breakpoints which allow us to catch references to data, to tracers which may record a whole sequence of actions prior or subsequent to a programmed event [WALZ80]. These hardware modules are in general conceived as add-on units to a standard section of a communication channel such as a bus, but their operations are limited by the visibility offered by the channel itself. Other powerful tools, the "In Circuit Emulators" (ICE's) have been developed to increase the visibility up to the internal registers of the processor [LIVE76].

4 DISTRIBUTED REAL-TIME SYSTEMS

The availability of microprocessors and the consequent decreasing cost of computer hardware has changed the attitude of designers towards problems that require great processing power. Rather than taking and adapting the whole problem to a large machine, it is now sufficiently cheap and easy to use a number of small machines specific for that problem [WATW82]. In other words, functions are distributed not only within the single microcomputer, but among many relatively independent machines. This DISTRIBUTED PROCESSING approach gives two main avantages:

- the big application problem is divided into smaller ones assigned to different computers and handled by separate software packages;

- the system can survive faults because other units can pick up the essential functions of the failed part.

The term "distributed processing" applies to a wide range of computing machinery. We shall here revue the usual taxonomy, pointing out the features that become relevant in real-time systems [ANDE75,ENSL77].

The parameters which classify distributed systems are:

- the control philosophy;

- the degree of coupling among processors;

- the connection topology;

- the physical distribution.

A discussion of system control choices has already been outlined in section 2. When all the units (in our case all the computers) are managed by a main processor the control is CENTRALIZED. If all units are equal and have the possibility for managing themselves without any asymmetry, the control is DISTRIBUTED. Centralized-control machines are easier to set up and rely on well-defined software design techniques. There is now a great deal of research activity on distributed control, mainly because the second goal outlined above, i.e. fault survival, is better achieved with this class of system.

From the point of view of processor coupling, there are two main possibilities:

- LOOSELY COUPLED systems, which communicate with each other through I/O ports. A computer network is a typical example of a loosely coupled distributed system.

- TIGHTLY COUPLED machines, where all processors can directly access some common memory. These systems are called MULTIPROCESSORS.

Multiprocessors are better suited for hard real-time on-line systems [ARNO82], where the main requirements are speed and processing power. Networks or other loosely coupled machines are widely used for less time-demanding applications, such as office automation, where they allow better exploitation of system resources.

The paths through which the processors can communicate can be described by connection topology. Some examples are given in figure 13.

The physical distribution specifies essentially how far the computer are from each other. If we consider only real-time systems, the distribution is usually limited because communication over long distances introduces long and sometimes unpredictable delays. The ranges in this case span from less than one meter up to several hundred meters. The former figure applies to tightly coupled machines, which require very high communication bandwidths; the latter is typical of office automation computer networks, or of plant automation systems.

Since the length of the interconnection impacts the speed and the reliability of the information transfer machines, which must exchange large amounts of data within short times are usually kept as close as

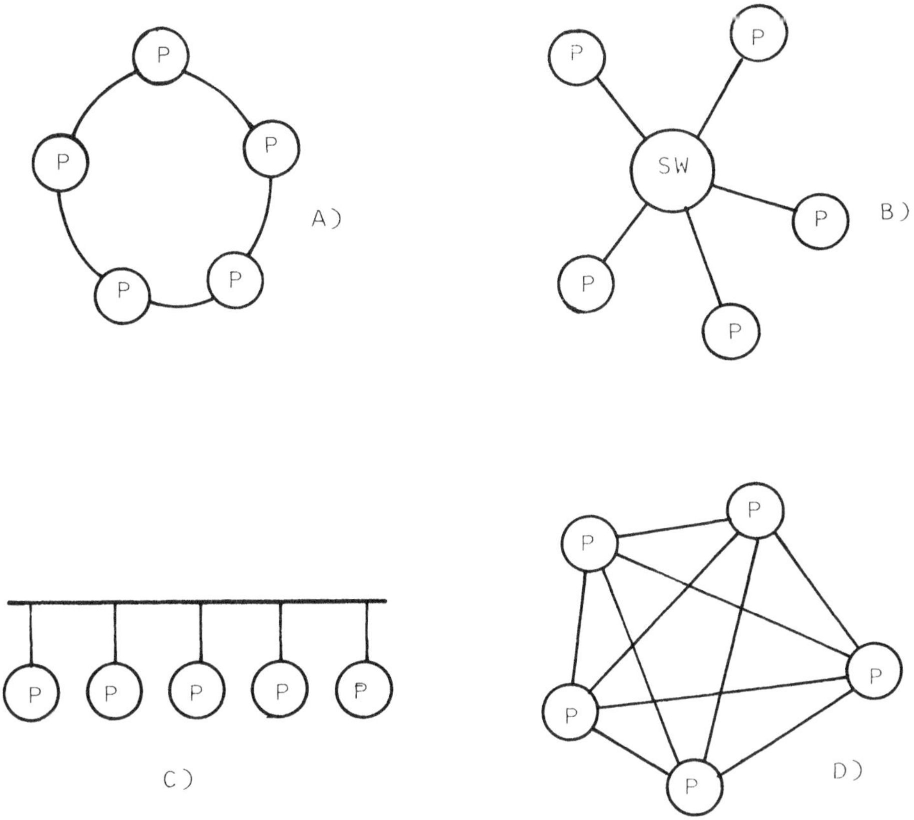

Fig.13 - Connection topologies of multiple-processor systems.
A) Ring
B) Star with central switch
C) Time-shared bus
D) Fully connected network

possible. For this reason the three parameters of distribution are related; hard real-time problems, requiring fast communication, are handled using tightly coupled machines, concentrated in a small area. The most generally used connection topologies are the global bus, the ring, and the star. When time requirements are less tight, as in the interaction with an operator, the loosely coupled networks offer better price/performance tradeoff, and better exploitation of expensive resources, as such as high quality printers, plotters, or large data bases.

5 REFERENCES

The references are classified by topics: survey, microprocessor architecture, software tools, special architectures and multiprocessors. Each reference may cover more than the topic related to the group. For each group of references we added some titles of books and proceedings of conferences, that are of specific interest.

The Proceeding of the International Conference on Computer Architecture and of the IFAC/IFIP symposia are recomended for further readings on the subject. Suggested keywords are: Real-time systems, process computer, process control systems, process control software, multiprocessor, distributed system architecture.

5.1 Survey and Miscellaneous References

[ENSL76] Enslow: "What does 'distributed processing' mean?", Distributed Systems INFOTECH, 1976.

[JENS76] E.D.Jensen: "Distributed processing in real-time environment", Distributed Systems INFOTECH, 1976.

[WEIS77] A.J.Weissberger: "Analysis of multiple-microprocessor system architectures", Computer Design, Vol.16 n.6, June 1977.

[RENN80] D.A.Rennels: "Modelling distributed function architectures", Distributed Systems INFOTECH, 1976.

[BUSI80] W.Busing: "Distributed process automation systems: requirements and interfacing", Process Automation n.2, 1980.

5.2 Microprocessor system design

[BOOK] J.B.Peatman: "Microcomputer based design", Mc Graw Hill, 1978.

[BOOK] J.K.Thurber, G.M.Masson: "Distributed Processor communication architecture", Lexington Books, 1979.

[BOOK] M.R.Skrovkov: "Mini and micro computer control in industrial processes", Van Nostrand, 1980.

[THUR72] K.J.Thurber et al.: "Computer interconnections - A systematic approach to the design of digital bussing structures", Proc. Fall Joint Computer Conference, Dec. 1972.

[LIVE76] B.Live et al.: "The in-circuit approach to the development of microcomputer based products", IEEE Proceedings, June 1976.

[LEVI78] J.V.Levy: "Busses, the skeleton of computer structures", Computer Engineering: a DEC view of Hardware System Design, Digital Press, 1978.

[WALZ80] H.V.Walz, R.Downing: "FASTBUS snoop diagnostic module", Nuclear Science and Nuclear Power System Symposium, Orlando, Florida, Nov. 1980.

[PLAT81] B.Plattner, J.Nievergelt: "Monitoring program execution, a survey", Computer, Nov. 1981.

5.3 Software Tools

[BOOK] P.B.Hansen: "The architecture of concurrent programs", Prentice Hall, 1977.

[BOOK] ed. Hasegawa: "Real-time programming", Proc.of IFAC/IFIP Aug. 1981, Pergamon Press.

[BOOK] ed. Novak: "Software for computer control", Proc.of IFAC/IFIP Symp., June 1979, Pergamon Press.

[DICK76] Dickey: "Distributed computer systems", Distributed Systems INFOTECH, 1976.

[FRIE76] Frieder: "Cooperative module architectures - interfaces, internals and inferences", Distributed Systems INFOTECH, 1976.

[BASS78] C.Bass: "PLZ : A family of System Programming Languages for Microprocessors", Computer, March 1978.

[BROW78] W.Brown: "Modular Programming in PL/M", Computer, March 1978.

[KAHN78] K.C.Kahn: "A small scale Operating System Foundation for Microprocessor Applications", IEEE Proceedings, Vol.66, Febr. 1978.

[LIND80] F.Van Der Linden, I.Wilson: "Real time executives for Microprocessors", Microprocessors and Microsystems, Vol.4, n.6, Aug. 1980.

[CRES80] S.Crespi Reghizzi, P.Corti, A.Dapra': "A survey of Microprocessor Languages", Computer, Jan. 1980.

[MORR80] J.Morris: "Real-Time process Control in the '80s", Microprocessors and Microsystems, Vol.4, n.9, Nov. 1980.

5.4 Multiprocessors

[BOOK] B.A.Bowen, R.J.A.Buhr: "The logical design of multiple-microprocessor systems", Prentice Hall ,1980.

[JENS74] E.D.Jensen: "A distributed function computer for Real-Time control", Proc.2nd Symposium on Computer Architecture, 1974.

[ANDE75] G.A.Anderson, E.D.Jensen: "Computer interconnection structures: taxonomy, characteristics and examples", ACM Computing Surveys, Vol.7, n.4, Dec. 1975.

[SIEW76] Sieworek, M.Barbacci: "Modularity and multiprocessor structures -some open problems in the construction and utilization of mini- and microprocessor networks", Distributed Systems INFOTECH, 1976.

[BAER76] Baer: "Modelling distributed function architectures", Distributed Systems INFOTECH, 1976.

[ENSL77] P.H.Enslow: "Multiprocessor Organization - A Survey", ACM Computing Surveys, Vol.9, n.1, March 1977.

[WATK77] D.C.J.Watkins: "Distributed processing - a logical approach to real-time, on-line systems", IEE Intern.Conf.on Distributed Computer Control Systems, 1977.

[LIPO78] G.J.Lipovsky, K.L.Doty: "Developments and directions in computer architecture", Computer, Aug. 1978.

[MOTT79] D.R.Mott: "A distributed computing architecture for real-time system control and information processing", Proc.1st Inter.Conf. on Distributed Computing Systems, Huntsville, Alabama,1979.

[SCHM80] G.Schmidt, W.Sendler: "A failure tolerant multi-microcomputer controller for process control application", Process Automation, n.2, 1980.

[BRIG81] F.A.Briggs, M.Dubois, K.Hwang: "Throughput analysis and configuration design of shared-resource multiprocessor system: PUMPS", Proc.8th ann. Symp. on Computer Architecture, 1981.

[ARNO82] R.G.Arnold, R.O.Berg, J.W.Thomas: "A Modular Approach to Real Time Supersystems", IEEE Transactions on Computers, Vol. C-31, n.5,May 1982.

CHAPTER 3

HARDWARE AND SOFTWARE ASPECTS OF CONTROL APPLICATIONS OF MICROPROCESSORS

Daniel Tabak
Dept. of Electrical and Computer Engineering
Ben Gurion University of The Negev
Beer Sheva, Israel

Abstract. The principles of hardware realization and the preparation of software for microprocessor-based digital controllers are presented. A general, representative block diagram of a microprocessor-based digital controller is given along with a description of its subsystems, their tasks and their properties. Problems of sampling period selection, timing organization and the quantization effect, are discussed. Alternatives in hardware realization are mentioned. Basic steps of software preparation, requirements for reliable software and language selection considerations are presented.

1. INTRODUCTION

Since the advent of the microprocessors in the early seventies, a considerable amount of research and development effort has been dedicated to explore their applicability in control. Now, at the end of 1982 there exists a vast amount of literature, reporting case studies, general design considerations and surveys of a variety of aspects of the implementation of microprocessors in different areas of control. Only a small sample of the existing literature is mentioned here (Auslander et al. 1978, Mayman and Tabak 1978, Neuman and Baradello 1979, House 1980, Tabak and Lipovski 1979, Tabak and Lipovski 1980, Bibbero 1977, Chen 1982a, b, Moroney et al. 1980, Farrar and Eidens 1980, Walter 1982, Nechmadi and Tabak 1981). From the control theory point of view there is no basic difference if a microcomputer or any other computer is used to implement the controller. The preliminary analysis and control design of the system would be the same. For linear systems we would use the z-transform and possibly the bilinear transform in the w- or r-domain, and eventually obtain a z- transformed transfer function of the digital controller D(z), to be placed in the control loop. The methods and techniques to be used in such an analysis have been well documented in textbooks, such as Kuo 1980, Franklin and Powell 1980, Cadzow and Martens 1970, and many others. These general aspects of digital control will not be discussed in this article. It will be assumed that the reader already has a block diagram of a control system, with a plant

S. G. Tzafestas (ed.), Microprocessors in Signal Processing, Measurement and Control, 33–44.

transfer function G(s), an established sampling period T, and a digital controller D(x). It will be assumed that the system performance satisfies the design specifications. This can be confirmed by simulation. The purpose of this article is to deal with the next step of the design: the implementation of the digital controller D(z) and its interfaces (interconnections) with the controlled process G(s), using microprocessor-based equipment. While making the next step of actual implementation, several practical problems, involving hardware and software, arise. These problems and some possible solutions will be discussed in this article. The hardware aspects - in section 2, the software aspects - in section 3. Conclusions will be given in section 4. It will be assumed that the reader is familiar with some basic notions of microcomputer architecture. There exists by now a large amount of literature on this topic, such as Leventhal 1978, Wakerly 1981.

2. HARDWARE ASPECTS

2.1 General System Structure

A block diagram of a typical microprocessor-based feedback control system is shown in Fig. 1 (Kuo 1980, Biehl and Ditzinger 1981). The feedback control system can be divided into two main parts:
(a) The analog part, consisting of the controlled process, represented by its transfer function G(s), a measurement system, which transmits the measurable process variables to the controller, and possibly an amplifier, in case the controller signals are too low to activate the controlled process.
(b) The digital, microprocessor-based controller system, consisting of a microcomputer, endowed with additional subsystems needed in digital control: A/D and D/A converters, a sample and hold unit and a timer, particularly geared to the sampling rate. The regular microcomputer components are: the microprocessor, which is a Central Processing Unit (CPU), a Read Only Memory (ROM), a Random Access Memory (RAM), and Input/Output (I/O) Ports.

The digital controller system and its interfaces with the analog part (the controlled process) will be the main topic of the discussion of this section. The digital controller subsystem is a special purpose microcomputer, whose main tasks are:
(1) Compare the actual feedback signals with the user supplied inputs and calculate the error signals (the differences), e.
(2) Calculate the control signals, u, based on the error values and the digital controller z-transformed transfer function D(z). In particular, a PID controller (Kuo 1980) may be implemented, or any other control scheme, supplied by the designer.
(3) Supply the interface between the analog signals of the controlled process and the discrete, binary-coded signals within the microcomputer.
(4) Perform, if necessary, additional tasks of data acquisition and processing, connected with the controlled process, but not necessarily connected with the feedback control loop.

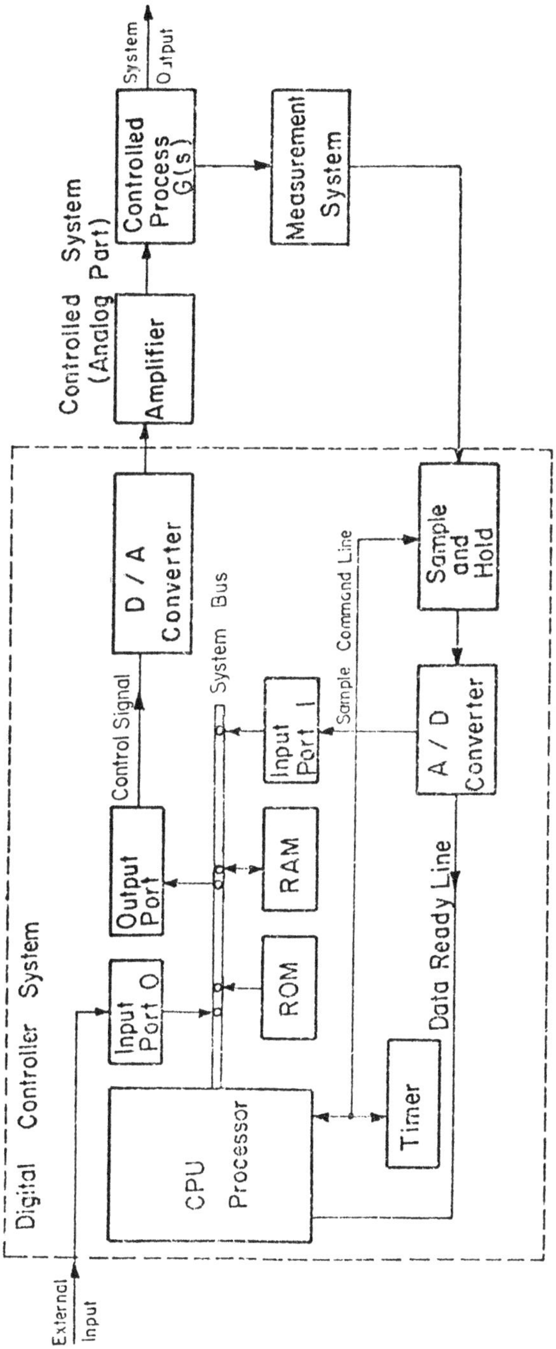

Fig.1 A Microprocessor - Based Control System

(5) In addition to the above, the microcomputer may be used, on a time-shared basis, for other computing tasks, unconnected with the control process.

The CPU (microprocessor) performs all of the calculations and supplies the control for the whole digital system. The ROM usually contains the fixed part of the controller program, while the RAM contains the variable data. In particular cases (Intel 8051, Motorola 6801), the microprocessor contains some ROM and RAM on the same chip, and sometimes, in very simple control schemes it may suffice, to the extent that no additional ROM or RAM chips will be necessary. In general, we will have a number of ROM and RAM chips, connected to the CPU through a data/address/control system bus (Fig. 1). On the same bus we have also I/O ports. In the system, shown in Fig. 1, we have a single output port to transmit the calculated control signal. We have two input ports: one for the feedback signals and the other for external inputs. In other systems, the number of I/O ports may be different, depending on the system's tasks and implementation. It should be noted that in general we may have a number of variables, shown in each path (line) in Fig. 1.

The measured feedback variables are sampled at the sampling frequency and held at a constant value for the duration of a sampling period by the sample and hold device (Kuo 1980), which is considered to be a part of the digital controller system. Subsequently, these signals are converted to digital, binary coded numbers by the Analog to Digital Converter (A/D Converter or ADC). The feedback signals, in their digital form, are fed into the CPU through Input Port 1 and the system bus. The control signals, calculated by the CPU are put out through the Output Port to the Digital to Analog Converter (D/A Converter or DAC) which converts them into an analog signal (voltage level). The details of the specific design of ADC or DAC are outside of the scope of this article; the interested reader is referred to the existing literature (Kuo 1980, Hoeschele 1968, Sheingold 1972, Schmid 1970). Input signals from the outside, transmitted through a keyboard or an outside communications bus, are transferred to the CPU through the Input Port 0 and the system bus. These input signals may represent any microcomputer input, needed for its continued operation. In particular, these inputs may constitute the desired values of the controlled variables, to be compared with the measured variables. The timer circuit controlled by the CPU is designed to fit the sampling period of the system. By issuing appropriate timing signals it determines the start of the next sampling period.

The details of the operation of some of the subsystems mentioned above and the problems associated with them, will be discussed in the subsequent paragraphs.

2.2 The Sampling Period

The selection of the sampling period, T, (or sampling frequency ω_s),

of the digitally controlled system, is conducted in two main stages:
(I) During the control loop design.
(II) A further refinement may be needed during the design of the digital controller.
Stage I is conducted according to well established principles of the design of discrete time systems (Kuo 1980, Franklin and Powell 1980, Cadzow and Martens 1970). This design would yield, among other things, the transfer function of a digital controller D(z), expressed in the z- domain, for a specific sampling period T. A formal, first step criterion for the selection of T in stage I is the sampling theorem, according to which, T should be no more than a half of the smallest period or time constant, inherent in the controlled process transfer function G(s) (or: the sampling frequency should be at least twice the highest frequency harmonic of G(s). A rule of thumb, given by Franklin and Powell 1980, states that the sampling frequency should be from 4 to 20 times the closed loop bandwidth of the system. This corresponds approximately to rules of thumb given by other authors.

When we start stage II, we already have a tentative value for T, based on the application of the above rule of thumb and the sampling theorem. At this stage, the first and foremost consideration is the detailed analysis of the computing tasks of the microcomputer. Particular attention should be given to the computing tasks which are to be performed periodically, during each sampling period (such as calculation of the error function and the realization of the digital controller transfer function). If the total computation time is T_c, we should always preserve the inequality $T_c < T$. The analysis, leading to the establishment of the computing time T_c, should also take into account the conversion times of the ADC and DAC, as well as the delay, introduced by the sample and hold circuit (Kuo 1980, Tsuchiya 1982, Tendulkar and Muller 1980). If $T_c > T$, the sampling period T should be appropriately increased.

At this stage, the following trade-off considerations should be taken into account. The larger T, the more adverse effect it has on the feedback system's stability (Kuo 1980). The larger T, the smaller are the stability margins of the system. This consideration forces the designer to keep T as small as possible, while preserving the inequality $T_c < T$. On the other hand, cost considerations force the designer to increase T, or to lower the sampling frequency (Franklin and Powell 1980). If T is large, a slower and therefore a less costly microcomputer can be used. Moreover, for a sufficiently large T, we can use ADC and DAC with a lower conversion speed, which will in turn decrease the cost. The final decision, taking into account the above trade-offs, should be made by the designer according to the particular properties, specifications and constraints, involved in the system under consideration. A general solution, which would fit all possible cases, can not be offered.

2.3 Timing

The microprocessor has to complete the calculation of the error function, the control output, all signal transfers and conversions within one sampling period. In addition, other tasks may be imposed on it by the user. A new sampling period can not start until all of the tasks of the current period are completed. Therefore, the organization of timing is an important feature of a digital control system.

Because of the importance of the timing, a special timer circuit is usually implemented (Fig. 1). The timer determines the start of the next sampling period by transmitting a pulse every T seconds (Kuo 1980). The pulse is used as follows:
(a) It performs an interrupt of the CPU. The currently running routine is stopped. A timer interrupt service routine is started, which transmits the calculated control signal, through the output port to the DAC and then to the amplifier.
(b) It is simultaneously transmitted through the Sample Command Line to the Sample and Hold circuit, causing the current measured signals to be sampled and held for one sampling period T. The value of the measured signals is then converted to a binary code by the ADC.
As soon as the ADC completes the conversion it sends a control signal on the Data Ready Line to the CPU, which reads in the converted signals through the Input Port 1. Within the same sampling period the external input (if any) is read in through the Input Port 0 and the next control output is calculated. The new control output is transmitted out upon the arrival of the next timer interrupt, at the beginning of a new sampling period.

An alternative to a hardware implemented timer would be to use timing loops, realized by software. Additional timing arrangements may be added, by software, within the sampling period. For instance, in case of multiple signal lines there may be crosstalk problems in both the ADC and DAC. Both ADC and DAC have a finite settling time, which has to be taken into account. A simple solution for the above, reported by Chen 1982b, is to use several identical IN (Input) or OUT (Output) instructions in a row, wherever applicable. In other cases, a timing modification can be achieved by using a number of NOP (No Operation) instructions. An alternative solution to minimize the output crosstalk is to use a dummy output (if an unused channel is available) line, prior to outputting the signal to the desired channel (Chen 1982b).

2.4 The Quantization Effect

The variables, dealt with by the digital controller (measured variables, errors, control variables, inputs) are represented in a discrete, binary code of a finite word length. We will denote the word length of the controller, excluding the sign bit, by n bits. The range of all variables can be scaled in such a manner that the variables can be considered to be positive numbers. In this case, the largest number, represented by an n-bit word (fixed point binary notation) is $2^n - 1$, and the smallest, nonzero number is 1. If on the other hand we change our scaling so that the number which was 2^n in the previous scaling, will now represent 1.0, the smallest nonzero number will now be 2^{-n} (or $1/2^n$). This number, 2^{-n}, represents the resolution of the n-bit microcomputer word. It also represents the weight of the Least Significant Bit (LSB) of the word with respect to its full scale (FS) value, which in this case is one (1.0). By increasing the word length, that is, by increasing the number of bits per word, n, we improve the general precision of number representation, by improving the resolution. However, it should be noted that in order to improve the resolution by increasing the number of bits, n, the full scale (FS) analog and digital signal values should be maintained the same (Kuo 1980).

By performing an A/D conversion we represent the analog signal values by their n-bit binary equivalents. In other words, we quantize the analog signals, with a quantization level q, equal to the resolution of the binary representation: $q = 2^{-n}$. Naturally, due to the quantization procedure, the value of the discrete signal, obtained after the A/D conversion, is different from the analog value. Thus, a quantization error, which is the difference between the values of the analog and digital signal, is introduced. It can be shown (Kuo 1980, Franklin and Powell 1980) that the maximal absolute value of the quantization error is q/2. In other words, the quantization error is between -q/2 and q/2.

Example. Suppose the FS is 1.0V and n = 4. The resolution is $1/2^4 = 1/16$; the quantization level, q = FS/16 = 1/16 = 0.0625V. The maximum absolute value of the quantization error: q/2 = 0.03125. The maximal 4-bit binary number, representable in this system is 1111, which is equivalent to 15/16 FS, or: 0.9375V. Therefore, the maximal value of an analog signal that can be converted, without exceeding the quantization error, is:

$$15/16 \text{ FS} + q/2 = 0.9375 + 0.03125 = 0.96875\text{V} \quad . \square$$

The practical effect of the quantization error is that all of the system variables and equations coefficients are represented within the digital part of the system, at values which are different (by the quantization error at most) from the actual values, as they appear in the analog part, or as they were represented in the preliminary analysis. In other words, once we implement the digital control system, we are really dealing with a system which is different from the one analyzed and designed in the previous stage. Since the system is now different, its dynamic behaviour is now different from the one, expected by the design. This phenomenon should be taken into account in a subsequent stage of the design. The quantization error can be represented as an external noise source of amplitude q/2. Such an external noise source can be introduced into the block diagram of the system and its influence on the system performance can then be evaluated (Kuo 1980).

Example. Consider a simple discrete-time feedback system, represented in Fig. 2. Assume that the input is zero, r(k) = 0. The steady

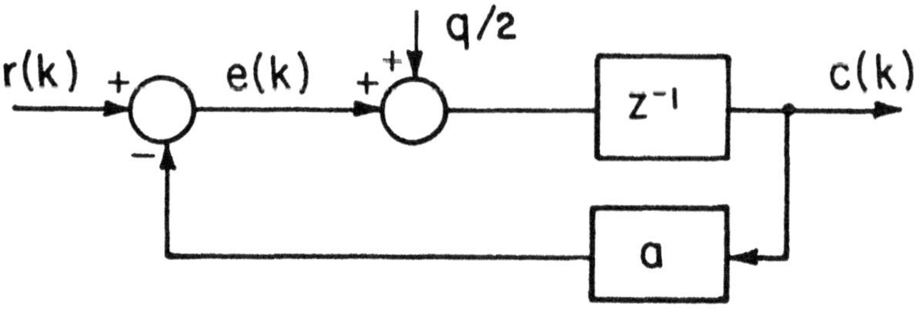

Fig. 2 A simple discrete - time system

state influence of the quantization error, q/2, represented as an external noise source, on the system output, c(k), is:

$$c_s = \lim_{k\to\infty} c(k) = \lim_{z\to 1} (1-z^{-1})\, C(z) = \lim_{z\to 1} \frac{z-1}{z} \frac{z^{-1}}{1+az^{-1}} \frac{qz/2}{z-1} =$$

$$= q/2(a+1)$$

Some c_s values for different values of parameter a:

a	1	1/2	1/3	1/10	0	-1/10	-1/3	-1/2	-9/10
c_s/q	0.25	0.33	0.375	0.45	0.50	0.55	0.75	1.0	5.0

Of course, as a approaches -1, the system approaches the limit of stability. In any case, such a model permits us to analyze the effects of the quantization error q/2. □

Another effect of the quantization is the introduction of an extra nonlinearity (of the quantizer type) into the feedback loop. The introduction of a nonlinearity may influence the stability of the system. In some cases it may cause oscillations in an otherwise stable system. The effects of this type of nonlinearity in systems using 8- and 16-bit microprocessor equipment, has been investigated by Maymon and Tabak 1978, by the describing function approach. For the class of systems investigated, the destabilizing effect was negligible.

2.5 Alternative System Configurations

The block diagram of a microprocessor-based control system should be regarded as a representative particular case. Many variations of this configuration are possible. For instance, the same microcomputer may serve several control loops. All of the subsystems of the digital controller, shown in Fig. 1, may be included within a single chip, such as the Intel 2920. A detailed design of a digital controller, including its software, implemented on a 2920 chip, was recently presented by Chen 1982b.

Another alternative, particularly when the controller tasks involve numerous, floating point calculations, would be to use a special numeric co-processor, along with the CPU microprocessor. For instance, one could use an Intel iAPX86 (8086) microprocessor, along with the floating point 8087 chip, thus forming an iAPX86/20 system. Many other digital controller configurations can be formed, depending on the system's needs and performance specifications.

3. SOFTWARE ASPECTS

The preparation of software for a digital control system is one of the most crucial tasks. The control system that we design and implement should be reliable, and naturally this calls for a reliable software as well. We can summarize the basic requirements for reliable software as follows (Lambrechs and Rodd 1982, MacLeod 1982, Nechmadi and Tabak 1982, Myers 1975, 1979) :

(a) The software should have clear, well documented specifications.
(b) The software should be well designed, legible, with clear and meaningful documentation.
(c) The software should be well structured; preferably with a tree structure.
(d) The software should be prepared in a completely modular structure.
(e) The software should be designed in a top-down fashion. This allows a systematic breakdown of the functional specifications into smaller units, until a level is reached in which each unit consists of a single indivisible function which can be defined as a module.

One of the important decisions to be made is whether to use Assembly or a High Level Language (HLL). The use of HLL yields more compact, cheaper, more legible and more reliable programs. On the other hand, once an HLL program is compiled it may not yield the shortest and the most efficient machine language version. As a result, more memory storage may be required, resulting in turn in a more expensive implementation. The rule of thumb in such a case should be, that whenever we have a relatively short and simple software task - it should be programmed directly and efficiently in assembly language. However, if the task is lengthy and complicated, the use of HLL is recommended. Although the initial investment may be high, it will pay off in the long run. Even the initial cost will be reduced, because HLL programming requires a considerably smaller effort. It frees the programmer from worries about memory organization, transfers of data among CPU registers, and other tasks, which appear in assembly language programming. Moreover, HLL programs are easier to debug, modify and to document.

The software preparation, for long and more elaborate programs, should go according to the following recommended steps (Nechmadi and Tabak 1982):
(a) Analysis of the software tasks and performance specifications.
(b) Establish modules, procedures and variables which will appear in the software package. Assign names to all of the above, in a manner which will express their nature and function. This will permit their easy identification in the future.
(c) Prepare a variables table. Every row in such a table should contain the variable's name, its nature and other pertinent data.
(d) Prepare a list of all of the procedures and their tasks for each module.
(e) Prepare detailed flow diagrams for each module.
(f) Prepare a list of variables which are to be used by several modules. This is important for both the debugging and the documentation of the software package.
(g) Complete the coding of all modules, placing comments abundantly all along the programs.

The software development for the microprocessor-based controller should be performed with the help of a microcomputer development system, such as the Intel MDS 231 (Leventhal 1978). A typical development system includes the following software subsystems:
(1) Monitor - a program which supervises the whole system, including the I/O. It organizes the memory and transfers commands to various programs.
(2) Editor - a program which organizes any text, which consists of a set of coded characters (such as ASC II code), into lines and blocks.
(3) Debugger (Simulator) - a program which analyzes the object code of the control program. It stops the object code in various control points (break points), advances the program in single steps and checks the contents of the registers and the memory.
(4) Compilers and/or Assemblers, depending on the computer languages implemented.
(5) Cross assembler - a program which translates programs written in an assembly language to the machine language of another (host) computer.
(6) In Circuit Emulator (ICE) - a mix of software and hardware which enables the user to debug, in real time, the written software, together with some parts or whole hardware subsystems.

4. CONCLUDING COMMENTS

The hardware and software aspects of the realization of microprocessor-based controllers have been surveyed and discussed. Many of the issues raised are generally applicable to computer-based control, and not necessarily to microcomputers alone. With the advent of modern and powerful 32-bit microprocessors, such as the Intel 43201, 43202 chips of the iAPX432 system, or the National Semiconductors 16032 chip of the NS16000 system, the differences between micro-, mini-, and mainframe computers, will gradually disappear. We will continue to discuss computer-based control; the general principles are still the same.

REFERENCES

Auslander, D. M., Takahashi, Y., and Tomizuka, M.: 1978, Proc. IEEE, 66, pp. 199-208.

Bibbero, R. J.: 1977, Microprocessors in Instruments and Control, Wiley, NY.

Biehl, G., and Ditzinger, A.: 1981, Microprocessing and Microprogramming, 7, pp. 326-333.

Cadzow, J. A., and Martens, H. R.: 1970, Discrete-Time and Computer Control Systems, Prentice-Hall, Englewood Cliffs, N.J.

Chen, C. S.: 1982a, Control Systems Magazine, 2, No. 1, pp. 17-19.

Chen, C. S.: 1982b, Control Systems Magazine, 2, No. 3, pp. 16-22.

Farrar, F. A., and Eidens, R. S.: 1980, IEEE Trans. on Automatic Control, AC-25, pp. 461-468.

Franklin, G. F., and Powell, J. D.: 1980, Digital control of Dynamic Systems, Addison Wesley, Reading, MA.

Hoeschele, D. F.: 1968, Analog to Digital/Digital to Analog Conversion Techniques, Wiley, NY.

House, C. H.: 1980, Computer, 13, No. 12, pp. 35-49.

Kuo, B. C.: 1980, Digital Control Systems, Holt, Rinehard and Winston, NY.

Lambrechs, J. S. D., and Rodd, M. G.: 1982, Highly Reliable Software for use in Fail-Safe Control, in Proc. of 3rd IFAC/IFIP Symp. on Software for Computer Control, Madrid, Oct. 1982, pp. 389-394.

Leventhal, L. A.: 1978, Introduction to Microprocessors, Prentice-Hall, Englewood Cliffs, NJ.

MacLeod, I. M.: 1982, The Application of Modern Software Engineering Techniques in the Development of a Process Control Package, in Proc. of 3rd IFAC/IFIP Symp. on Software for Computer Control, Madrid, Oct. 1982, pp. 395-399.

Maymon, E., and Tabak, D.: 1978, Int. J. Systems Sc., 9, pp. 827-835.

Moroney, P., Willsky, A. S., and Haupt, P. K.: 1980, IEEE Trans. on Automatic Control, AC-25, pp. 621-630.

Myers, G. J.: 1975, Reliable Software Through Composite Design, Petrocelli/Charter, NY.

Myers, G. J.: 1979, The Art of Software Testing, Wiley, NY.

Nechmadi, M., and Tabak, D.: 1981, Microcomputer-Based Laser Control and Synchronization System, in Proc. IFAC 8th Triennial World Congress, Kyoto, Japan, Aug. 1981, Vol. XI, pp. 70-76.

Nechmadi, M., and Tabak, D.: 1982, Software for a Microcomputer Controlled Laser Synchronization System, in Proc. of 3rd IFAC/IFIP Symp. on Software for Computer Control, Madrid, Oct. 1982, pp. 495-499.
Neumann, C. P., and Baradello, C. S.: 1979, IEEE Trans. on Systems, Man and Cybernetics, SMC-9, pp. 856-860.
Schmid, D.: 1970, Electronic Analog/Digital Conversion, Van Nostrand Rheinhold, NY.
Sheingold, D. M. (ed): 1972, Analog/Digital Conversion Handbook, Analog Devices Inc., Norwood, MA.
Tabak, D., and Lipovski, G. J.: 1979, Microcomputers in Control, in Proc. 1979 JACC, Denver, CO, June 1979, pp. 30-33.
Tabak, D. and Lipovski, G. J.: 1980, IEEE Trans. on Computers, C-29, pp. 180-190.
Tendulkar, G.A., and Muller, B.: 1980, Microprocessor Architecture for Industrial Controller, in Sami, M., Thompson, L. and Mezzalira, L. (eds), Microprocessor Systems, North Holland, Amsterdam, pp. 233-242.
Tsuchiya, T.: 1982, IEEE Trans. on Automatic Control, AC-27, pp. 295-306.
Wakerly, J. F.: 1981, Microcomputer Architecture and Programming, Wiley, NY.
Walter, C.: 1982, Functional Programming for Discrete Process Control, in Proc. of 3rd IFAC/IFIP Symp. on Software for Computer control, Madrid, Oct. 1982, pp. 377-382.

Part II

SIGNAL PROCESSING

CHAPTER 4

DESIGN AND IMPLEMENTATION OF THE DIGITAL FILTER VIA THE SIGNAL PROCESSOR

C. S. Chen
Electrical Engineering Department
The University of Akron
Akron, Ohio 44325, USA

ABSTRACT

Basic IIR and FIR digital filters design techniques are reviewed, with emphasis on various analog-to-digital conversion methods for deriving IIR filter transfer functions, as well as windowing and frequency sampling methods for FIR filters design. The computational algorithm of filtering is then realized in recursive and nonrecursive network structures for hardware implementation. The recently developed INTEL 2920 signal processor is used for demonstrating the implementation of the four basic operations of digital filtering: input/output, add/subtract, multiply and unit delay. Key features of other similar processors are also listed.

I. INTRODUCTION

A digital filter can be viewed as a computational algorithm implemented using finite-precision arithmetics. It processes a sequence of numbers called input and produces an output sequence with the desired properties called for by the algorithm. The design of the digital filter involves the following three procedures: first, the desired filter characteristics are specified and next, a relation is derived relating the input-output (I/O) sequences that meet the desired filter specifications at least approximately and finally the I/O relation is realized using finite-precision arithmetics.

In this chapter, we start off with the IIR (infinite-impulse-response) and FIR (finite-impulse-response) filter design. In IIR filter design, we assume that the analog filter transfer function satisfying the desired specification is already available, various analog-to-digital transformation techniques are then described to derive the digital filter transfer function. This approach is taken because the analog filter design methods are well established. For FIR filter design, windowing method and frequency sampling method are covered.

S. G. Tzafestas (ed.), Microprocessors in Signal Processing, Measurement and Control, 47–73.

After the digital transfer function is derived, different realization techniques of the transfer function are described, the effects of quantization error and the finite-word length are discussed.

The computational algorithm that realizes a given digital transfer function can be implemented on a digital processor. In Section V, we illustrate the use of a signal processor - Intel 2920 in implementing the digital filter. Intel 2920 can be considered as a special purpose microcomputer for signal processing. Since it samples and digitizes analog input signals, processes the sampled values in the digital processor and outputs analog signals, it is also known as analog processor. Other commercially available products such as signal processing peripheral S2811 of American Microsystems Inc., signal processing interface μpD7720 of NEC and signal processing computer TMS 3020 of TI can also be used for implementing the digital filter. The listing of key features of those devices is given in Appendix.

II. IIR FILTER DESIGN

A filter the impulse response of which has infinite duration is called infinite-impulse-response (IIR) filter. On the other hand, if the impulse response is of finite duration, the filter is called finite-impulse-response (FIR) filter.

The design of IIR digital filter can be easily accomplished by first designing prototype analog filter meeting a set of given specifications and then converting the analog filter transfer function to digital transfer function. This approach takes full advantage of the established and well documented analog filter synthesis procedures [1,2,3]. There are several techniques available for the conversion and each method will result in an approximate digital transfer function with different frequency characteristics, step responses and impulse responses. A particular method is usually chosen on the basis of the desired frequency characteristics, impulse response, and/or step response.

In the following we shall describe methods which can be used to convert an analog filter transfer function to a digital transfer function.

II.1 Impulse-Invariance Method

For an analog filter with transfer function $G_a(s)$, the impulse response $g_a(t)$ is the inverse Laplace transform of $G_a(s)$, that is

$$g_a(t) = L^{-1}[G_a(s)]$$

By this method we are to find a digital transfer function $G(z)$ such that the unit-sample response $g(KT)$ matches $g_a(KT)$.

If a transfer function $G_a(s)$ is written as

$$G_a(s) = \sum_{\ell=1}^{n} \frac{A_\ell}{s-s_\ell} \tag{1}$$

where s_ℓ represents one of the n poles of $G_a(s)$. Then the impulse response of the continuous system is

$$g_a(t) = \sum_{\ell=1}^{n} A_\ell \exp(s_\ell t)U(t)$$

since we want to match the unit-sample response g(kT) to $g_a(t)$ at t=kT, k=0, 1, 2,..., we have

$$g(kT) = g_a(kT) = \sum_{\ell=1}^{n} A_\ell \exp(s_\ell kT)U(kT)$$

$$= \sum_{\ell=1}^{n} A_\ell (\exp(s_\ell T))^k U(kT)$$

The digital transfer function G(z) is obtained by taking Z-transformation of g(kT),

$$G(z) = \sum_{\ell=1}^{n} \frac{A_\ell}{1-\exp(s_\ell T)z^{-1}} \tag{2}$$

The following comments could be made concerning this method:

1. A pole $s=s_\ell$ in $G_a(s)$ is transformed to a pole $\exp(s_\ell T)$ in G(z). Thus, if $G_a(s)$ is the transfer function of a stable filter in the s-domain, the poles of G(z) will be all within the unit circle and G(z) represents the transfer function of a stable digital filter.

2. The coefficients in the partial-fraction expansion of $G_a(s)$ and G(z) as shown in eqs. (1) and (2) are all equal.

3. Whereas pole at $s=s_\ell$ in the s-domain are mapped into the z-plane by the relation $z=\exp(s_\ell T)$, zero in the z-plane depends on the coefficients of the partial fraction expansion and the poles.

4. The DC gains $G_a(o)$ and G(1) of the analog filter $G_a(s)$ and the

digital filter G(z) are in general different.

5. The sequence g(kT) could be considered as the sampled values of $g_a(t)$ and since the original filter frequency response $G(z)\big]_{z=\exp(jwT)}$ is related to the analog filter frequency response $G_a(jw)$ by the equation [4,5,6]

$$G(z)\Big]_{z=\exp(jwT)} = \frac{1}{T}\sum_{m=-\infty}^{\infty} G_a(jw + j\,\frac{2m\pi}{T}) \qquad (3)$$

The choice of the sampling period T must be made to minimize the aliasing effect. Thus, this method is applicable to low pass filters only and is not suitable for high-pass and/or band-stop filters.

II.2 Step-Invariance Method

In this method, one seeks a z-domain transfer function G(z) such that its step response $S_d(kT)$ matches the continuous step response Sa(t) at the discrete sampling instants kT, for k=0,1,2,... That is

$$S_d(kT) = Sa(t),\ t=kT \text{ and } k = 0,1,2,\ldots$$

To derive such a function G(z), assume again that $G_a(s)$ can be expressed as

$$G_a(s) = \sum_{\ell=1}^{N} \frac{A_\ell}{s-s_\ell} \qquad \text{its step response is}$$

$$Sa(t) = L^{-1}\left(\frac{1}{s}G_a(s)\right) = L^{-1}\left(\sum_{\ell=1}^{N}\frac{A_\ell}{s(s-s_\ell)}\right)$$

$$= \sum_{\ell=1}^{N}\frac{1}{s_\ell}L^{-1}\left(\frac{A_\ell}{s-s_\ell} - \frac{A_\ell}{s}\right) = \sum_{=1}^{N}\frac{A_\ell}{s_\ell}(\exp(s_\ell t)-1)U(t)$$

and $$S_d(kT) = S_a(kT) = \sum_{\ell=1}^{N}\frac{A_\ell}{s_\ell}(\exp(s_\ell kT)-1)U(kT)$$

Therefore, the Z-transform of the unit-step response is

$$S_d(z) = Z(S_d(kT) = \sum_{\ell=1}^{N}\frac{A_\ell}{s_\ell}\left(\frac{1}{1-\exp(s_\ell T)z^{-1}} - \frac{1}{1-z^{-1}}\right)$$

$$= \sum_{\ell=1}^{N} \frac{A_\ell}{s_\ell} \left(\frac{z^{-1} \quad (\exp(s_\ell T)-1)}{(1-z^{-1})(1-\exp(s_\ell T)z^{-1})}\right)$$

Since $Z(\text{unit step})\cdot G(z) = S_d(z)$

$$\left(\frac{1}{1-z^{-1}}\right) G(z) = \sum_{\ell=1}^{N} \frac{A_\ell}{s_\ell} \left(\frac{z^{-1}(\exp(s_\ell T)-1)}{(1-z^{-1})\ (1-\exp(s_\ell T)z^{-1})}\right)$$

thus

$$G(z) = \sum_{\ell=1}^{N} \frac{A_\ell}{s_\ell} \left(\frac{\exp(s_\ell T)-1}{z-\exp(s_\ell T)}\right) \tag{4}$$

The remarks on this method are:

1. A pole $s=s_\ell$ of $G_a(s)$ is transformed to a pole $\exp(s_\ell T)$ of $G(z)$. Thus, if $G_a(s)$ is the transfer function of a stable filter in the s-domain, the poles of $G(z)$ will be all within the unit circle, and $G(z)$ represents the transfer function of a stable discrete filter.

2. The DC gains $G_a(0)$ and $G(1)$ of the analog filter and the digital filter are the same, for from equations (1) and (4),

$$G_a(0) = G(1) = \sum_{\ell=1}^{N} - \left(\frac{A_\ell}{s_\ell}\right).$$

3. As in the case of impulse-invariance method, the choice of the sampling period T must be made to minimize the aliasing effect.

4. This is a useful method if one intends to retain the desirable step response of the analog filters.

II.3. Backward Difference Method

Consider a transfer function $G_a(s)$ expressed as the ratio of two polynominals

$$G_a(s) = \frac{Ya(s)}{X\ (s)} = \frac{\sum_{k=0}^{M} d_k s^k}{\sum_{k=0}^{N} C_k s^k}$$

where $X(s)$ and $Ya(s)$ are the Laplace transforms of the input and output of the filter.

In time domain, the input-output of the system is related by the following differential equation,

$$\sum_{K=0}^{N} C_k \frac{d^k ya(t)}{dt^k} = \sum_{K=0}^{M} d_k \frac{d^k x(t)}{dt^k} \quad (5)$$

Suppose the first derivative is approximated by the first backward difference $\nabla^{(1)}$, called the first order shift operator,

$$\left.\frac{dya(t)}{dt}\right]_{t=nT} \rightarrow \nabla^{(1)} (y(n)) = \frac{y(nT)-y(\overline{n-1}T)}{T} \quad (6)$$

where $y(nT) = y_a(nT)$.

Approximations to the higher-order derivatives are obtained by the repeated application of the first order backward difference,

$$\left.\frac{d^k ya(t)}{dt^k}\right]_{t=nT} = \frac{d}{dt}\left.\left(\frac{d^{k-1}}{dt^{k-1}} y_a(t)\right)\right]_{t=nT}$$

$$\rightarrow \nabla^{(k)}(y(n)) = \nabla^{(1)}(\nabla^{(k-1)}(y(n)) \quad (7)$$

and the original differential equation is approximated by the following equation,

$$\sum_{K=0}^{N} C_k \nabla^{(k)}(y(n)) = \sum_{K=0}^{M} d_k \nabla^{(k)}(x(n)) \quad (8)$$

Denote

$$\nabla^{(0)}(y(n)) = y(n)$$

and note the Z-transforms of $\nabla^{(1)}(y(n))$ and $\nabla^{(k)}(y(n))$ are,

$$Z\,(\nabla^{(1)}(y(n)) = \frac{1-z^{-1}}{T}\, Y(z)$$

and $$Z(\nabla^{(k)}(y(n)) = \left(\frac{1-z^{-1}}{T}\right)^k Y(z)$$

Taking the Z-transform of the difference eq. (8), we have

$$G(z) = \frac{Y(z)}{X(z)} = \frac{\sum_{k=0}^{M} d_k \left(\frac{1-z^{-1}}{T}\right)^k}{\sum_{K=0}^{N} C_k \left(\frac{1-z^{-1}}{T}\right)^k} \quad (9)$$

Comparing $G_a(s)$ with $G(z)$, we can conclude that the digital transfer function $G(z)$ can be obtained from $G_a(s)$ by the substitution of variable,

$$s \text{ by } \frac{1-z^{-1}}{T}, \text{ or } \frac{1}{s} \text{ by } \frac{T}{1-z^{-1}} = \frac{Tz}{z-1} \tag{10}$$

The substitution of the variable $\frac{1}{s}$ by $\frac{T}{1-z^{-1}}$, as indicated in eq. (10) can be interpreted in another way. Consider the integration of a function $r(t)$. The integration of $r(t)$ from 0 to t gives rise to $x_a(t)$

$$X_a(t) = \int_0^t r(\tau)d\tau$$

If $r(\tau)$ is sampled with an interval of time T, and if t is equal to kT, $x_a(t)$ can be written as

$$x(kT) = \int_0^{kT} r(\tau)d\tau = \int_0^{(kT-T)} r(\tau)d\tau + \int_{KT-T}^{kT} r(\tau)d\tau$$

$$= x(kT-T) + \int_{kT-T}^{kT} r(\tau)d\tau \tag{11}$$

In the above integration, if the second term on the right hand side is approximated by a rectangle of width T and amplitude $r(kT)$, then the integral is approximated by the following difference equation,

$$x(kT) = x(kT-T) + TR(kT) \tag{12}$$

The Z-transform of the above difference equation is

$$X(z) = z^{-1}X(z) + TR(z)$$

or

$$\frac{X(z)}{R(z)} = \frac{T}{1-z^{-1}} = \frac{Tz}{z-1} \tag{13}$$

The transfer function shown in eq. (13) is the same as that in eq. (10). Thus the backward difference method can be interpreted as a rectangular approximation to the integration.

II.4. Bilinear Transformation Method

If the second term on the right side of eq. (11) is approximated by a trapezoid with width T and two sides of r(kT-T) and r(kT), eq. (11) then is

$$X(kT) = X(kT-T) + \frac{T}{2}[r(kT-T) + r(kT)]$$

and its Z transform is

$$X(z) = z^{-1} X(z) + \frac{T}{2}(z^{-1} + 1) R(z)$$

or

$$\frac{X(z)}{R(z)} = \frac{T}{2}\left(\frac{z+1}{z-1}\right) \quad (14)$$

which is the trapezoidal approximation to $\frac{1}{s}$, and the transformation $\frac{1}{s} \rightarrow \frac{T}{2}\left(\frac{z+1}{z-1}\right)$ is called a Bilinear transformation. This transformation has the following interesting properties.

1. It maps the entire imaginary axis in the s-plane to the unit circle in the z-plane, the left side of s-plane to the interior of the unit circle.

2. A stable analog filter with poles in the left side of s-plane results in a stable digital filter with poles within the unit circle in the z-plane.

3. Frequency Warping Phenomenon. The frequency is distorted in the transformation. To see this let $s=j\omega$ and $z=\exp(j\omega T)$ = $\exp(j\Omega)$, where $\Omega = \omega T$ is interpreted as digital angular frequency, then

$$s = jw = \frac{2}{T}\left(\frac{z-1}{z+1}\right) = \frac{2}{T}\left(\frac{\exp(j\Omega) - 1}{\exp(j\Omega) + 1}\right) = j\left(\frac{2}{T}\right)\tan(\Omega/_2)$$

$$w = \left(\frac{2}{T}\right)\tan(\Omega/_2) \quad \text{or} \quad \Omega = 2\tan^{-1}\left(\frac{wT}{2}\right) \quad (15)$$

Eq. (15) shows the analog frequency is distorted in the transformation, especially at the high end of the frequency scale. Therefore, the design of the digital filters using this transformation is useful only in cases the frequency distortion can be tolerated or compensated.

II.5. Pole-Zero Mapping Method

Consider a continuous-time transfer function $G_a(s)$ of the form,

$$G_a(s) = \frac{A\sum_{n=1}^{N} (s-s_{on})}{\sum_{m=1}^{M} (s-s_{pm})} \tag{16}$$

The heuristic rules for finding the approximate discrete transfer function G(z) by the pole-zero mapping technique are:

1. All poles of $G_a(s)$ are mapped by $z = \exp(sT)$. For example, a pole at $s = s_{pm}$ of $G_a(s)$ is mapped onto $z = \exp(s_{pm}T)$ of the discrete transfer function G(z).
2. All finite zeros at $s = s_{on}$ of $G_a(s)$ are also mapped by $z = \exp(s_{on}T)$ as in step 1.
3. All zeros of $G_a(s)$ at infinity are mapped in G(z) to the point $z = -1$. For example, $G_a(s)$ of equation (16) has (M-N) zero at infinity, therefore, G(z) should have (M-N) zeros at $z = -1$.
4. The gain of G(z) is selected to match the gain of $G_a(s)$ at some critical frequency. In many applications, the critical frequency is taken as DC, and hence, the gain of G(z) is selected so that

$$G_a(s)\Big]_{s=0} = G(z)\Big]_{z=1}$$

II.6. Digital Frequency Transformation

Other frequency selective filters such as high-pass, band-pass and band-stop filters can be designed by applying appropriate frequency transformation to low-pass digital filters [5,7,8,9,10].

II.7. Computer-Aided Design Method

The digital filter design through the analog-to-digital transformation method is most convenient if the design formulas and/or tables exist for the prototype analog filters, such as Butterworth, Chebyshev or Elliptic filters. However, if a filter is to be designed to match a set of arbitrary frequency specifications, computer-aided design methods should be considered [5,11,12,13,21-27].

III. FIR FILTER DESIGN

III.1. Fir Filter Characteristics

IIR filters designed through analog-to-digital conversion method essentially simulate digitally the prototype analog filter. Although this approach has the advantage that the closed-form design formula for the prototype analog filters are readily available, it also has some disadvantages. One of the disadvantages is that the filters so designed usually have nonlinear phase characteristics and thus the filters are not suitable for situations where linear phase is essential, such as in the case of speech processing application; the other disadvantage is that one can't take full advantage of the computational speed of fast Fourier transfer (FFT) in its implementation.

Therefore, it is important to consider the FIR filter design problem. This class of filters has all its poles located at z=0 and its frequency selective characteristics is attained through proper placement of zeros in the z-plane. It will be shown as an example that exact linear phase filter can be obtained. Efficient realizations of FIR filter exists as both recursive and nonrecursive structures.

III.2. Windowing Design Technique

Consider the problem of designing a digital filter with the frequency response $H(e^{jwT})$, where T is the sampling period. Since $H(e^{jwT})$ is a periodic function of period $\frac{2\Pi}{T}$, it can be expanded in Fourier series

as
$$H(e^{jwT}) = \sum_{-\infty}^{\infty} h_k e^{-jwkT} \tag{17}$$

where
$$h_k = \frac{T}{2\Pi} \int_0^{\frac{2\Pi}{T}} H(e^{jwT})\, e^{jwkT}\, dw \tag{18}$$

The h_k coefficients are recognized as the impulse response (weighting sequence) of the filters. Eq. (17) is an infinite series which means that the filter is an IIR filter.

In order to yield an FIR approximation to the specified $H(e^{jwT})$, the infinite series of eq. (17) is truncated at $k = \pm M$ through windowing. Among the popular windows in common use are rectangular, Hamming and Kaiser windows.

The desired window characteristics are generally: (1) small width of main lobe containing as much of the total energy as possible and (2) side lobes decrease in energy rapidly.

The advantages and disadvantages of various windows are discussed fully elsewhere [5,14].

III.3. Frequency Sampling Technique

The discrete Fourier transform (DFT) coefficients of an FIR filter are related to its N weighting sequence h_n as

$$H_k = \sum_{n=0}^{N-1} h_n \exp(-j\frac{2\Pi}{N} nk) \tag{17a}$$

$$h_n = \frac{1}{N}\sum_{k=0}^{N-1} H_k \exp(j\frac{2\Pi}{N} nk) \tag{18a}$$

The DFT samples, H_k for a FIR sequence can be regarded as samples of the filters Z-transform evaluation at N points equally spaced around the unit circle, i.e.

$$H_k = H(z)\Big]_{z = \exp(j\frac{2\Pi}{N} nk)}$$

Thus the Z-transform of a FIR filter is related to its DFT coefficients as:

$$H(z) = \frac{(1-z^{-N})}{N}\sum_{k=0}^{N-1} \frac{H_k}{1-z^{-1}\exp(j\frac{2\Pi}{N}k)} \tag{19}$$

Thus, from the samples of the frequency response of a FIR filter H_k, k = 0,1.2,...,N-1, the transfer function is obtained as shown in eq.(19).

If we let z = exp(jwT) in eq. (19), we have the frequency response function

$$H(\exp(jwT)) = \frac{1-\exp(-jwNT))}{N}\sum_{k=0}^{N-1}\frac{H_k}{1-\exp(-jwT)\exp(j\frac{2\Pi}{N}k)}$$

$$= \frac{\exp(-j(N-1)WT/2)}{N}\sum_{k=0}^{N-1} H_k \exp(j(N-1)k\Pi/N)\frac{\sin(N(wT-2k\Pi/N)/2)}{\sin((wT-2k\Pi/N)/2)} \tag{20}$$

Eq. (20) indicates that the frequency response evaluated at the sampling frequency $W_k = 2\Pi k/(NT)$ is equal to H_k for k=0,1,2...,N-1. Therefore, a digital filter designed using this method is called a frequency sampling filter. At points other than the sampling frequencies, the frequency response could be obtained through interpolation as indicated in eq. (20).

III.4. Linear Phase FIR Filter

Since the filter weighting sequence h_k is constrained to be real, the magnitude of its frequency response is a symmetric function and the phase an antisymmetric function, that is

$$||H(\exp(jwT))|| = ||H(\exp(-jwT))||, \; 0<wT<\Pi$$

$$\Theta(wT) = -\Theta(-wT), \; \Theta(wt) = \underline{/H(\exp(jwT))}$$

If we desire $\Theta(wT)$ to be linear phase, $\Theta(wT)$ is of the form

$$\Theta(wT) = -\beta wT, \; -\Pi<wT<\Pi$$

where β is a constant phase delay in samples.

It is shown [5] that to satisfy the linear phase condition, β and h_k are,

$$\beta = (N-1)/_2, \; h_k = h_{N-1-k} \tag{21}$$

eq. (21) shows that for a given value of N, there exists only one value of β for which linear phase could be attained. Furthermore, for odd value of N, the filter delay is an integer number of samples. For an even N, the delay is a half integer of sample. Eq. (21) also shows that the filter weighting sequence is symmetrical with respect to the midpoint of the data.

IV. DIGITAL FILTER REALIZATION STRUCTURE

IV.1. Recursive and Nonrecursive Realization

Once the digital filter transfer function has been derived following the procedures in the previous sections, it is next converted to a computational algorithm. The computational algorithm consisting of three basic operations of add/subtract, multiply and unit delay of sampling interval can be represented by a variety of network structures. The implementation of the various forms of network structures takes different numbers of arithmetic operations and requires different amounts of memory storage unit. In addition, as the effect of truncation error due to digital hardware finite word length also depends on the form of the realization, one should naturally select the realization which is least sensitive to error and which requires least number of arithmetic operations and storage memories.

A digital transfer function can be realized recursively and nonrecursively. Let $[x_k]$, $[h_k]$ and $[y_k]$ be the input sequence, weighting sequence and output sequence of the filter. In recursive realization, the current output is a function of past outputs, as well as current and past input samples, that is,

$$y_k = F(y_{k-1}, y_{k-2}, \ldots, x_k, x_{k-1}, \ldots) \qquad (22)$$

On the other hand, in nonrecursive realization, the current output is a function of current and past input samples only,

$$y_k = F(x_k, x_{k-1}, \ldots) \qquad (23)$$

Both IIR and FIR filters can be realized either recursively or nonrecursively. However, the nonrecursive realization of IIR filter is not effective, and is omitted from discussion.

IV.2. Recursive Realization of IIR Filter

We assume that the filter transfer function is physically realizable. The physical realizability implies that no filter output can precede its input signal. Suppose the transfer function is the ratio of two polynomials in z,

$$G(z) = \frac{b_o z^m + b_1 z^{m-1} + \ldots + b_{m-1} z + b_m}{z^n + a_1 z^{n-1} + \ldots + a_{n-1} z + a_n} \qquad (24)$$

the physical realizability requires that the highest power of the denominator n must not be less than the highest power of the numerator m, or $n \geq m$[4].

Since z^{-1} represents one unit of sample delay in digital implementation of the transfer function, the transfer function is often more conveniently expressed as the ratio of two polynomials in z^{-1} rather than z. Dividing both the numerator and the denominator of eq. (24) by z^n, it is converted to

$$G(z) = \frac{b_o z^{m-n} + b_1 z^{m-n-1} + \ldots + b_{m-1} z^{1-n} + b_m z^{-n}}{1 + a_1 z^{-1} + \ldots + a_{n-1} z^{1-n} + a_n z^{-n}} \qquad (25)$$

The realization of the transfer function in eq. (25) can be carried out in many different techniques. Three of the more well-known decomposition methods are described in the following.

1. Direct Form Decomposition

If R(z) and Y(z) are the Z-transforms of input and output sequences r(k) and y(k), eq. (25) represents the following relation,

$$\frac{Y(z)}{R(z)} = \frac{b_o z^{m-n} + b_1 z^{m-n-1} + \ldots + b_{m-1} z^{1-n} + b_m z^{-n}}{1 + a_1 z^{-1} + \ldots + a_{n-1} z^{1-n} + a_n z^{-n}} \tag{26}$$

if the Z-transform of an intermediate variable $X_1(z)$ is multiplied into the denominator and numerator of eq. (26), we have

$$R(z) = (1 + a_1 z^{-1} + \ldots + a_{n-1} z^{1-n} + a_n z^{-n})\ X_1(z) \tag{27}$$

$$Y(z) = (b_o z^{m-n} + b_1 z^{m-n-1} + \ldots + b_{m-1} z^{1-n} + b_m z^{-n}) X_1(z) \tag{28}$$

or the equivalent difference equations,

$$x_1(k) = r(k) - \sum_{i=1}^{n} a_i x_1(k-i) \tag{29}$$

$$y(k) = \sum_{j=o}^{m} b_j x_1(k+m-n-j) \tag{30}$$

Eqs. (29) and (30) are realized in the following direct from realization, shown in Fig. 1.

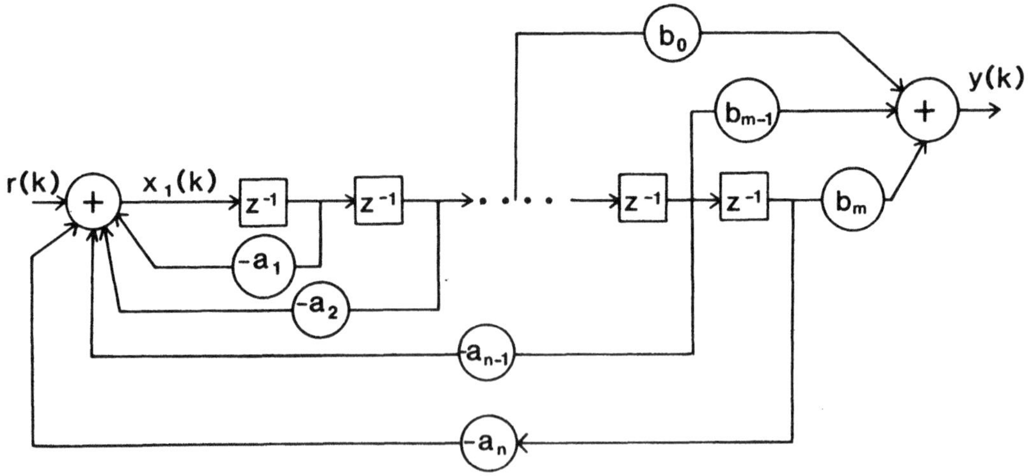

Fig. 1. Direct Form Decomposition

The implementation of this block diagram requires (n+m+1) multiplications, (n+m) additions/subtractions, and n memory storage units. Inaccurate implementation of any coefficient due to finite word length will affect all the poles/zeros of the filter and thus its frequency response. Thus the direct form decomposition is not recommended for use in implementing high-order filters on small word length digital processors.

2. Cascade Form Decomposition

If the filter transfer function can be factored as the product of a number of first order and second order transfer functions as

$$G(z) = \left(\frac{N_1(z)}{D_1(z)}\right)\left(\frac{N_2(z)}{D_2(z)}\right)\cdots\left(\frac{N_n(z)}{D_n(z)}\right) \tag{31}$$

where each factor $\frac{N_i(z)}{D_i(z)}$, $i = 1, 2, \ldots, n$ is either a first order transfer function in the case of simple real pole/zero or a second order transfer function in the case of complex conjugate pair of poles/zeros.

G(z) then is represented by the cascade connections of network structures of first and second order transfer functions. For example, the cascade form decomposition for the filter transfer function G(z),

$$G(z) = \left(\frac{b_o z}{z+a_1}\right)\left(\frac{z(z+b_1)}{z^2+a_2z+a_3}\right), \quad a_2^2-4a_3<o \tag{32}$$

is shown in Fig. 2.

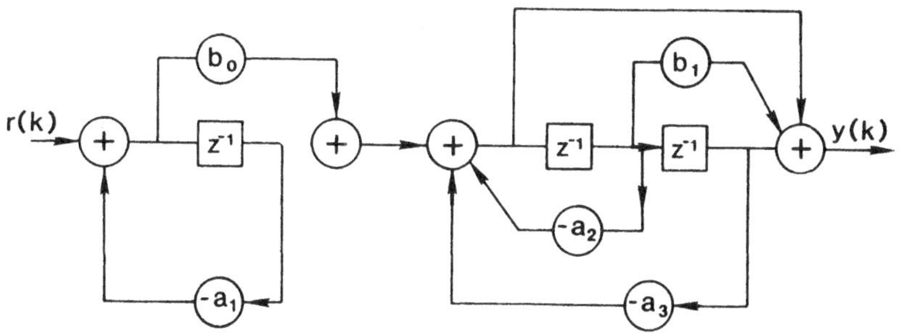

Fig. 2. Cascade Form Decomposition

The advantage of this decomposition over the direct form is that the error of a coefficient implementation in one section affects only the poles and zeros in that section.

3. Parallel Form Decomposition

The filter transfer function can also be realized in parallel form if its denominator can be factored out as the product of simpler forms and then by partial fraction expansion, G(z) is written as

$$G(z) = \sum_{i=1}^{L} \frac{\alpha_i z}{z+p_i} + \sum_{j=1}^{M} \frac{c_j z+d_j}{z^2+a_j z+b_j} \tag{33}$$

where $a_j^2-4b_j<o$.

The realization of eq. (33) is then the parallel connections of network structures of first and second order transfer functions. As an example, the parallel form decomposition for the filter transfer function G(z),

$$G(z) = \frac{\alpha_1 z}{z+p_1} + \frac{c_1 z+d_1}{z^2+a_1 z+b_1} \tag{34}$$

is shown in Fig. 3.

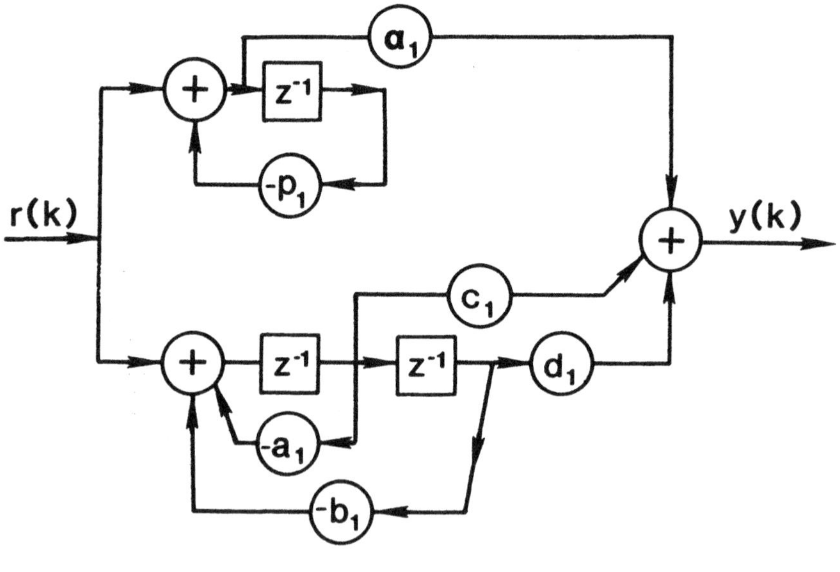

Fig. 3. Parallel Form Decomposition

The advantage of this decomposition is that the error of coefficient implementation in one section affects only the poles and zeros in that section.

IV.3. Effect of Finite Word Length

In implementing a filter on a digital hardware, there will be error in the digitization of the input signal and the error introduced due to the inaccurate filter coefficient implementation because of the finite word length nature of the digital device.

If the filter coefficients are not implemented exactly, the poles and zeros of the filter may not be the same as the desired poles and zeros. The movement of the poles and zeros will cause the frequency response of the implemented filter to deviate from the designed frequency response. Thus it is important that the frequency response of the filter after the implementation be checked to make sure the frequency response does meet the original design specification.

The effect of truncation error is highly dependent on the filter realization structure. However it is demonstrated [15] that direct form realization of high-order transfer functions is more sensitive to truncation error and its use is not recommended if the bit length for filter implementation is small.

In the direct form realization the variation of one coefficient due to truncation will affect the locations of all the poles and/or zeros of the transfer function. In a cascade or a parallel form realization the variation of one coefficient affects only the poles/zeros in one section and the poles/zeros in other sections are not affected.

IV.4. Nonrecursive Realization of FIR Filters

If $[h_m]$, m=0,1,2,... N-1 are N weight sequences of the filter, the filter output y_k is related to the input sequence $[X_n]$ as

$$y_k = \sum_{m=0}^{N-1} h_m x_{k-m}$$

A realization of the equation is depicted in Fig. 4.

IV.5. Recursive Realization of FIR Filters

The Z-transform of the FIR filter is

$$H(z) = \sum_{n=0}^{N-1} h_n z^{-n} \tag{35}$$

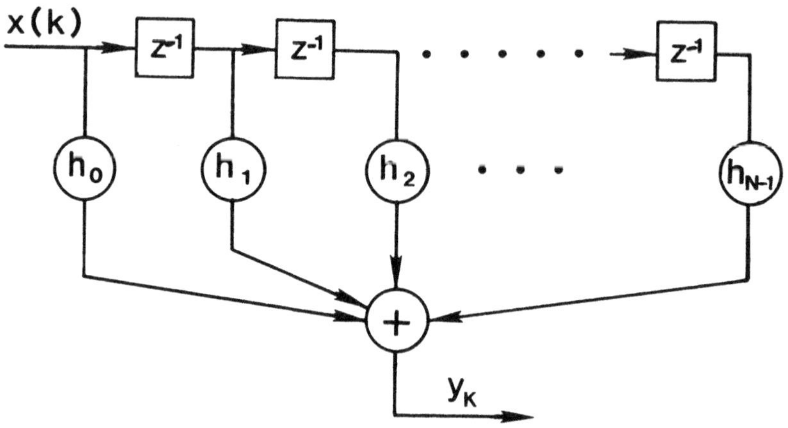

Fig. 4. Nonrecursive Realization

and $[h_k]$ is related to DFT coefficients $[H_k]$ as in eq. (18a). Substitute eq. (18a) into eq. (35) we have

$$H(z) = \sum_{n=0}^{N-1} \{\frac{1}{N} \sum_{k=0}^{N-1} H_k \exp(j\frac{2\pi}{N}nk)\} z^{-n}$$

$$= \frac{(1-z^{-N})}{N} \sum_{k=0}^{N-1} \frac{H_k}{1-\exp(j\frac{2\pi}{N}k)z^{-1}} \tag{36}$$

Since each section in eq. (36) is complex, it is necessary to combine the two sections with complex conjugate poles $z = \exp(j\frac{2\Pi}{N}k)$ and $\exp(j\frac{2\Pi}{N}(N-k))$ to give a second order equation, thus eq. (36) becomes

$$H(z) = \begin{cases} (1-z^{-N}) \left(\frac{Ho/N}{1-z^{-1}} + \sum_{k=1}^{\frac{N-1}{2}} \frac{\alpha_K - \beta_K z^{-1}}{1-2\cos(\frac{2\pi}{N}k)z^{-1}+z^{-2}}\right), & N = \text{odd} \\ (1-z^{-N}) \left(\frac{Ho/N}{1-z^{-1}} + \frac{H_{N/2}/N}{1+z^{-1}} + \sum_{k=1}^{\frac{N}{2}-1} \frac{\alpha_K - \beta_K z^{-1}}{1-2\cos(\frac{2\pi}{N}k)z^{-1}+z^{-2}}\right), & N = \text{even} \end{cases} \tag{37}$$

where

$$\alpha_K = 2REAL(H_k)/N,\ \beta_K = 2(REAL(H_k)/N \cos(\frac{2\pi}{N} k)+IM(H_k)/N \sin(\frac{2\pi}{N}K))$$

Eq. (37) can be then realized in recursive structure.

V. DIGITAL FILTER IMPLEMENTATION

V.1. Implementation of Filter on Signal Processor

The implementation of the filter network realization consists of four operations--input/output, add/subtract, multiply and a unit delay. The mechanization of these four operations are described below. The use of Intel 2920 signal processor is illustrated. The fundamental architecture and the instruction set as well as the key features of other similar processors are included in Appendix.

Consider the block diagram of a second-order digital filter shown in Fig. 5, X(K) and Y4(K) are

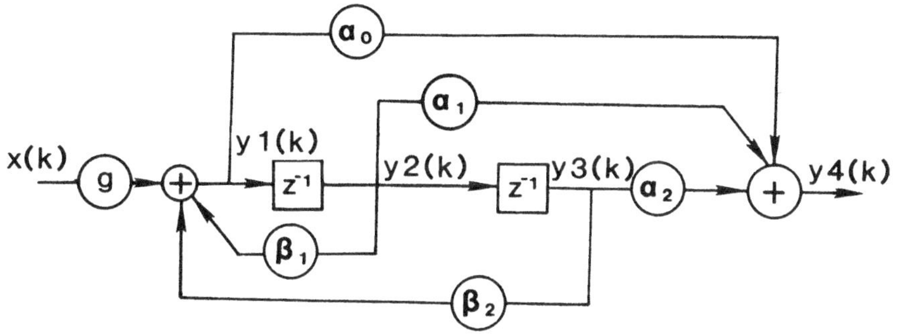

Fig. 5. Block Diagram of the Example

the input and output signals of the filter, g, α's, and β's are the filter coefficients and z^{-1} represents unit delay, that is, delay of one sample interval.

The difference equations the digital processor must evaluate are

$$Y1(K) = \beta_1 * Y2(K) + \beta_2 * Y3(K) + g * X(K)$$

$$Y4(K) = \alpha_0{*}Y1(K)+\alpha_1{*}Y2(K)+\alpha_2{*}Y3(K)$$

$$Y3(K+1)= Y2(K)$$

$$Y2(K+1)= Y1(K)$$

The incremental time is understood to be one sampling period. For g=0.015625, α_0=1, α_1=1.75, α_2=0.875, β_1=1.5, and β_2=-0.75 the step-by-step program development on Intel 2920 signal processor of the above block diagram is described below:

1. Input and Analog-to-Digital (A/D) Conversion

(a) The input channel consists of four analog sampling switches. Since the acquisition time should be equal to six or more time constants [16], approximately six INK instructions would be required for sampling the input signal at Channel K, where K=0, 1, 2 and 3 channel.

(b) For A/D conversion, CVTS instruction first sets the sign bit and CVTK instruction is used to determine the value of the Kth digital-to-analog register (DAR) bit starting with the most significant bit 7. NOPS (no operation) equivalent to 1.2 μsec. must follow each CVT instruction to allow the DAR to settle. Thus if input signal at channel 0 is to be sampled and converted, the input and conversion sequence using a 5MHz clock are

```
IN0        CVT5
IN0        NOP
IN0        NOP
IN0        CVT4
IN0        NOP
IN0        NOP
CVTS       CVT3
NOP        NOP
NOP        NOP
CVT7       CVT2
NOP        NOP
NOP        NOP
CVT6       CVT1
NOP        NOP
NOP ──┐    NOP
      └──► CVT0
```

2. Add/Subtract and Multiply -- implementing coefficients

The filter coefficients can be approximated as closely as the RAM word-length permits. Each ADD or SUB instruction can be thought of as adding to or subtracting it from the destination operand. Intel 2920 does not have multiplier, and multiplication of a variable (operand) is converted to a series of additions and subtraction. For example, the equation Y1(K)=1.5*Y2(K)-0.75*Y3(K)+0.015625*X(K) is implemented by the following sequence of the instruction. Any statement following ";" symbol is the comment statement and is ignored by the processor. X(K) is the sampled and digitized input, stored in DAR.

```
; to program Y1(K)=1.5*Y2(K)-0.75*Y3(K)+0.015625*X(K)
LDA Y1,Y2        ; Y1  ← Y2
ADD Y1,Y2,R01    ; Y1  ← Y1+0.5*Y2=1.5*Y2
SUB Y1,Y3,R01    ; Y1  ← Y1-0.5*Y3=1.5*Y2-0.5*Y3
SUB Y1,Y3,R02    ; Y1  ← Y1-0.25*Y3=1.5*Y2-0.75*Y3
ADD Y1,DAR,R06   ; Y1  ← Y1+(1/2)^6*DAR=1.5*Y2-0.75*Y3
                 ;+0.015625*X
```

Similarly, the equation Y4(K) = Y1(K) + 1.75*Y2(K)+0.875*Y3(K) can be executed by the following sequence of instruction.

```
LDA Y4,Y1        ; Y4  ← Y1
ADD Y4,Y2,L01    ; shift Y2 to the left one bit then add it to Y4.
                   Y4  ← Y4+2*Y2=Y1+2*Y2
SUB Y4,Y2,R02    ; Y4  ← Y4-0.25*Y2=Y1+1.75*Y2
ADD Y4,Y3,R01    ; Y4  ← Y4+0.5*Y3=Y1+1.75*Y2+0.5*Y3
ADD Y4,Y3,R02    ; Y4  ← Y4+0.25*Y3=Y1+1.75*Y2+0.75*Y3
ADD Y4,Y3,R03    ; Y4  ← Y4+0.125*Y3=Y1+1.75*Y2+0.875*Y3
```

3. Unit Delay Operation

The two unit-delays in the block diagram realize the following difference equations

Y3(K+1)=Y2(K)
Y2(K+1)=Y1(K)

These two equations are executed by the following sequence of instructions.

```
LDA Y3,Y2        ; Y3  ←  Y2, equivalently Y3(K+1)=Y2(K)
                 ; for the next program pass.
LDA Y2,Y1        ; Y2  ←  Y1, or Y2(K+1)=Y1(K) for the
                 ; next program pass.
```

4. Digital-to-Analog (D/A) Conversion and Output

Each of the eight analog output channels includes an individual sample-and-hold circuit demultiplexed from a common, buffered D/A converter output. The following rules must be observed for outputting [16,17]:(a) no outputting should be executed while writing to the DAR,

(b) A sequence of NOPs equal to 6 μsec. should be used to settle the D/A converter, (c) A sequence of OUTK (K=0,1,2,...,7 channel) instruction equal to 1 μsec. should be used to settle the sample and hold output. Thus for a 0.8 sec. instruction cycle time, the suggested output sequence is

```
LDA DAR,Y4      ; to load Y4 into DAR for outputting
NOP             ; to settle the D/A converter
NOP
NOP
NOP
NOP
NOP
NOP
NOP
OUT0            ; output DAR to channel 0
OUT0
OUT0
```

V.2. Sampling Period and Submultiple Sampling

It takes four clock cycles for Intel 2920 to execute one instruction, and the sampling period is the time it takes to complete one full cycle of filter calculations. Thus, for a five MHz clock and the maximum program length of 192 instructions, the sampling period is 0.1536 msec.

In implementing a very low frequence filter, it may be convenient to reduce the sampling rate or equivalently to increase the sampling period. As Intel 2920 does not have branching instruction, the sampling rate could be reduced by using the conditional load operation triggered by a counter reinitiated at a submultiple of the program sampling rate. And the filter calculation is updated once every N program cycles while the counter runs at 1/N of the sampling rate. A program for submultiple sampling is given in section 5.6.6 of reference [17].

REFERENCES

[1] Guillemin, E.A.
Synthesis of Passive Networks; John Wiley & Sons, Inc., New York, 1957.

[2] Ghausi, M.S., and Laker, K. R.
Modern Filter Design; Prentice-Hall Inc., 1981.

[3] Van Valkenburg, M. E.
Analog Filter Design; Holt, Rinehart & Winston, 1982.

[4] Kuo, Benjamin C.
Digital Control Systems; Holt, Rinehart & Winston, 1980.

[5] Oppenheim, Alan V. and Schafer, R.W.
Digital Signal Processing; Prentice-Hall, Inc., 1975.

[6] Tretter, Steven A.
Introduction to Discrete-time Signal Processing, John Wiley & Sons, 1976.

[7] Constantinides, A. G.
"Frequency Transformations for Digital Filters,"
Elec. Lett., Vol. 3, No. 11, Nov. 1967, pp. 487-489.

[8] Constantinides, A. G.
"Frequency Transformations for Digital Filters,"
Elec. Lett., Vol. 4, No. 7, Apr. 1968, pp. 115-116.

[9] Constantinides, A. G.
"Special Transformations for Digital Filters,"
Proc. IEEE, Vol. 117, No. 8, Aug. 1970, pp. 1585-1590.

[10] Jong, M. T.
Methods of Discrete Signal and System Analysis, McGraw-Hill Book Company, 1982.

[11] Steiglitz, K.
"Computer-Aided Design of Recursive Digital Filters," IEEE Transactions on Audio Electroacoust., Vol. AU-18, June 1970.

[12] Rabiner, L. R., and K. Steiglitz
"The design of Wideband Recursive and Nonrecursive Digital Differentiators," IEEE Transactions on Audio Electroacoustics. Vol. Au-18, June 1970, pp. 204-209.

[13] Peczky, A. G.
"Synthesis of Recursive Digital Filters Using the Minimum p Error Criterion," IEEE Trans. on Audio Electroacoust., vol. AU-20, Oct. 1972, pp. 257-263.

[14] Antoniou, A.
Digital Fitlers Analysis and Design, McGraw-Hill Book Co., 1979.

[15] Chen, C. T.
One-Dimensional Digital Signal Processing, Marcel Dekker Inc., 1979.

[16] 2920-16 Signal Processor Data Sheet, Intel Corp., 1981.

[17] The 2920 Analog Signal Processor Design Handbook, Intel Corp., 1980.

[18] Signal Processing Computer Functional Specification, Texas Instruments Inc., July 1981.

[19] S2811 Signal Processing Peripheral Advanced Product Description, American Microsystems, Inc., January 1980.

[20] μPD7720 Signal Processing Interface Technical Manual, NEC Electronics U.S.A. Inc., 1982.

[21] Charalambous, C.
"Minimax Optimization of Recursive Digital Filters Using Recent Minimax Results," IEEE Trans. on Acoustics, Speech and Signal Processing, Vol. ASSP-23, No. 4, Aug. 1975, pp. 333-345.

[22] Chan, Andrew K. and Charles K. Chui
"Nonrecursive Digital Filter Designs by Means of Korovkin Kernels," IEEE Trans. on Acoustics, Speech and Signal Processing, Vol. ASSP-27 No. 3, June 1979, pp. 218-222.

[23] Shenoi, K. et al
"On Least- Squares Design of Recursive Digital Filters," IEEE Trans. on Acoustics, Speech and Signal Processing, Vol. ASSP-24, No. 4, Aug. 1976, pp. 337-339.

[24] Gray, Augustine H., Jr. and John D. Markel
"A Computer Program for Designing Digital Elliptic Filters," IEEE Trans. on Acoustics, Speech and Signal Processing, Vol. ASSP-24 No. 6, Dec. 1976, pp. 529-538.

[25] Steiglitz, Kenneth
"Optimal Design of FIR Digital Filters with Monotone Passband Response," IEEE Trans. on Acoustics, Speech and Signal Processing, Vol. ASSP-27, No. 6, Dec. 1979, pp. 643-649.

[26] Programs for Digital Signal Processing, edited by IEEE Digital Signal Processing Committee, IEEE Press, 1979.

[27] Szentirmai, G. (ed.)
Computer-Aided Filter Design, IEEE Press, 1973.

[28] Chen, C. S.
"Application of One-Chip Signal Processor in Digital Controller Implementation," IEEE Control Systems Magazine, Vol. 2, No. 3, September 1982, pp. 16-22.

APPENDIX

The architecture, input/output, data conversion and the instruction set of the Intel 2920 signal processor is briefly described first, then the key features of other manufacturers' similar products are listed. The list includes, besides Intel 2920, signal processing peripheral S2811 of American Microsystems Inc., Signal processing interface μPD7720 of NEC and signal processing computer TMS 3020 of Texas Instruments, Inc. Only brief information about the products are provided here, the interested readers are urged to consult the manufacturers' product brochure for details [18-20].

I. Intel 2920 Signal Processor

The Intel 2920 can be viewed as one-chip special purpose microcomputer for real time signal processing and digital control applications.

I.1 Architecture

The architecture diagram of 2920 chip is shown in Fig. 6. There are three distinct sections: an analog I/O section, digital processing unit, and a program memory unit.

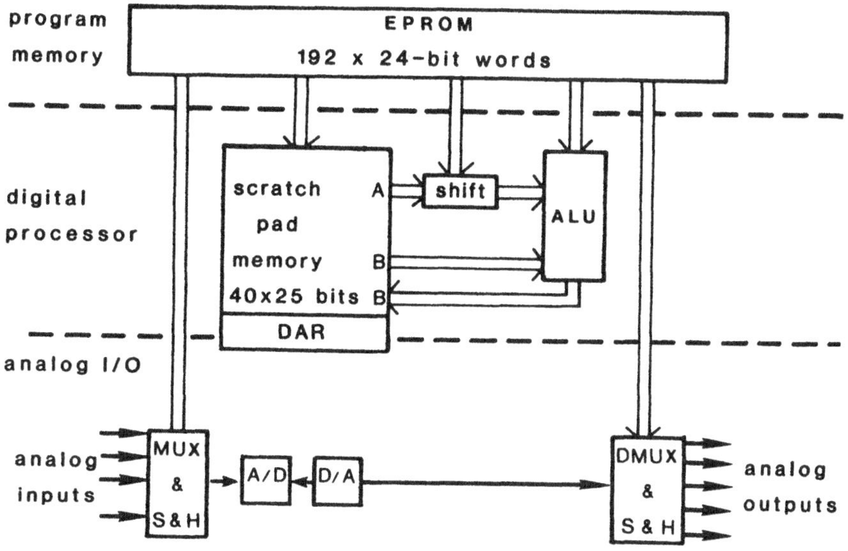

Fig. 6. 2920 Chip Diagram

(1) The incoming analog signal is input to one of the four multiplexed analog input channels, converted into digital form through a 9-bit A/D converter and stored in the digital-analog register (DAR). Output signals are placed in the demultiplexer which then outputs the signals to the eight output analog channels.

(2) There are scratch pad memory, arithmetic logic unit (ALU) and a scale shifter in the digital processing unit. The scratch pad memory is organized as 40 words of 25 bits each accessed through A and B ports. The scaler is located at the exit of the A port; mutliplying by two is a shift of one bit to the left, division by two a shift of one bit to the right. There are 16 constant codes; each code contains some multiple of eighths, from -8/8, -7/8, ... to + 7/8.

(3) The program memory section contains an erasable, programmable read-only memory (EPROM). The EPROM is arranged as 192 words of 24 bits, each word corresponding to one instruction.

I. 2 Instructions.

The instructions are stored in consecutive memory locations and are executed sequentially. A 2920 statement, equivalent to one PROM word, contains five basic fields: (1) ALU operation, (2) destination address, (3) source address, (4) optional scaler shift and (5) an analog control command. More detailed information could be found in [17,28].

II. Signal Processing Microcomputers/Microprocessors

The key features of the commercially available signal processing microcomputers/microprocessors are listed in the table below. The direct comparisons among different devices may not be meaningful because of different architectures. The readers are urged to refer to the manufacturer's product brochures [18-20].

	Intel 2920	AMI S2811	NEC μPD7720	TI TMS3020
instruction cycle (nsec)	400	300	250	200
data converter	on-chip	separate	separate	separate
instruction ROM (bits)	192x24	256x17	512x23	1.5Kx16
RAM capacity (bits)	40x25	256x16	128x16	144x16
input/output	4 inputs 8 outputs	8-bit data bus	8-bit data bus	8-bit data bus

	Intel 2920	AMI S2811	NEC μPD7720	TI TMS3020
multiplier	---	12 bit x 12 bit	16 bit x 16 bit	16 bit x 16 bit
data-word length (bits)	25	16	16	32
branching instruction	NO	YES	YES	YES

CHAPTER 5

MICROCOMPUTER-BASED WAVEFORM ANALYZERS AND THEIR APPLICATIONS

R. D. Jackson and E.J. Prendergast*
Engineering Department
Cambridge University
Cambridge, England

ABSTRACT.

The application area for waveform analysis instruments is considerably extended by the use of digitised signals and the signal processing capabilities of microcomputers. Wide band signals require digital to analogue converters with fast conversion rates, and techniques are available which achieve conversion times to 8-bit accuracy as low as a few nanoseconds. Various constraints are imposed on the signal processing algorithms by the quantisation of the analogue input signals. Two applications selected from a wide range are semiconductor device testing and signal spectral analysis.

1.INTRODUCTION.

Applications for waveform analysis are widespread, electrical engineering examples vary from semiconductor device testing to on-line fault monitoring in large turbo-alternators. The combination of a means of converting an analogue waveform into digital form and the compact computing power of a microprocessor provides a powerful instrument both for the acquisition and storage and for the analysis of time-varying quantities. Such systems are particularly well suited to the acquisition and analysis of very fast transient events and other signals hard to capture using a normal oscilloscope. Their bandwidth can be very wide as analogue to eight-bit digital converters are now available with conversion times of a few nanoseconds. With present generation microprocessors such conversion rates require some intermediate data storage between converter and processor memory.

The block diagram of a typical instrument is shown in Figure 1, and a variety of instruments are commercially available. This Figure also shows a direct view oscilloscope which continuously displays the digitised waveforms via a digital to analogue converter and provides a

*Present address:Bell Telephone Labs., Allentown, Pennsylvania, USA.

S. G. Tzafestas (ed.), Microprocessors in Signal Processing, Measurement and Control, 75–98.

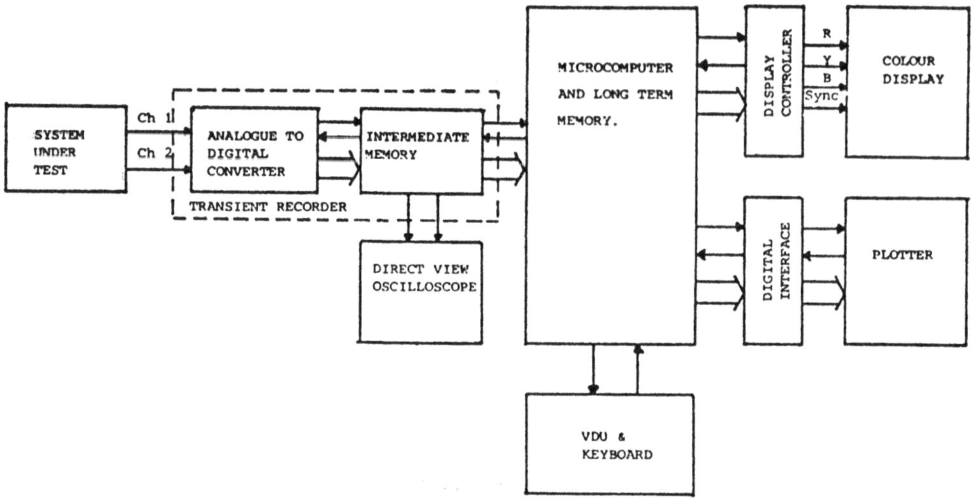

Figure 1. Waveform analysis system.

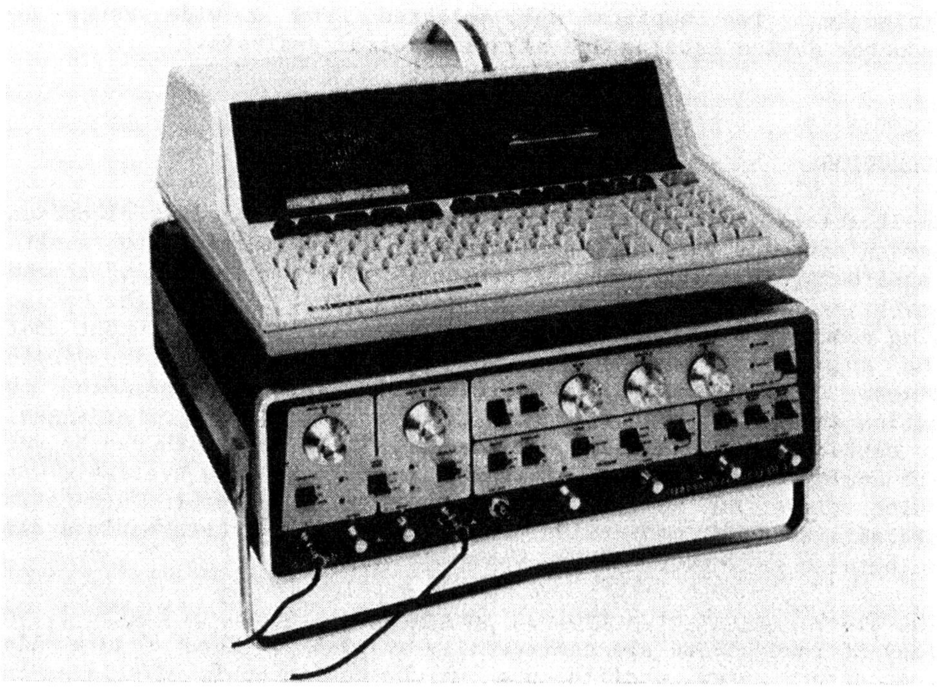

Figure 2. Datalab 985 Transient Recorder.
(Reproduced by permission of Data Laboratories Ltd.)

useful check on the recorded signals. An example, combining the Hewlett-Packard HP-85 microcomputer and the Datalab 912 waveform digitiser, is shown in Figure 2. The use of standard data bus protocols, as used by the HP-85, allows the straightforward addition of further devices as peripherals, such as a counter/timer or colour plotter.

Applications for waveform analysers can be divided into two main classes; those in which continuous operation in 'real-time' is required, as in a fault-monitoring mode, and those operating on 'time-windowed' sets of data performing such tasks as rise-time measurements on transistor pulse waveforms. This chapter is concerned mainly with the second class, that is, measurements following the sequence given below:

(a) Data acquisition and temporary storage,
(b) Display and measurement of stored data,
(c) Computation on data and storage in long-term memory.

The easy availability of computing power leads to a blurring of these two classes by the use of parallel processing. In such a system successive arrays of data may be acquired and handed to a processor, leaving the acquisition device free to continue. The processor can operate on the data array by performing, for example, a Fourier transform, and hand the result to a further link in the chain for display.

An important aspect of acquisition and analysis systems is the interaction between instrument and user. The use of general purpose microcomputers allows the user to develop programs for specialised tasks, and to fit such programs into an overall framework of standard routines for data acquisition and transfer. A system, based on a 20MHz, 4096 8-bit word, waveform digitiser and a Cromemco CS-3 computer, is described later in this chapter.

Computations such as scaling,offsetting, multiplication and Fourier transforms are well within the capabilities of the microprocessor, although considerable care has to be taken due to the nature of the input data. There are a number of important factors to be considered in the use of a series of digital numbers to represent an analogue waveform. First, the input signal waveform is sampled at successive points in time. These samples are normally equally spaced in time, although parts of a set of recorded data may use different time intervals. (see section 2.2). The sampling rate imposes a limit on the bandwidth of signal that can usefully be sampled. Second, the digital representation of each sampled value is a finite-length word and hence is either a rounded or truncated representation of the original value. Third, the sampling and analogue conversion processes are subject to error, causing, for example, an irregular sampling rate or non-monotonic conversion.

The sampling process is illustrated in Figure 3. The sample-hold

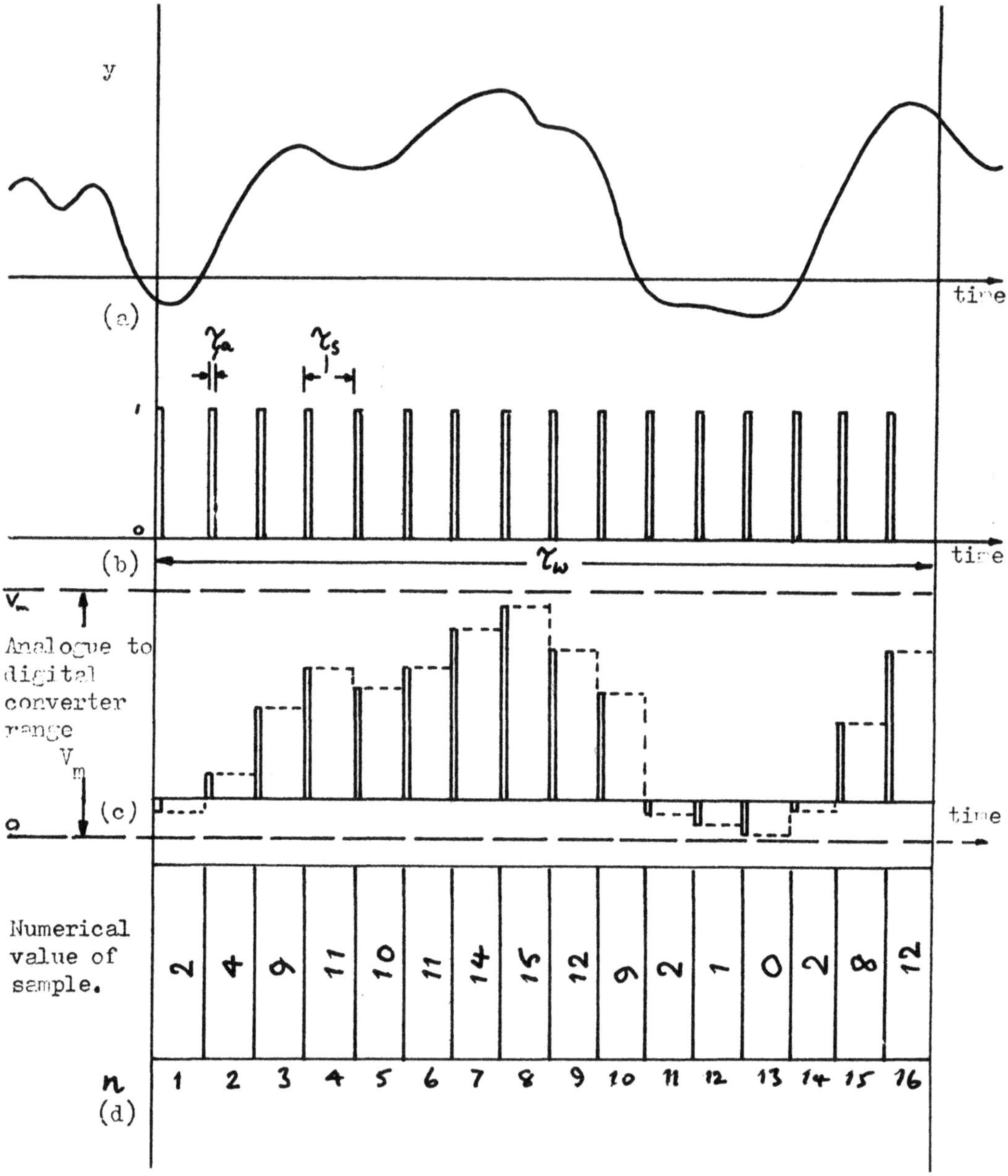

Figure 3. Sampling and windowing waveforms.

circuit preceding the analogue-to-digital converter (see Figure 1) samples the input waveform at intervals τ_s with a sampling aperture τ_a over a time-window of τ_w. This process can be represented by taking the product of the input waveform and the waveform of Figure 3(b). The resulting pulse train is digitised, with a consequent truncation or rounding error, depending on the form of converter, leading to a series of numbers. Each number then represents the value of the waveform samples, with a conversion error, known as quantisation error, due to its finite length.

Figure 4 shows some examples of signals recorded in the investigation of an electronic circuit. The converter data word length is 8-bit, and the quantising effect is clearly visible.

Section 2 of this chapter describes the individual components that make up a complete waveform analysis instrument, and section 3 describes numerical algorithms that may be used on the recorded data to derive information additional to that found from direct measurement on a time-varying waveform. Finally section 4 describes two applications in some detail and shows the place of these instruments in the general area of data acquisition and analysis.

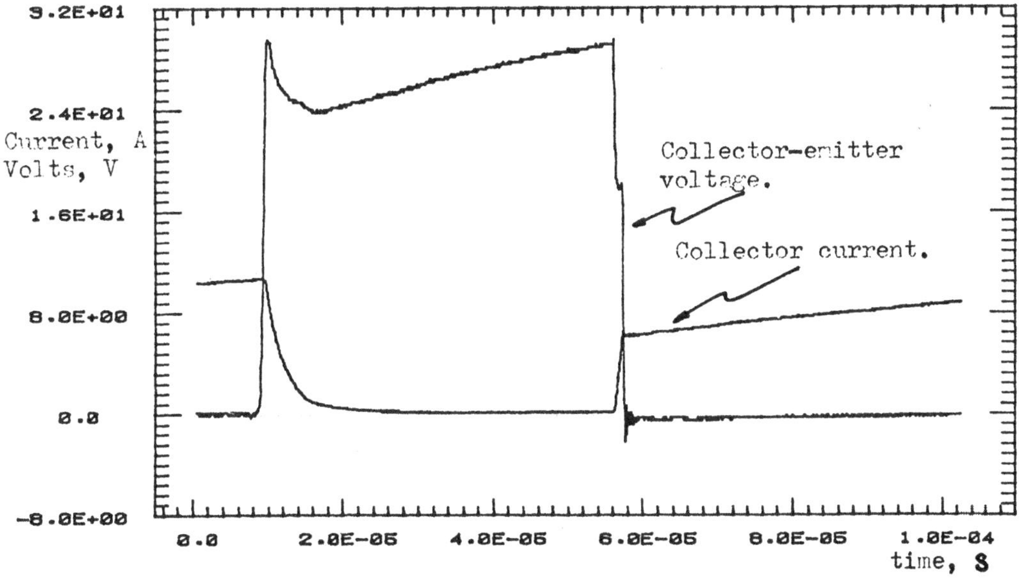

Figure 4. Transistor circuit transient voltage and current waveforms.

2. SYSTEM COMPONENTS.

2.1 Analogue to digital converters.

It is convenient here to consider briefly the consequences of the sampling and rounding processes referred to in section 1. Consider the original input waveform to be y(t) and the value at the start of the nth sampling window period to be $y(n\tau_s)$. The discrete version will then be given by $y_d(n\tau_s)$, where $y_d(n\tau_s)=y(n\tau_s)+e(n\tau_s)$, and $e(n\tau_s)$ represents the conversion error. The requirement that the original input y(t) shall be obtainable from the sampled values $y(n\tau_s)$ imposes a restriction on sampling interval τ_s in terms of the maximum frequency component f_m in y(t). The effect of the sampling process on the frequency spectrum of y(t) is to produce a series of repetitions of the original spectrum each centred on the sampling frequency, $1/\tau_s$, or its harmonics. (Stanley, 1975) The Nyquist sampling theorem requires $f_m < 1/\tau_s - f_m$, or $f_m < 1/2\tau_s$ so that the repeated spectra may be separated by the use of an ideal low-pass filter. The input signal may then be recovered without distortion. In practice, this restriction is too severe without substantial post-processing of the recorded signal, and a more reasonable restriction is that the original waveform shall be closely represented by linear interpolation between the sampled values. This requirement means that the amplitude/frequency characteristic of the function which describes linear interpolation, ($\sin^2(\pi f\tau_s)/(\pi f\tau_s)^2$), shall be sufficiently small at the frequency $1/\tau_s - f_m$, and $f_m < 1/5\tau_s$ is a more reasonable limit.

The analogue-to-digital converter following the sampling circuits will convert a signal within a given range, say 0 to V_m volts, into its digital equivalent, with resolution $\Delta y=V_m/2^N$, where N is the number of bits. Ideally, the digital output of the converter will represent the input value rounded to N-bit accuracy, that is, $e(n\tau_s) < 0.5\Delta y$. Many forms of converter will directly produce a truncated digital equivalent of the input value, that is, $e(n\tau_s) < \Delta y$ (which could be offset by $0.5\Delta y$).

Some form of input signal conditioning to scale and offset the input waveform is needed to ensure the best use of the converter; namely, that its input signal should fill the range 0 - V_m as far as possible. The input circuits required are similar to those used on conventional oscilloscopes. The input circuit bandwidth should be such that signals of frequency f_m or less are not significantly attenuated. The trigger facilities are an important part of the input circuits and the exact trigger point on the waveform needs to be precisely definable.

The bandwidth limitation imposed by the sampling rate has been discussed above. A further, and often more severe, restriction, on the input signal bandwidth, is due to the aperture uncertainty imposed by the sampling window τ_a (see Figure 3). The requirement here is that the input signal should not change by more than the voltage equivalent to the least significant bit of the digital output during the time τ_a.

For a sinusoidal signal between the limits 0 and V_m, at frequency f_m, the limit is: $f_m < 1/(2^N . \pi\tau_a)$.

Converters with conversion times down to around 1us, and word lengths of 8, 10 or 12 bits are readily and cheaply available in monolithic form. Such converters normally employ the successive approximation method (Hnatek,1976).

The block-diagram of a simple 8-bit successive-approximation converter is shown in Figure 5(a). The input signal is compared with the analogue equivalent of the digital number held in the successive-approximation register and the result used to adjust that number to a closer approximation of the input. The testing starts with the most significant bit by using the word 10000000, and the bit is set to zero if the analogue equivalent is greater than the input. After testing each bit in sequence the successive-approximation register will ideally contain a digital number equal to $y(t_1) - e(nt_1)$,where $y(t_1)$ is the input and $e(t_1) < \Delta y$. The method is serial, and faster conversion may be achieved using parallel methods in which the whole digital number is found in a single comparison.

Conversion times in the range 1us to 10ns can be achieved using parallel comparator arrays (Shoreys, 1982). The method is illustrated in Figure 5(b), and requires 2^N-1 comparators for N-bit conversion. Such arrays are now becoming available in monolithic form (Saul et al, 1980) enabling, for example, 8-bit converters to operate with 10ns conversion times. Very high speed bipolar circuits are used, and the packages described in reference 4 contain 4-bit converters (16 comparator arrays) together with decoding logic.

Combinations of the serial and parallel conversion methods are used to obtain conversion times in the range 1us to 50ns without the large comparator arrays necessary for a full parallel converter. One such method is illustrated for an 8-bit converter in Figure 5(c), using a parallel/serial/parallel principle.

The input signal is first digitised to 4-bit accuracy using a parallel converter to form the four most significant bits of the output. This number is then converted to analogue form, and subtracted from the original input. The result may then be digitised in a second 4-bit parallel converter to obtain a complete 8-bit output. Thus 30 comparators only are needed for full 8-bit conversion. A more accurate procedure is illustrated in Figure 5(c) and uses a 5-bit converter for the difference value. The full scale range of the second converter is thus twice that of the subtractor output. If the subtractor output is arranged to be centred in the middle of the range of the 5-bit converter, errors accruing from earlier stages which result in positive or negative offsets are not out of the range of the second converter but utilise otherwise unused comparators. Overrange of the second converter may also be used to correct the output of the first converter. The waveform digitiser shown in Figure 2 uses this method, and its

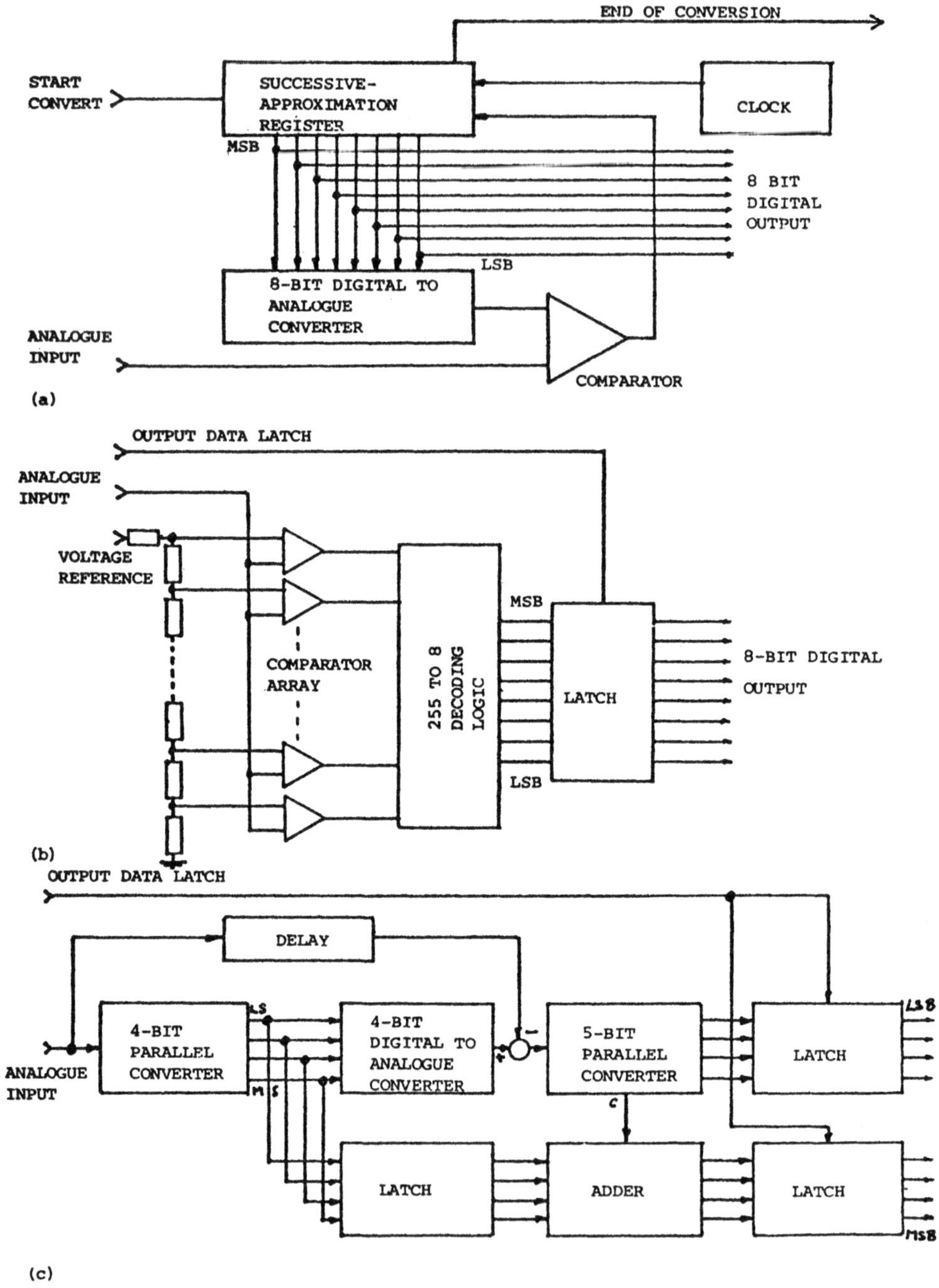

Figure 5. (a) Successive-approximation digital to analogue converter.
(b) Parallel comparator array converter.
(c) Parallel-series converter.

specification is given below:

Signal input:
Input range: 100mV to 20V, ±1%
Bandwidth: dc to 6MHz.

Analogue to digital converters:
Conversion rate: 20MHz maximum, simultaneous sampling on two inputs.
Resolution: 8-bits.
Aperture time: 30ps.

Memory: 8Kbyte static RAM, 2 channel 4096 x 8-bits, 1 channel 8192 x 8-bits.

Timebases: A and B, sample period 50ns to 20ms on each.

Digital input/output options: IEEE 488, RS232C, parallel TTL interface.

Table 1. Specification of Datalab 912 recorder.

A conversion technique capable of times in the 5-10 ns region is used by Tektronix (Robin and Ramirez, 1980). This converter employs an electron beam tube with a semiconductor target. The beam is focussed into a narrow ribbon, and the target consists of an array of diodes, 8x256 for an 8-bit converter, in the form of a Gray code pattern. The beam width spans the 8 columns of the array, and vertical movement is controlled by Y deflection plates over the 256 rows. Thus beam deflection can be digitised to 8-bit accuracy, and the method is essentially a means of digitising directly an oscilloscope display. The output is a rounded equivalent of the analogue input.

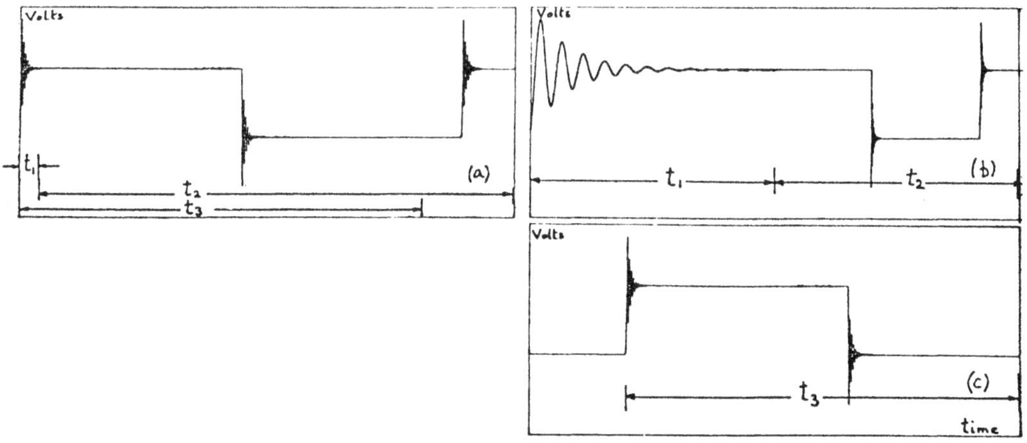

Figure 6. Signal acquisition using expanded and delayed timebases.

The time-base and triggering circuitry form an important part of the converter. For example, if a waveform is sampled continuously and stored in a limited memory array, which is successively overwritten, information can be presented for a limited period prior to the triggering signal. Typical acquisition possibilities are discussed in conjunction with Figure 6.

Figure 6(a) shows a sequence of oscillatory pulses, and the trigger circuits of the digitiser are assumed to operate on the initial rising edge. Figure 6(b) shows the digitiser output for a dual rate time base over the intervals t_1 and t_2 , while Figure 6(c) shows the 'pre-trigger' mode in which the memory is continuously overwritten, and the time-base arranged to show only t_3 following the trigger.

A convenient assessment of converter performance can be made by digitising a sinewave signal to full-scale. If then the record is passed to the controlling computer, an idealised sinewave may be fitted using a least squares fit. The result may be quantised to the resolution of the original signal, N-bits say, and the rms error which would have resulted had an ideal analogue to digital converter been used can be calculated. The effective number of bits of the converter is then defined as:

$$\text{Effective bits} = N - \log_2 (\text{Actual rms error})/(\text{Ideal rms error})$$

This measure is affected by converter and amplifier non-linearities, aperture uncertainty, and noise.

2.2 Memories.

The intermediate or cache memory is required to store the recorded data in binary form until the controlling computer transfers the information into its own memory, or it is overwritten. The speed requirement for the cache memory is therefore severe. This is particularly so for the Tektronix electron-beam digitiser,with its minimum conversion time of 5ns. One solution to the problem is to provide slower memory and multiplex the input data into a number of rows, for example, 200ns memory would require 4 rows to satisfy a 50ns digitising time.(Data Laboratories Ltd.)

Large, dynamic shift registers can provide a suitable memory system, and a block diagram for a 4-bit memory is shown in Figure 7. The system is designed to operate at a maximum frequency of 20MHz,and uses 4-way multiplexing combined with an array of dynamic, 5MHz, MOS shift registers.

The data from the analogue to digital converter is latched into a set of 4-bit latches, each latch taking every fourth four-bit word. The latches are strobed by a four-bit ring counter clocked at 20MHz so that each is latched in turn. The shift registers are each configured in a recirculating path with two-input multiplexers at the input. These multiplexers accept data from the latches or recirculate the existing shift register contents depending on the operating mode. The final stage of each shift register is fed into a multiplexing circuit which ensures that the individual data bits from each register are interlaced into a coherent order for the readout.

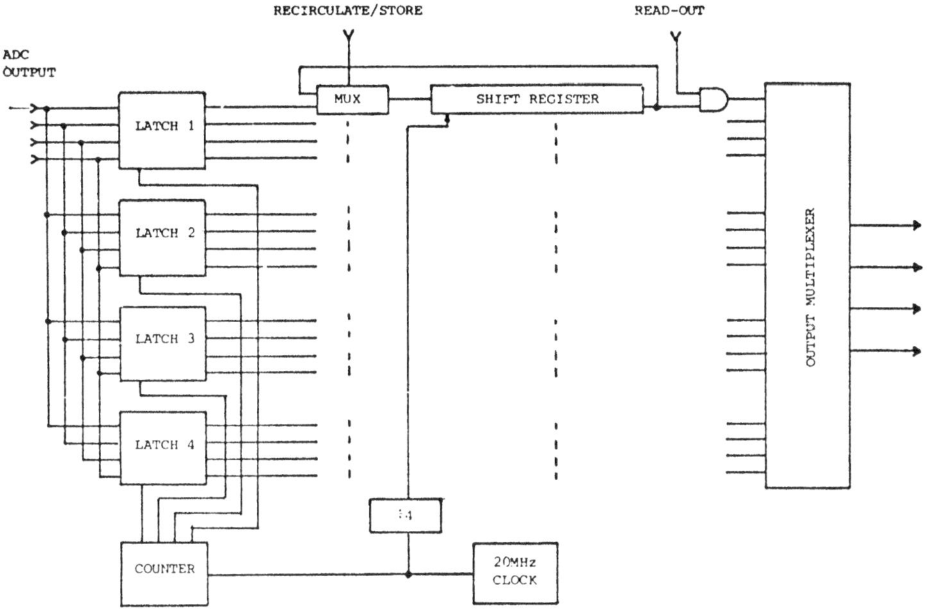

Figure 7. Block diagram of cache memory circuits.

The data output from this circuit may then be routed at a lower rate to the digital interface and transferred under handshake control to the microcomputer. If faster data transfer is required, a direct memory access interface may be implemented. A possible form for the handshake sequence is described in section 2.3. In addition, the output data may be converted to analogue form and used to provide a continuous readout via an oscilloscope, or, at a slower rate, a single trace readout to an analogue plotter.

The availability of large-scale cheap memory allows a very large quantity of information to be stored. For example, the Hewlett-Packard 5180 recorder has a memory of 16,384 10-bit words which can be segmented into 1,2,4,8,16 or 32 equal blocks.

Long-term memory is normally associated with the controlling computer in the form of floppy or hard disc units, although certain stand-alone recorders contain floppy-disc storage systems. The long-term storage requires careful labelling and indexing and some form of database program is required for large numbers of records.

2.3 Microprocessors.

The microcomputer block shown in Figure 1 represents the main signal processing and data handling capability of the waveform analyser. It may be a general purpose computer with the capability of program development in a low or high level language, or a more dedicated system allowing the use only of, say, BASIC. Its tasks are: to supervise the analogue to digital converter; to control the transfer of data from intermediate to processor memory, and to long-term memory if required; to process the data on command; to control the display, printing and plotting peripherals available to the instrument.

The main hardware requirement is for a fast data transfer between cache memory and processor memory. Such a transfer may be achieved under handshake control using a standard parallel port and its associated control lines. At least two control lines are necessary, one indicating data valid, (DAV=1, say, for valid data), and the other data accepted, (DAC=1, say, for data accepted). The timing sequence for a transfer of data words between the digitiser cache memory (or any other peripheral) acting as 'talker' and controlling DAV, and the processor memory acting as 'listener' and controlling DAC is summarised below:

1. Talker sets DAV=0, and puts Word 0 on the data bus.

2. Talker waits for DAC to become 0, and then sets DAV=1. DAV stays at 1 until DAC is set to 1.

3. Listener accepts Word 0, and sets DAC=1. DAC remains at 1 until DAV is reset to 0.

The sequence is then repeated with Word 1 and subsequent data words. Data may be transferred at rates of a few Kbytes/second using straightforward handshake procedures and standard processor input/output instructions. Faster rates, in the region of 1 Mbyte/second, may be achieved using direct memory access interfaces.

The IEEE-488 bus definitions form a standard for control functions, timing, connectors and cables for bidirectional parallel data transfer. (Peatman, 1977) The bus connections are implemented with three handshake lines instead of the two described above, and use 8-bit data words. The 'front panel' controls (for voltage and time scaling etc) on a waveform digitiser with an IEEE standard interface may be set either directly by the user or indirectly via the controlling computer. The digitiser then becomes an addressable 'listener' on the bus; this facility finds important applications in automatic test equipment.

For many applications the combination of an 8-bit processor, such as the Z-80, and an 8-bit analogue to digital converter will provide fast transfer rates and adequate converter resolution. In addition, it is important that long-term storage is available, and that transfer to and from long-term memory is fast.

The speed of computation, particularly for signal processing applications, is also an important factor. An 8-bit processor, relying mainly on software arithmetic routines for any floating point calculations, may have an unacceptably slow response for a sequence of calculations on stored data. The response may be significantly improved by the use of special purpose arithmetic processors (Croson et al, 1981). For example, a comparison between FORTRAN software routines and a hardware arithmetic card based on the AMD 9511 processor gave the following times for a benchmark of 10,000 arithmetic operations:

Function.	9511 processor.	Software.
SIN	13s	102s
COS	14s	114s
EXP	16s	109s

Operations such as multiplication do not show such significant improvements as the transfer of data betweeen MPU and 9511 package forms a significant part of the arithmetic operation.

A major improvement in calculation time over that of an 8-bit processor can be achieved by using a 16-bit processor such as the Motorola 68000. The 16-bit wide data-bus also allows direct connection to higher resolution converters, and the very full instruction set is well suited to signal processing algorithms.

The software forms a major part of the overall system, and determines the 'user friendliness' of the instrument. It should be organised to provide prompts to the user, and helpful error messages. A possible scheme for a framework of standard routines is outlined below. It has been implemented in a program written in FORTRAN for a Z-80 based Cromemco CS-3 microcomputer system with floppy-disk storage linked to a Datalab 922 waveform digitiser. (Prendergast and Jackson, 1980) The compiled program occupies about 40Kbytes of memory and allows fast interactive control of recorded data.

The cache memory in the waveform digitiser contains 4096 8-bit words, so a standard data file consists of 4096 8-bit words together with floating point numbers representing scaling factors and offset values. The program uses two working data files, a primary file and a secondary file; the primary file can originate from disk or from direct transfer from the digitising recorder, the secondary file can originate only from disk. The waveform is displayed using 256x256 point colour graphics, thus each level of digitised voltage can be displayed, but only every 16th time point in the original data set. System control is exercised via an alphanumeric VDU, and analogue input is provided for controlling a cursor on the graphics monitor. The cursor is constrained to move along the waveform. Different colours are used to distinguish separate waveforms, or parts of the same waveform, and to separate waveform and cursor. Most of the facilities can, however, be satisfactorily implemented in single colour graphics.

The following basic menu of operations is provided by the program:

Restart - generate a new primary working data file, either from disk or via direct transfer from the record
The data is displayed with the cursor, and x and y cursor coordinates displayed on the VDU.

Pan - move a 256 point window across the working file such that each data point is displayed on the monitor. The movement of the window may be stopped to display a section of interest.

Sector - a section of the display, of selectable length, is chosen and indicated by a colour change, and then displayed as the full picture on the monitor. The section may be established as the working file, with each point replaced by a line at the same voltage level. This approximation is preferable to linear interpolation, which may lead to misleading results.

Save data - store the present working file on disk.

Plot data - plot the working file on a digital plotter.

Set cursor - reset the coordinates describing cursor position, that is, change stored offset values.

Restore working file - display working file again, following pan or sector commands.

All these options allow the user to examine and measure the stored waveform. A further menu can be called to specify certain computations to be performed, using one or two data files. Typical computations are:

Multiply - take the point-by-point product of two files.

Divide - take the point by point quotient of two files (with care over dividing by zero).

d/dt - calculate and display the slope of the original waveform.

Integrate - calculate the area under the waveform between points specified by using the cursor.

Fourier transform - form the Fourier transform of a single file, using the fast Fourier transform algorithm.

Correlate - form the cross-correlation function of two waveforms (or the auto-correlation of a single waveform).

These last two computations require considerable care if the results are not be, at the least, misleading. They are considered in detail in section 3.

2.4 Displays.

A basic premise in this description of waveform analysers is that the recorded waveform should be displayed with adequate resolution, and that the instrument operator should be able to make measurements on the displayed waveform. Some possible interations between user and display have been described in the section 2.3. The form of the display should be compatible with the resolution of the waveform digitiser, or some means of expanding the time or voltage axes of the display must be provided if information is not to be lost. A satisfactory display is formed by a VDU with full graphics capability; an alternative solution is the use of a separate graphics display monitor. The scale and offset factors of the recorded data should be displayed, together with a title and additional identification information (see Figure 2). For example, trigger levels, signal maxima and derived information such as rise-time, may all be presented to the user. Standard graphics software for axes and axis scales can be used with advantage in the presentation of data.

Colour can form an important part of display presentation, as indicated in the previous section, although adding considerably to system cost.

3.SIGNAL PROCESSING ALGORITHMS.

3.1 Data types and number storage.

The optimisation of data storage is desirable to minimise memory requirements and maximise data flow rates. Equally spaced sampled data representing part or all of an acquired signal can be efficiently stored using fixed-point numbers to define relative magnitudes. In addition, signal reconstruction requires the storage of x-axis and y-axis scaling factors and x-axis and y-axis offset values. These four additional quantities need to be stored in a floating-point format with sufficient significant digits to avoid further quantisation errors in the reconstructed signal. For the waveform digitiser described in section 2 only 4112 bytes of storage are required for a waveform consisting of 4096 equally spaced samples. Consequently, data flow rates are high and in this system the movement of a sampled waveform using parallel transfers with full handshaking takes a maximum of 10 seconds and typically less than a second. The worst case is that of generating and addressing a new floppy-disk file.

Storing sampled data in this manner can often speed computations. Many waveform manipulations simply involve changes to offset and scaling parameters. Further, computations with fixed-point numbers are considerably faster than the equivalent floating-point calculations. Intermediate results can be stored in a fixed-point format, and normalised to the standard format for display or storage after the final result has been calculated. However, since the length of a computer word is limited not all computations can be performed in this way and it is often necessary to convert the quantised signals into a string of floating-point numbers prior to processing.

Errors which result from sampling are described in section 2.1. Further errors will occur when computations are performed on waveforms. Round-off errors occur in the storage of computed waveforms. Truncation errors occur because of limited range arithmetic. Simplifications to mathematical algorithms often result in errors which are difficult to estimate. Some errors can be eliminated through careful programming, for example, rewriting ill-conditioned formulas. However, it is impossible to eliminate all errors which result from processing and it is important to be able to estimate the bound for the magnitude of error.

The propagation of errors is dependent on the operations being performed. For example, in addition and subtraction the bounds for the absolute errors in the results are equal to the sum of the bounds for the absolute errors of the operands. However, in multiplication and division the bounds for the relative errors in the results are equal to 1.5 multiplied by the sum of the bounds for the relative errors in the operands, for relative errors less than unity. No general rule exists for error propagation and it is necessary to plan each calculation carefully. The sensitivity of output waveform variations can be experimentally checked by systematically perturbing the inputs.

3.2 General purpose algorithms.

The computational capabilities of microprocessors can be used to multiply, divide, add, subtract, differentiate and integrate waveforms. The above calculations are straightforward; however, particular care must be taken in ratio-taking and derivative calculations to avoid amplifying roundoff noise. Averaging a series of repetitive waveforms will improve resolution if truncation is statistically random.

In addition to the above calculations, microprocessors are sufficiently powerful to perform Fourier transforms (Cooley and Tukey, 1965), and auto- and cross-correlations. The discrete Fourier transform (DFT) for a signal y(t), sampled at fixed intervals τ_S , is defined as follows:

$$Y_n(nf_o) = (1/N) \sum_{m=0}^{N-1} y(m\ \tau_S).\exp(-jm2\pi\ n/N)$$

where $f_o = 1/N\tau_S$ and $N\tau_S = \tau_\omega$, the time window.

The DFT assumes that it is operating on a periodic waveform and produces frequency coefficients at discrete frequencies only. The fast Fourier transform (FFT) algorithm (Brigham, 1974) restricts the number of samples to integer powers of 2. However, it is considerably faster than a straightforward application of the DFT. Special precautions are necessary in the use of the FFT algorithm: the main sources of trouble are spectral overlap, leakage, and the discrete form of the frequency output. Spectral overlap, or aliasing, will occur if a signal contains frequencies above half the sampling frequency and it may be necessary to filter a signal prior to sampling. Leakage can occur owing to truncation in the time domain. Terminating a signal after a finite number of terms is equivalent to multiplying the signal by a square-wave function in the time domain (see Figure 3). The resulting transform is the convolution of the desired spectrum with a sinc function. This distortion can be reduced by using alternative window functions. A simple example is the Hanning window function (Stanley, 1975).

$$w(t) = \tfrac{1}{2}(1 - \cos(2\pi t/\tau_\omega) \quad \text{for } 0 < t < \tau_\omega$$
$$= 0 \text{ elsewhere.}$$

This particular window function also alleviates the effect of leakage associated aliasing. Finally, because of the discrete nature of the FFT result, it is possible that major peaks lie between discrete transform points and will not be detected. This effect is commonly termed the 'picket-fence' effect.

Correlation is performed by taking the inverse transform of the conjugate multiplication of two complex spectra of the time signals to be correlated. This calculation requires a double use of the FFT algorithm for auto-correlation and a triple use for cross-correlation. Precautions are necessary in performing these operations. Convolving two nonperiodic signals results in a function with a width approximately equal to the sum of the widths of the two functions being convolved. Since time signals are assumed to be periodic, the resulting convolution is forced to 'wrap-around' itself. (Stanley, 1975)

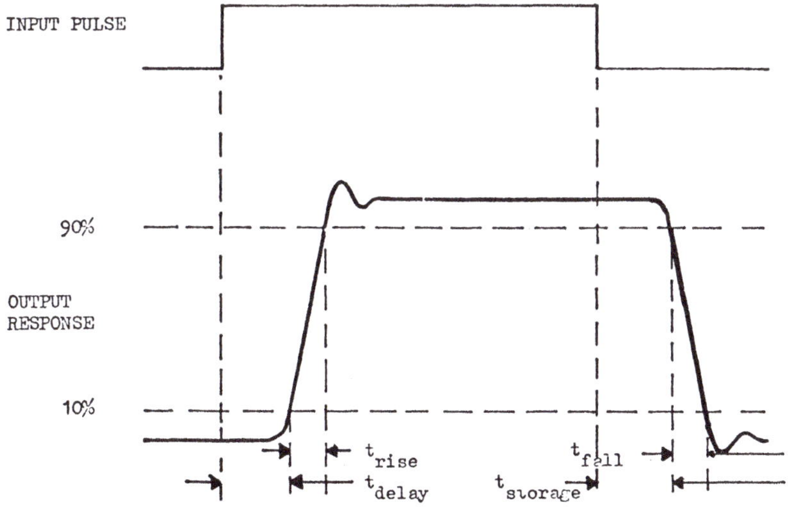

Figure 8. Typical waveforms showing switching parameters to be calculated.

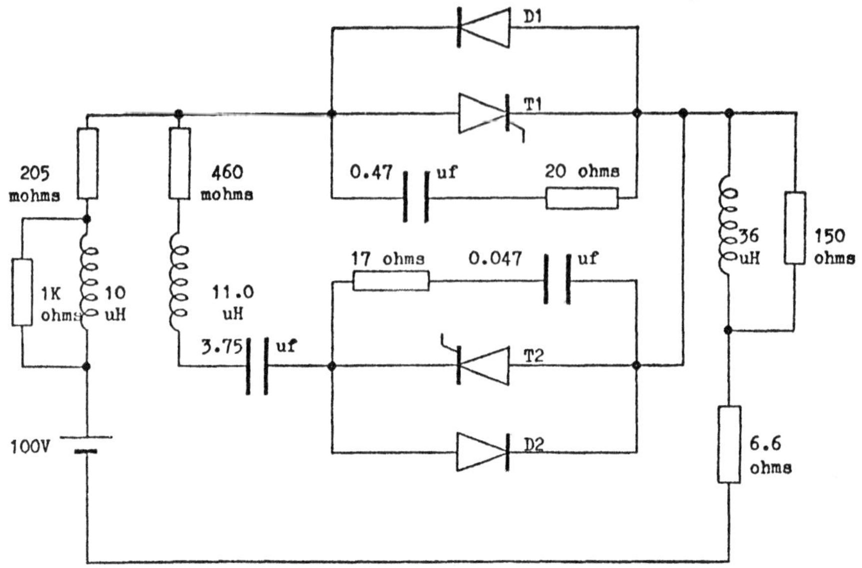

Figure 9. Circuit diagram of thyristor chopper.

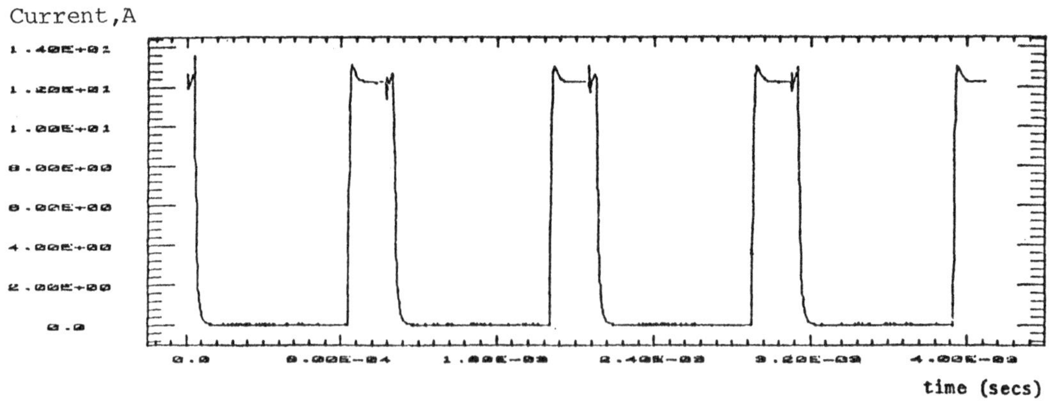

Figure 10. Waveforms in thyristor chopper circuit.

3.3 Special-purpose algorithms.

Microcomputer-based waveform analysers are particularly versatile in permitting the user to develop special purpose algorithms for acquired waveforms. Figure 8 shows a typical example, from which the user may make measurements of rise-time, fall-time, delay-time and storage time. For this simple problem, the user must write a routine to determine the maximum and minimum levels of the waveform. Then the 10% and 90% points can be computed and the relevant time-spans determined. For noisy repetitive waveforms averaging algorithms are worth using prior to parameter extraction.

4. APPLICATIONS.

4.1 Power electronics.

The performance of circuits employing high power semiconductor devices is critically dependent on transient switching characteristics. The waveform analyser described in section 2 has been used for a detailed examination of the switching behaviour of the d.c. chopper circuit shown in Figure 9. In particular, it was used for parameter extraction and compact model development for a power semiconductor circuit simulator. (Prendergast and Jackson, 1979)

During the operation of this chopper circuit load current flows through the primary thyristor T1 and is current-impulse commutated by gating the auxiliary thyristor T2. Gating pulses are applied with rise-times of 200ns. An example of load current flow is shown in Figure 10. Figure 11 demonstrates the versatility of the waveform analyser for comparing circuit simulations made on an IBM 370/165 with measured responses. This Figure shows detailed switching behaviour during the turn-on period of the main thyristor T1.

Many types of semiconductor circuits can be examined in this way. Voltage levels, current levels and timing parameters may vary with different device types but the acquisition, processing, displaying and storage of data is essentially the same. To illustrate this point, the switching behaviour of power transistors for the circuit shown in Figure 12 is examined. In particular, the waveform analyser was employed in a study of the commutation of current from D1 to Q2. During the commutation transient the collector voltage on Q1 changes from a low negative value to nearly Vcc. Recorded waveforms are shown in Figure 13, along with a waveform depicting instantaneous power, calculated by the waveform analyser.

4.2 Signal spectrum measurements.

The waveform analyser was used to determine the amplitude spectra of waveforms arising in the application of electromagnetic fields for the stimulation of bone growth and repair. (Hinsenkamp, 1982) Considerable care is necessary in the use of the FFT algorithm to avoid

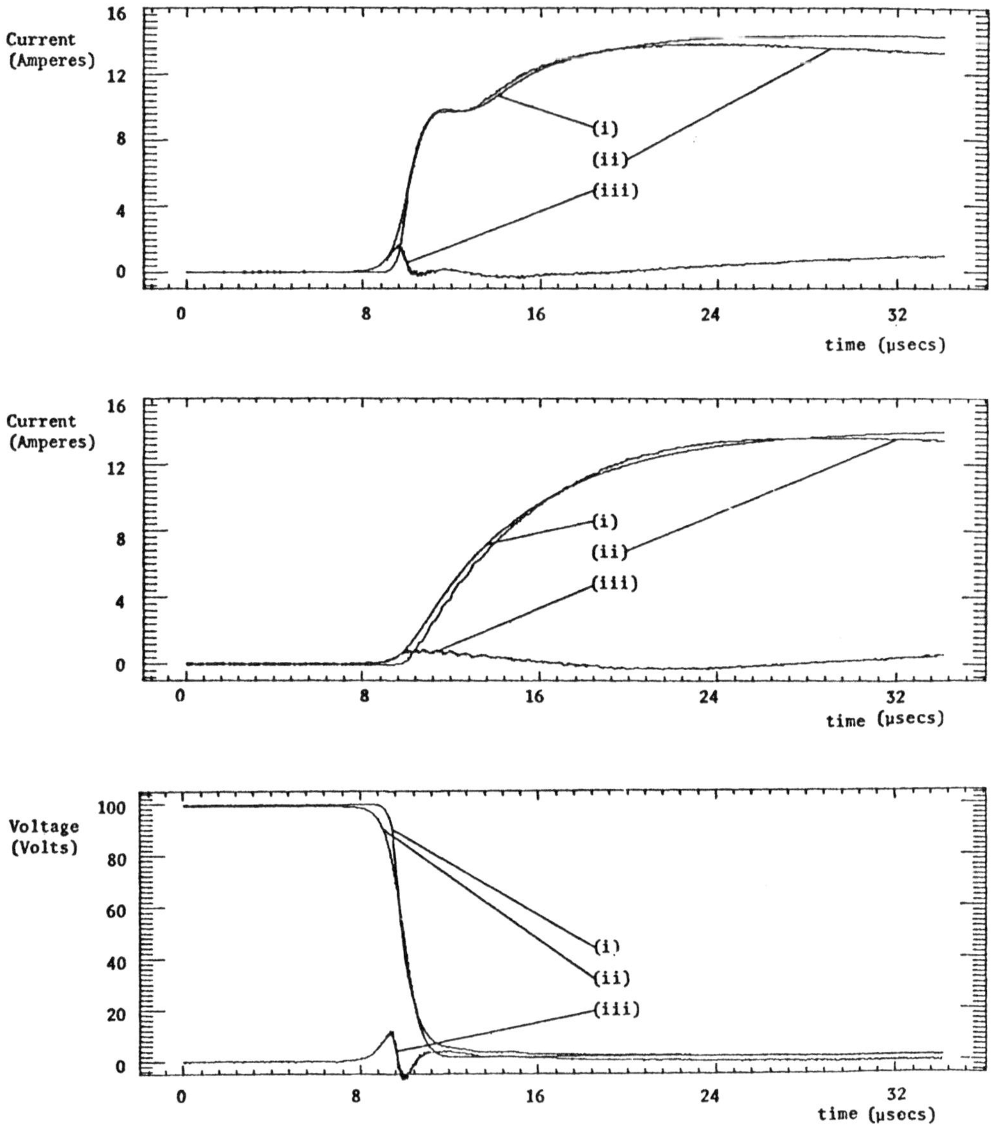

Figure 11. Comparison between (i) simulated and (ii) measured response during the turn-on period of thyristor T1, and showing (iii) error.
(a) Thyristor T1 current.
(b) Load current.
(c) Voltage across thyristor T1.

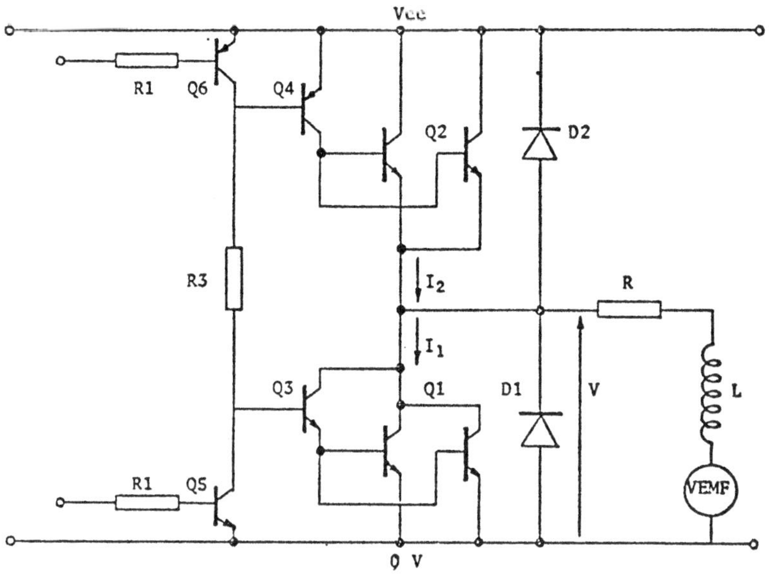

Figure 12. Circuit diagram of power transistor chopper.

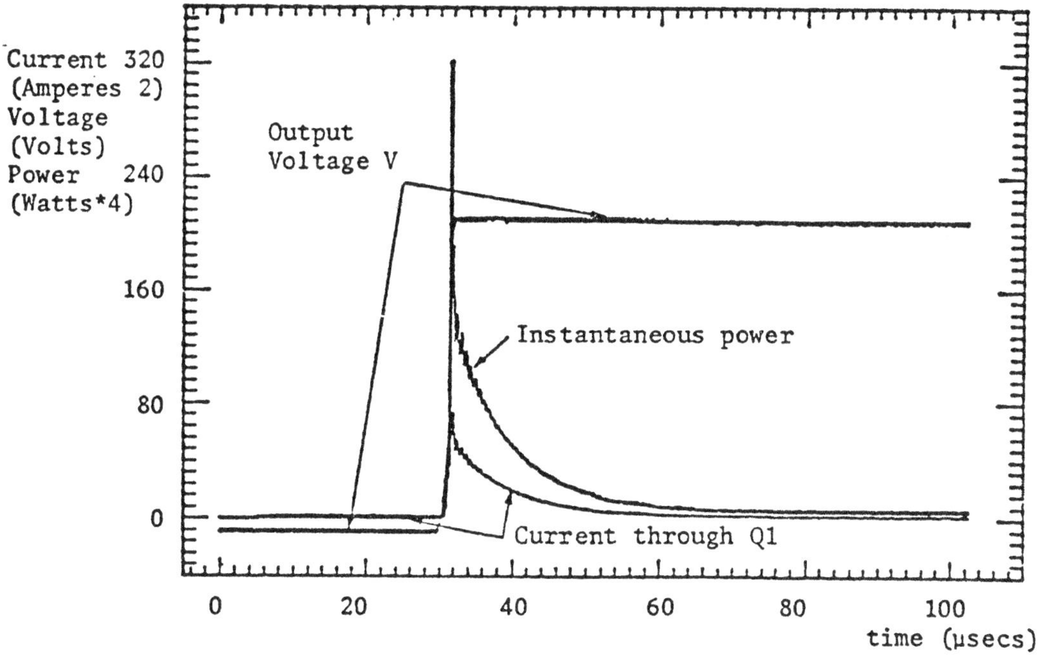

Figure 13. Waveforms in power transistor circuit.

misleading results. The general form of the waveforms under study, representing the voltage output from a search coil in the field, is sketched in Figure 14. It consists of a series of pulses, of width t_1 (5-20us), and repetition frequency approximately 40kHz, occurring for time t_2 (approximately 5ms) with repetition period t_3 (approximately 67ms). The nature of the spectrum is indicated in Figure 14(b). The overall envelope is the spectrum of an individual pulse, the pulse repetition leads to a series of discrete samples of this envelope, each convolved with a sinc function itself sampled at the low repetition frequency. It is not practicable to show the fine detail of the 15Hz frequency lines in an overall spectrum, and the spectrum must be developed by a combination of measurement and analysis.

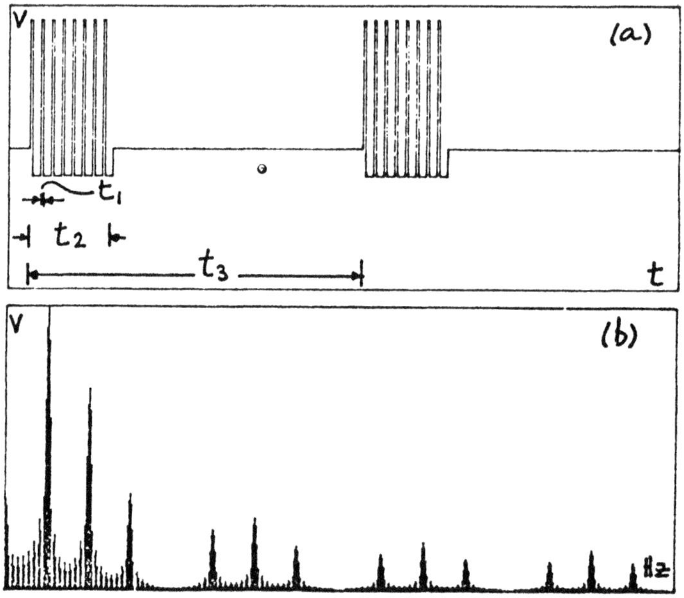

Figure 14. (a) Voltage output from search coil in stimulating field.
(b) General form of Fourier transform of (a).

Figure 15 shows some steps in the evaluation of these spectra. Figure 15(a) shows that 4096 samples is barely adequate to capture the pulses in period t2. There are about 10 samples per pulse, but the lack of resolution is evident in the low frequency ripple apparently modulating the pulse height. For a correct view of the spectral envelope, therefore, it is necessary to take the FFT of an individual pulse. Figures 15(b) and (d) show the individual pulses from two sets of apparatus, with sampling period 50ns, and Figures 15(c) and (e) the results of performing a 256 point FFT on each. Since the functions are zero at n=1 and n=4096, no window function is necessary. The frequency range is 1.25MHz, and the frequency resolution is 4.88kHz, set by the time window τ_ω . The complete spectrum can therefore be developed from these envelopes, and the appropriate sinc functions.

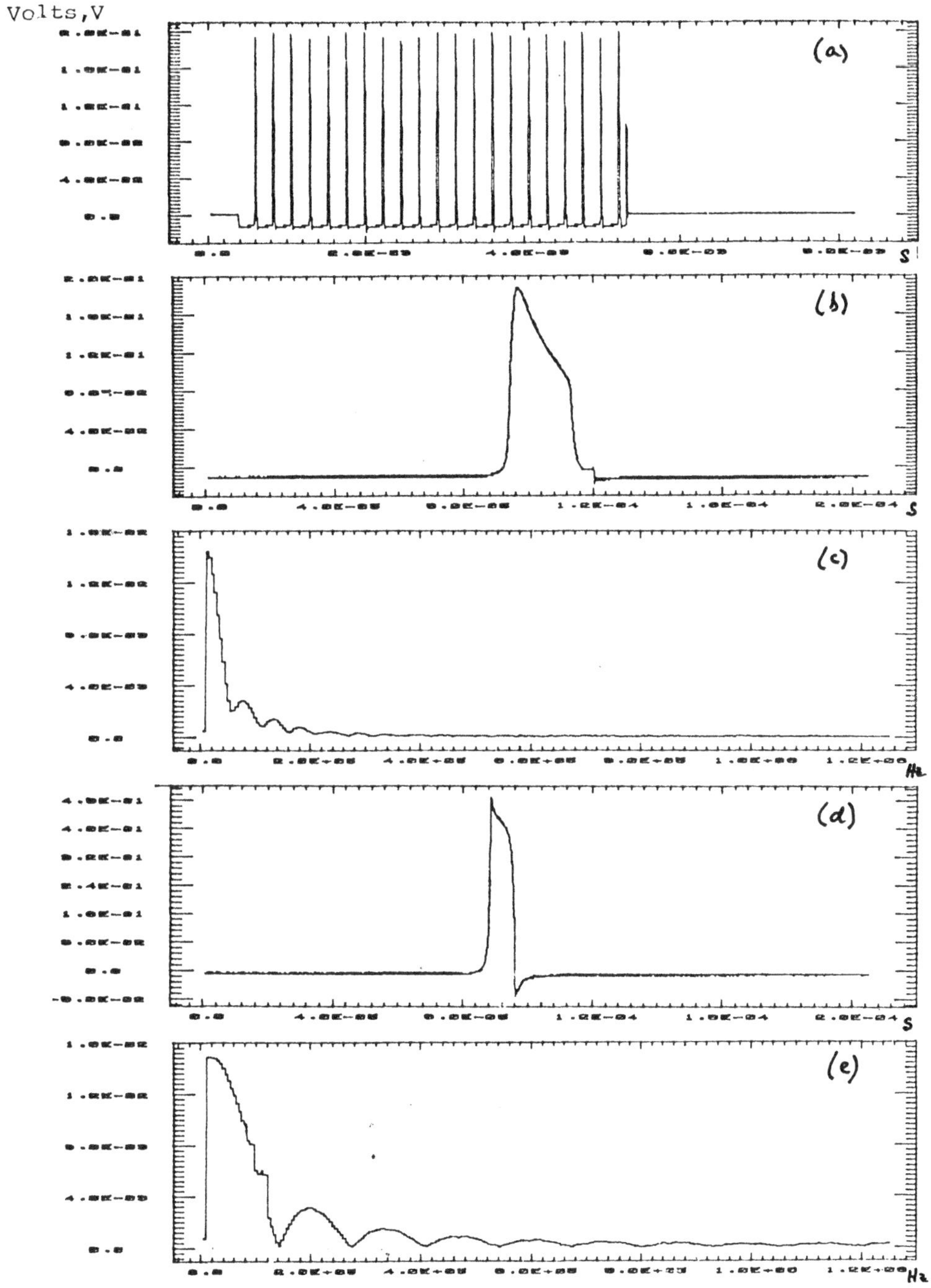

Figure 15. Waveforms and FFTs from bone stimulation investigation.
(a) Burst pulse waveform.
(b) Single wide pulse waveform. (c) 256 point FFT from waveform (b).
(d) Single narrow pulse waveform. (e) 256 point FFT from waveform (d).

5. REFERENCES..

1. Stanley,W.D.;'Digital signal processing', Prentice-Hall, 1975.

2. Hnatek,E.;'A handbook of analogue to digital converters', Wiley,1976.

3. Shoreys,F.;'New approaches to high-speed high-resolution analogue-to-digital conversion', IEE Electronics and Power, Feb. 1982, pp 175-179.

4. Saul,P.H., Fairgrieve,A. and Fryers,A.J.;'Monolithic components for 100MHz data conversion', IEEE Journal of Solid-State Circuits, SC-15, (3), June 1980, pp 286-290.

5. Robin,N.A. and Ramirez, B,; 'Capture fast waveforms accurately with a 2-channel programmable digitiser', Electronic Design, Feb. 1980, pp 50-55.

6. DL 920/922 Transient recorder technical manual, Data Laboratories Ltd.

7. Peatman,J.B.;'Microcomputer-based design', McGraw-Hill, 1977.

8. Croson,E.B.,Carlin,F.H. and Howard,J.A.;'Integrated arithmetic processing unit enhances processor execution times', Computer Design, April 1981, pp 163-171.

9. Prendergast,E.J. and Jackson, R.D.;'Microcomputer-based power semiconductor waveform analyser', Proc IEE, 127, 1980, pp 363-367.

10. Cooley, J.W., and Tukey, J.W.;'An algorithm for the machine calculation of complex Fourier series', Math. Comput., 19, 1965, pp 297-301.

11. Brigham, E.O.;'The fast Fourier transform', Prentice-Hall, N.J., 1974.

12. Prendergast, E.J., and Jackson, R.D.;'Computer-aided design of thyristor circuits; the sparse-network tableau approach', Proc IEE, 126, 1979, pp 369-374.

13. Hinsenkamp, M.,and Burny, F. (Ed); 'Electromagnetic stimulation of bone growth and repair', Supplement 196, Acta Orthopaedica Scandinavia, Vol 53, 1982.

CHAPTER 6

MICROCOMPUTER-BASED INTERACTIVE IMAGE PROCESSING

Fredrick M. Cady
Montana State University
Bozeman, Montana 59717,USA

Robert M. Hodgson
University of Canterbury
Christchurch, New Zealand

ABSTRACT

An analysis of interactive image processing systems is presented. The hardware and software of a microcomputer-based system is described. Typical applications are given. The system design of a multispectral remote sensing system is given and the designs of two new advanced microcomputer-based systems are shown.

1. INTRODUCTION

Microprocessors and other large scale integrated circuit devices are having a profound effect in many fields of activity, including digital image processing. The latter is normally performed with expensive, large scale equipment but the recently developed solid-state imaging array, combined with the microprocessor as a computing element, can lead to low cost but dramatically effective image processing equipment.

A digital image processing system can be subdivided into three parts including (1) the digitization of the image, (2) the performance of an image processing calculation or transformation, and (3) the display of the processed image. Rapid technological advances are occurring in each of these areas. Input transducers for image processing are now being fabricated as two-dimensional arrays of sensors using integrated circuit technology. For example, 800 by 800 element charge coupled device imaging arrays will be used in the large space telescope. Charge coupled devices may have dynamic ranges of up to 1000:1, possess quantum efficiencies as high as 50% and, when operated

S. G. Tzafestas (ed.), Microprocessors in Signal Processing, Measurement and Control, 99–120.

at a low temperature, allow very faint images to be recorded. The processing of images can be done with microprocessor computing elements. These large scale integrated circuits now have up to 500,000 transistors per chip. The processed image is usually displayed with conventional television techniques. Color displays are used either to distinguish between gray levels (false color) or as a means of presenting multispectral imagery.

In this chapter digital image processing is approached from the viewpoint of using simple image processing equipment in an interactive manner. Section 2 reviews interactive image processing systems, illustrated by an image chain which stretches from the original object, through the sensor and the digital processing stage, and finally to the display of the processed image. Section 3 presents a stand-alone microprocessor and charge coupled device (CCD) interactive image processing system that has been developed in our laboratory. The equipment is simple and inexpensive (less than $10,000) but has been remarkably effective in several areas. In Section 4 we describe three advanced microcomputer-based image processing systems under development at the present time.

2. INTERACTIVE IMAGE PROCESSING SYSTEMS

An interactive image processing system, as illustrated in Figure 1, consists of three major subsystems. These are the image conversion system which converts the two-dimensional image to an array of digital words, the digital processing section which performs some transformation of the picture data, and the image display system which presents the processed image for viewing by the operator of the image processing system.

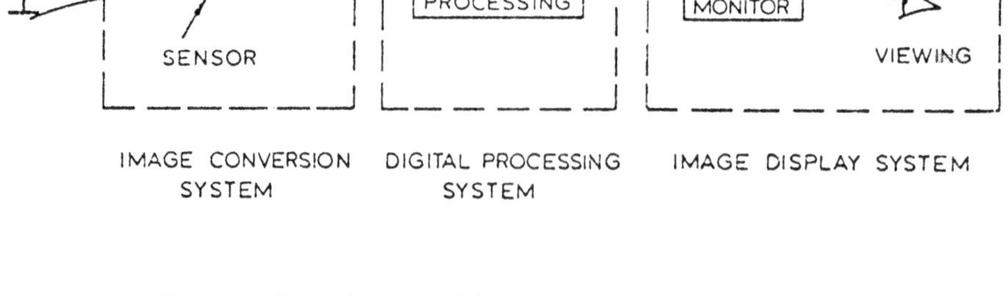

Figure 1. Interactive image processing system

There are many factors affecting the imaging process which must be considered in the design or the analysis of an imaging system. A methodology for this is known as image chain analysis (Booth and Schroeder 1977). The following sections deal with each of the three major blocks of figure 1 and analyze the components within the blocks.

2.1. Image Conversion System

The two-dimensional distribution of intensities over an object must be converted to a digital representation before it can be processed by a digital computer. The intensity of the image is detected by a sensor, or an array of sensors, to give an appropriate analog signal. The detection process requires that the two-dimensional distribution be spatially quantized into picture elements, or pixels. In addition to the spatial quantization, the intensity is also quantized when the analog signal produced by the detector(s) is converted to a digital value.

The image conversion system analyst must consider the scene, its illumination, the transmission path and imaging conditions, the optics, detector, and A/D converter. The object of interest can be characterized by attributes such as reflectivity and size. The optics and the detector must be matched to the conditions established by the scene, illumination, path, and imaging conditions. The optical components used in the imaging system depend to a large extent on the physical attributes of the imaging situation. The aperture determines the intensity of illumination 'seen' by the detector and can be adjusted for differing conditions. The focal length determines the field-of-view of each pixel. The physical image is a distribution of measurable physical properties. Attributes which can be used to describe this physical measurement include the spectral characteristics, intensity, contrast, dynamic range, and spatial frequency content.

The conversion of an image into an analog signal is accomplished by a photoelectric or photochemical mechanism. Solid-state detectors are the prime interest here although many photoemissive cathode devices, such as television vidicon tubes, are in use. Solid-state sensors are becoming widely available and are suitable for many digital image processing applications. They can be made smaller and more robust because they do not require a vacuum or high voltages to operate. They can be mass produced and are potentially less expensive than photocathode devices. Important characteristics of solid-state imaging devices include:

1. Quantum efficiency. Quantum efficiency is a measure of the sensitivity of the device and is expressed as the ratio of the number of effective photons, i.e. those producing observable events, to the number of incident photons. The variability in quantum efficiency as a function of wavelength of the incident radiation is known as the spectral responsivity.
2. Sensitivity. The sensitivity of a solid-state imaging array is

defined as the minimum input flux which gives a detectable output. Sensitivity is therefore related to the noise performance and the responsivity or gain of the detector.

3. Noise. There are many noise mechanisms-photon noise, shot noise and thermal noise-which operate in photoelectric devices. In addition, dark signal and fixed pattern noise act to obscure the desired signal.

4. Dark signal. The dark signal is the component of the output signal which is not generated by incident radiation or by any readout process. It is thermally generated and is a strong function of temperature, doubling every 7°C (Amelio and Dyk 1975).

5. Fixed pattern noise. Fixed pattern noise describes the fixed nature of defective picture elements in array sensors.

6. Readout noise. In some array devices, the process of transferring charge from the storage sites generates noise. These noise mechanisms are usually attributable to capacitive effects but also include the trapping of charge carriers in surface energy states for later release into other signal charge packets (Barbe 1975).

7. Dynamic range. The dynamic range of a device is defined as the ratio of the output signal when an input causes saturation to the output signal with no input radiation. The dynamic range is related to the average noise performance of the device.

8. Signal-to-noise ratio. The noise performance of different devices is usually compared by the signal-to-noise ratio, which gives, for a fixed illumination, the signal available compared to all sources of fluctuating noise.

9. Linearity. Linearity expresses the deviation away from a 1:1 relationship between the change in input radiation and the corresponding change in output signal.

10. Geometrical effects. Geometrical effects include the fixed pattern noise described above, distortions and aberrations introduced by the lens, and non-linear scanning mechanisms in scanning type sensors.

11. Modulation transfer function. The modulation transfer function (MTF) expresses the relative sensitivity of the device to spatial frequencies.

The imaging detectors convert radiation received from the source into an analog signal. The information must now be quantized and converted to a digital representation. This is performed by an analog-to-digital conversion. The input signal from the sensor is first processed by a conditioning element. This provides buffering and isolation, amplifies the signal, converts the polarity to that required by the sample-and-hold unit and bandlimits the input signal. The specification of an analog-to-digital converter for use in an imaging system is similar to that used in any other system. The designer must consider the accuracy, aperture time, conversion time and resolution required. In many imaging cases, high speed conversion is achieved using a sample-and-hold circuit and a parallel A/D Converter.

2.2. Digital Processing System

Image processing terminology tends to be used in slightly different ways by different authors. Green (1977) suggests that there are two broad categories of image processing. These are the subjective enhancement and the quantitative enhancement of images. Subjective enhancement is designed to re-display image data for viewing by a human observer and is usually performed interactively to allow an optimal interpretation of the image. Quantitative enhancement is based on some mathematical model and is not normally an interactive procedure. Quantitative enhancement techniques include geometrical transformations to correct for optical distortion in the picture recording process and the removal of defects caused by the input transducer. Others (Huang 1975; Andrews and Hunt 1977) distinguish between the terms restoration and enhancement. Restoration refers to processing which is designed to reconstruct an image that has been degraded by some process. Enhancement is an attempt to improve the appearance of the image for human viewing. The term reconstruction also appears in the literature (Mersereau and Openheim 1974) and is used to describe the production of an image from data which is generated or gathered in a transform plane, e.g. the Fourier transform plane. Examples of this kind of image reconstruction are found in computer aided tomography (CAT) scanners and synthetic aperture imaging systems.

A combination of these similar views and terms seems appropriate. It is convenient to consider techniques grouped according to their suitability for use in interactive image processing. Some techniques may be used to restore an image interactively, particularly when the degradation process is not well understood and parameters must be varied to obtain the 'best' image. Other techniques may be used in a batch mode of operation. This is often the case when the results of a processing step, e.g. the geometrical rectification of non-linear scanning distortions, are well understood and must be applied before other processing is undertaken.

In general, the total time between the start of processing and when the operator views the image on a display determines whether or not an image processing procedure is used interactively. The major part of this delay in interactive image processing is the time spent processing the image by the computer. In batch mode operations, the delay is predominately the time spent by the program waiting in the job queue. Elapsed time, along with the amount of storage required, can be used to determine if image processing procedures can be performed interactively, and if the equipment upon which the processing will be carried out is suitable.

We now consider digital image processing but restrict ourselves to reviewing the analysis of image processing procedures in general. The reason for this is twofold. First, many excellent reviews of image processing procedures have been published (Stockham 1972; Andrews 1974; Huang 1975; Andrews 1977; Andrews and Hunt 1977; Castleman 1979).

Secondly, a digital image processing procedure can be analyzed to determine whether or not it is suitable for use in interactive image processing. This suitability can be expressed in terms of the time which an operator can wait for results and still be effective in processing images or solving problems.

The analysis of the processing step in interactive image processing systems must consider the algorithm and the accuracy of the process. An important prerequisite for the development of an effective algorithm is a complete understanding of the problem to be solved. This helps ensure that the algorithm will perform as required. Program design techniques such as top down design and structured programming emphasize problem analysis and design before writing programs. Producing efficient algorithms requires an insight into the way an algorithm works. Unconventional approaches can produce highly efficient algorithm implementations. The 'brute force and ignorance' method can also be interminable. Bentley (1979) cites the example of a sorting problem requiring 138 hours of machine time for a naive comparison algorithm and 4 seconds when a hashing technique is used.

In image processing applications, the use of look-up tables can significantly reduce the execution time of programs. However, one must balance the time required to calculate the index into the table against the time to calculate the value itself. The use of in-line code (macros) instead of subroutine calls can speed up the operation. An analysis of the accuracy with which calculations are made can sometimes lead to a significant increase in the execution speed. Integer arithmetic is faster than floating point arithmetic when performed in software, and so changing from a floating point mathematics package to an integer package will allow the program to operate faster.

We now ask, "What is a reasonable processing time limit for interactive image processing?" An interactive processing procedure must be completed within some maximum elapsed time to allow effective user interaction. Comparisons of problem solving performance show the superiority of interactive (time-shared) systems over batch operations, especially where heuristic approaches to problem solving are used (Hansen 1976). The effect of system response time on the performance of users has been investigated and an attempt has been made in some cases to quantify the user's reaction, or 'feeling', toward the operation of the system when system response time has been altered (Goodman and Spence 1978). The man-machine interface has been investigated with the idea of improving certain parameters in the interface to improve the performance of the user. Miller (1977) concludes that various factors such as the output data rate and the flexibility of the command processor affect the performance of the user, and further careful research is needed to adequately define the man-machine interface.

The answer to the question posed above, however, requires an investigation of the psychology of problem solving. Miller (1968) indicates that during a problem solving activity we become aware of response delays greater than two seconds and that delays longer than this are inhibiting and become intolerable during creative activity. However, he also suggests that delays of up to fifteen seconds are tolerable when the user recognizes that a complex activity is required of the computer. Furthermore, it seems likely that when dealing with complex problems such as image processing, an operator may need to think for periods greater than two seconds. Our own experience with image processing procedures indicates that this is true. Therefore we propose that the result of an interactive image processing procedure should be available for viewing by the operator in less than fifteen seconds.

The processing time of a computer is generally specified by the time the central processor unit takes to perform some representative instruction. In some cases, particularly in the competitive microcomputer field, the absolute shortest instruction is specified. This tends to be misleading when attempting to evaluate image processing procedures. Most algorithms characteristically repeat some basic calculation or operation many times. For example, a Fast Fourier Transform (FFT) is characterized by the number of multiplications performed. Typically, these multiplications are performed using floating point arithmetic, and so an appropriate metric for the comparison of computers is the time taken to perform a floating point multiplication.

After the processing is finished, the data is output either to a display or back to a bulk storage medium to await further processing. In most systems, the rate at which data can be transferred to a display is limited by the speed at which the main storage memory can operate. The basic unit in this operation is the memory cycle time which gives the time for a read or write cycle. Memory cycle times can be as low as 10 to 20 ns for very high speed memory, but a more typical time for semiconductor random access memory would be in the range of 100 to 250 ns. The time to transfer data from the main processor memory to a display memory will be the number of bytes times the memory cycle time. This is the fastest time that is possible and is achievable only when using special direct memory access (DMA) techniques. In general, however, the total time required is much less than the 2 to 15 seconds limit.

The pictorial data are usually held in some type of bulk storage. In most image processing systems, the bulk storage medium is a disk unit. The fastest type of disk unit for retrieval of data is the fixed head disk which is commonly found on medium-to-large computer installations. Other popular types of disk storage, especially in smaller microcomputer-based installations, are the floppy disk and the Winchester disk. Magnetic tape is also used in many instances. The storage requirement of the image may sometimes determine the type of bulk storage used. For example, with the large amount of data required for a single Landsat scene (30 megabytes), magnetic tape is invariably used.

2.3. Image Display

The display forms the last link in the image chain. The information must now be of an analog nature and a digital-to-analog converter is included. Display devices can be classed into two general categories. These are devices in which the display is of a transient nature, such as a video display with a continual refresh, and display devices which store the image for an extended time, such as a storage cathode ray tube. Hardcopy of the image is usually produced by a photographic process, but other techniques such as the production of images on thermosensitive or electrostatic paper are also used.

The display device should faithfully recreate the image, at least to the extent of recreating what the human viewer would perceive if the image processing chain were not present. The display exhibits characteristics which relate to the image such as dynamic range, contrast ratio, and luminance. In addition, a display has the operating characteristics of geometrical fidelity, flicker, resolution, and persistance which should be considered.

There are few display devices commercially available which are suitable for interactive image processing work. An effective device must allow the operator to view the results of the processing step within the time limits established for effective interactive problem solving. Therefore, photographic reproduction techniques are eliminated. The television type refreshed display is one of the most popular devices used today. The ease with which full color displays are generated contributes to this popularity, and many systems are offered with color displays. New electronic devices, such as flat panel and electroluminescence displays, are being developed.

The presently available methods for the generation of hard copy images are even less satisfactory than display methods. To date, most hard copy devices available are used in graphics systems where a binary (black and white) copy is produced. An inexpensive, fast, high resolution, hard copy device for gray scale images does not exist.

Hard copy of images are commonly produced photographically. These methods produce images with better resolution and dynamic range than other methods, such as line printer overprinting. There are two techniques which use film as the hard copy medium. In the first, a computer controlled light beam is swept over a film plate. The intensity of the light beam is modulated by the computer to correspond to the intensity of the image. Commercial instruments, such as the Optronics C-4300 Colorwrite machine, produce high quality photographic films of digitized images. Color reproductions are produced by successively writing to the film with three different color beams. Positive transparencies or negatives for color prints can be produced easily. An alternative photographic method is to simply photograph the display monitor. This procedure means the photograph is limited to the same resolution and dynamic range as the display device. Nevertheless, useful photographs can be produced.

3. A MICROPROCESSOR-BASED INTERACTIVE IMAGE PROCESSING SYSTEM

A microcomputer-based image processing system has been constructed at the University of Canterbury, New Zealand. This system, known as the University of Canterbury Image Processing System (UCIPS), saw its first application in the evaluation of a charge coupled device (CCD) as an imaging array. The hardware was constructed so that the system was reasonably portable in order to test the premise that a microcomputer controlled solid-state imaging system could be operated for remote sensing applications in a light aircraft (Hodgson et al. 1979). It soon became clear that the imaging system was very useful and had many potential applications in stand-alone interactive operation.

This section describes the hardware and software which has proved to be so useful and versatile. It begins with a system overview and a description of the component parts. The system used for remote sensing is described and an example of images obtained during the flight trials are given. The software developed for use in interactive image processing is described and a number of applications where the system has been used are given.

3.1 Image Processing System Hardware Overview

Four major hardware subsystems are identified in the system diagram, Figure 2. These are (1) the CCD camera and its associated driving circuitry, direct memory access controller, and memory, (2) the visual display store interrogator (VDSI) and CCIR standard television display, (3) the Intel 8080 microcomputer system with CRT terminal, graphics input joystick and fast line printer and (4) the bulk storage system consisting of a dual drive floppy disk and a digital cassette recorder. The architecture of the system is that of a standard computer with central address, data and control buses through which all devices communicate. In operation, a program which controls the capture and display of information from the CCD camera is loaded into memory from the floppy disk unit. The repetitive process of capturing valid frames of data from the CCD and transferring them to the VDSI continues at about six pictures per second until a key is depressed on the CRT console device. This freezes the last captured picture on the display. The operator of the system then has an option of storing and/or processing the captured picture.

The CCD camera subsystem consists of the CCD and its optical components and drive circuitry, a Peltier effect cooling device, analog-to-digital converter, and direct memory access controller. The charge coupled device integrated circuit is mounted in the focal plane of the lens of a camera. The 3mm x 4mm format of the active area makes it most suitable for 8 mm optics. The thermoelectric cooling device allows the temperature of the CCD to be controlled in the range of -12°C

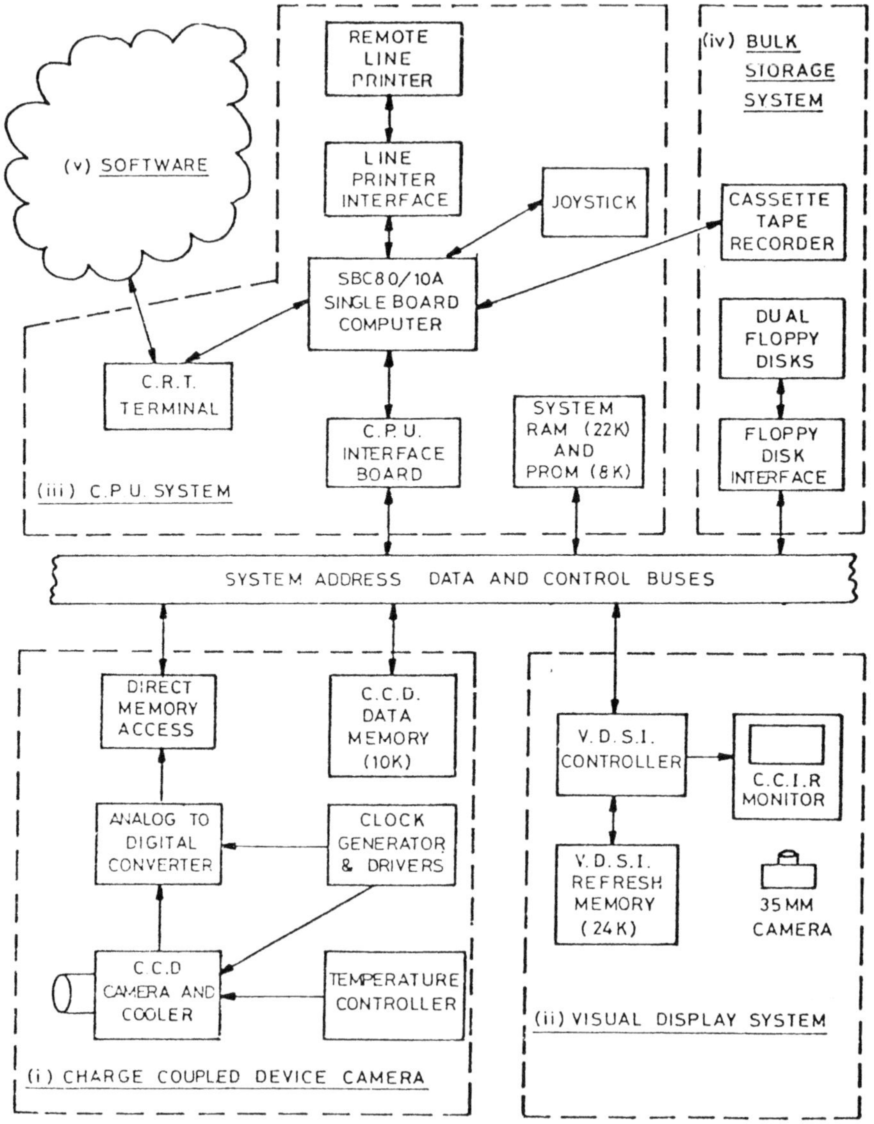

Figure 2. University of Canterbury Image Processing System.

to +30°C (Hodgson et al. 1979). The clock generator produces the required clocking signals for the CCD and generates a sample signal for the analog-to-digital converter. Clock driver/regulators allow the various clocking signals to be independently adjusted for optimum performance of the CCD.

The visual display store interrogator provides two basic functions. The first of these is to sequentially read data from the VDSI refresh memory and to convert these data bytes to a television compatible video signal. The display format is chosen to match the sensor that is being used and can be switched to display a centered picture of 128 x 128 or 100 x 100 pixels. The video is blanked outside the chosen display area except in an area of 16 x 100 pixels reserved at the bottom of the display for a title (see Figure 3). When generating the display each line of CCD information is repeated as four lines on the television monitor. This serves to enlarge the display and preserves the original 4x3 aspect ratio of the CCD imaging array. Twenty-four K bytes of refresh memory have been allocated to the VDSI memory space. This allows the storage of two complete images in memory. A starting address is programmed under software control into the VDSI address generator which allows either of the two images to be displayed. The second basic function of the VDSI controller is to control the access of the microcomputer to the memory dedicated to the display. The controller has been designed so that the microcomputer has absolute priority of access over the display address generator. When this occurs, the display is automatically blanked.

Figure 3. 100 by 100 CCD image displayed by the VDSI

The central processor unit subsystem utilizes a single board computer (Intel SBC80/10A) and contains random access memory, read only memory, and input/output ports. Communications with the operator of the imaging system are via a CRT display unit operating at 9600 Baud. This unit has a set of user programmable function keys which have proved to be very useful in image processing. A line printer interface has been constructed which connects the microcomputer to a remote line printer. The interface allows a number of microcomputer development systems to share a line printer on a single coaxial line. One of the input/output ports on the single board computer is used to interface a joystick to the system. This joystick allows the operator to identify any pixels displayed on the television monitor.

An image processing system requires a storage device for images. The type of bulk storage chosen depends somewhat on the size of the input device being used. A very large sensor, say 400 by 400 pixels will require much more storage than a smaller sensor such as the 100 by 100 array. The University of Canterbury system uses two types of bulk storage. The most commonly used device is the dual floppy disk unit, which stores 21 pictures from the 100 by 100 CCD camera. A digital cassette was chosen for the remote sensing system described in section 3.4 and will store 30 pictures on a single side of a 100 meter cassette tape.

3.2. Image Processing System Software

The software on an image processing system falls into two categories. The first, operating system software, includes programs which are necessary to interact with the floppy disk storage system. The operating system also includes the software development tools to develop and debug a user's software. The second category is the user's image processing software. This software is dedicated to the manipulation and processing of images and is particular to the image processing requirements. The image processing software can be considered to be an operating system because, like the floppy disk operating system, it allows the operator to interact with the floppy disk both for loading and executing programs and for storing and retrieving images. In image processing applications, a software package which allows a choice of image processing routines is generally developed. A program called PIKKY has been written to accomplish these tasks.

PIKKY contains three basic modules. These are (1) the command input and decoding section, (2) the floppy disk handling routines, and (3) the image processing routines inherent in the operating system. In addition, a suite of user generated programs has been developed. A fixed suite of image processing procedures is not sufficiently flexible to allow the investigation of different aspects of image processing. Therefore, an essential element of the picture processing system is its capability to allow a large number of image processing routines to be run under its control.

An image processing command, written as a command line, is entered into the picture processing system through an ASCII serial keyboard. Editing of the command line is very important and the rub-out key is permitted and line echo facilities are provided. The command line is terminated by a carriage return and held in a text buffer in system random access memory. It is therefore available to other user generated software. Command decoding is achieved by matching the command string in the text buffer with a valid command table. Control is transferred to the required routine and returned to the input command processor when the routine is finished. If the command processor fails to find a match for the command line in the command table, it assumes that the command line specifies a processing routine found on the floppy disk unit. It then loads and executes the required routine. Parameters may be passed to any of the processing routines through the command line buffer.

A number of image processing routines are available as standard options within the picture processing system. See Appendix A. These routines form the basis of many image processing procedures, and may be accessed and executed in any order at the whim of the operator. In addition, an often repeated sequence of routines may be defined as a macro or batch string and executed thereafter by a single keystroke.

One of the most powerful features of the interactive picture processing system is its facility to access and run any program which is stored in the floppy disk system. This means that picture processing is not limited to the routines inherent within PIKKY. Other software procedures can be developed and run without requiring the re-compilation of the picture processing software each time a new processing algorithm is to be tested. The concept of external, or user generated, routines is very useful. Many special purpose routines can be developed and tried without interfering with the basic operation of the main picture processing system program. See Appendix A for a list of typical modules developed by PIKKY users.

When developing a system to allow this type of operation, the following details should be considered:

1. A special-purpose routine must either be loaded into the maching at a fixed location or it must carry with it an automatic starting address for use by the software which loads the machine code from the disk.
2. Dedicated control buffers must not be overwritten. Examples of such buffers used in PIKKY include a command line text buffer, the floppy disk system parameter buffer, and a batch mode buffer.
3. The picture buffers must not be overwritten.
4. The user module must return to a fixed point in the main image processing program.

Another technique which has proved to be useful is to allocate space at the beginning of the main processing software for use as a vector table. This table contains either addresses of desired routines or jump instructions which allow a user to access modules within the main program. In PIKKY, a jump table is set aside to allow the user to access the floppy disk routines and other useful internal modules such as the pixel flashing routines.

3.3. Evaluation of Image Processing Software

Several metrics by which routines may be evaluated for their suitability in interactive image processing are defined in section 2.2. The image processing routines within PIKKY and the user generated routines are interactive routines because that is the way they are used, and although the limitations of using a microcomputer in image processing have not been investigated thoroughly, we have performed complicated and useful processing in realistic interactive processing times.

Appendix A shows the execution time and size of a number of useful image processing routines. Most routines are executed in less than one second, much less than the operator's response time. Note that all routines listed have been written in assembler language which is faster and more compact than most high level languages.

3.4. Applications of the University Of Canterbury Image Processing System

The UCIPS was designed to be portable so that flight trials could test that a sysem could be used in a light aircraft to provide calibration for Landsat data. During this development, the system was found to be very useful in its own right as an image processing machine and a number of application areas, in addition to the remote sensing system, can be mentioned.

The UCIPS was developed as part of a program to construct a system to provide local calibration of Landsat satellite imagery (Hodgson et al. 1979). In this application, a multispectral CCD camera system is to be flown over selected test areas simultaneously with the overflight of the Landsat satellite. Data collected by the aircraft system would be used to calibrate the data obtained from the satellite. A single channel test system based on UCIPS has been the outcome of this hardware development (Hodgson et al. 1981). The hardware of the remote sensing system is essentially the same as that in Figure 2 except that equipment in the UCIPS not specifically needed in the flying system is deleted. The floppy disk storage unit is replaced by a digital cassette recorder to reduce size and weight. The CRT terminal and keyboard were considered too bulky and difficult to use in an aircraft. Therefore, a control head was designed to provide operator input during flight. It controls all operations of the tape recorder, allows the selection of one of several modes of camera operation, and generates a run and survey code to identify each picture taken with the system. The initial flight trials of the system were held in August 1979. A number of photographs were taken and examples can be seen in figure 4 (Hodgson et al. 1981).

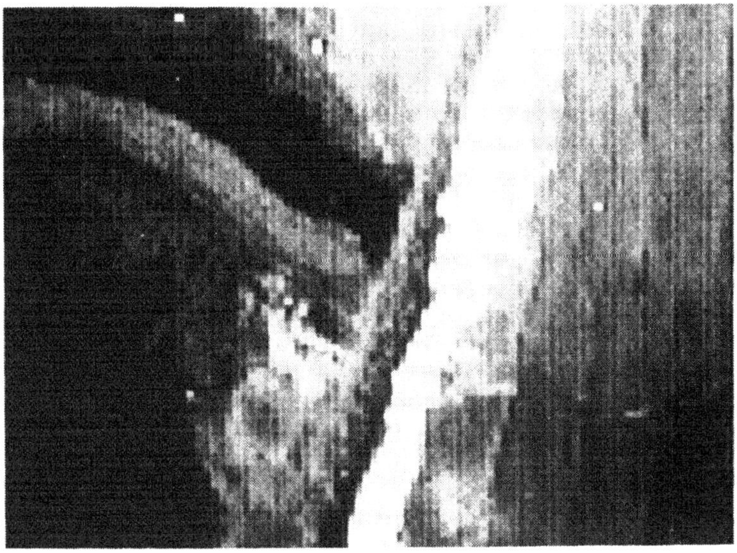

Figure 4a. CCD derived image of a coastal region.

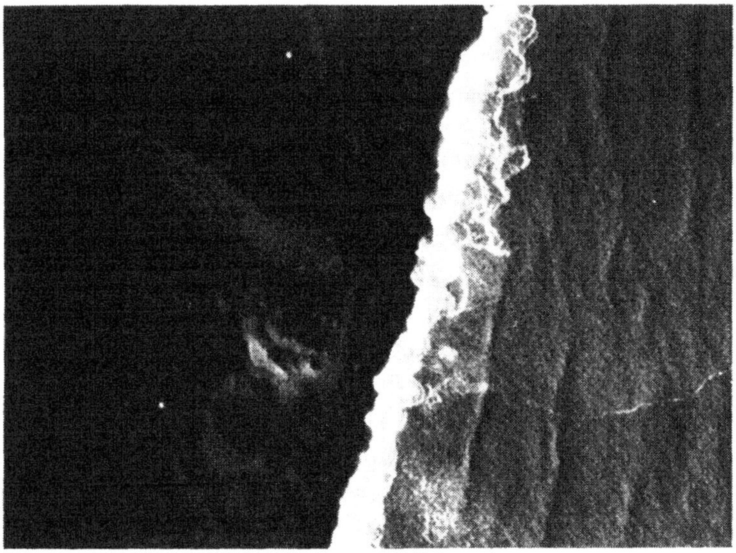

Figure 4b. A conventional photograph of the same region (Landsat Band 5).

A project to investigate the *in vivo* detection of multiple pregnancies in sheep is being sponsored by the Auckland Industrial Development Division of the New Zealand Department of Scientific and Industrial Research. They hope that early detection will reduce lambing mortality rate. We were able to show, using a sample x-radiograph film, that digital edge enhancement techniques could provide an image which was superior to the original and would allow an operator to determine the state of the ewe.

A research program investigating the restoration of images severely degraded by atmospheric turbulence has been ongoing at the University of Canterbury. An integral part of the research is an optics laboratory in which simulated astronomical images are generated. The CCD image processing system allows low light level images to be captured. Simple but remarkably effective image processing techniques have been developed using the system (Bates and Cady 1980; Cady and Bates 1980; Cady 1980).

A microscope fitting allows the CCD camera to be attached to any standard microscope. Initial tests of this attachment with the system have shown that digital processing can be very effective in enhancing photomicrographs. This use of the system has not been persued but possible applications include the determination of growth ring spacing in tree core samples and the calculation of open-to-closed area in very fine mesh screen. Each of these applications makes use of the precise physical construction of the CCD which allows accurate measurements to be made.

4. ADVANCED MICROCOMPUTER-BASED IMAGE PROCESSING SYSTEMS

Three new microcomputer-based image processing systems are now described. A four-channel multispectral imaging system has been developed from the single channel remote sensing system described in section 3.4 (Hodgson and Cady 1982). Two systems incorporating newer sixteen bit microprocessors are under development. The first is a High Resolution Image Processing System (HIP) at the University of Canterbury (Hodgson and McNeill 1982). A second, similar system is under construction at Montana State University.

4.1. A Multispectral Sensing System for Remote Sensing

A multispectral fast area digitizing scanner, dubbed Multi-FADS, is shown in figure 5. The operation of this system is similar to the single channel system described in section 3.4. Dual port memory boards provide frame storage for the four CCD image channels. Each of these boards is mapped into the same memory space and the microcomputer accesses them via a programmable bank-switching interface. The generation of CCD clock signals is performed under software control. The drive waveforms for two-dimensional solid-state array sensors can be generated by gating four basic waveforms with a clock signal. The CCD clock generator memory is programmed with data which describes the states of the four basic waveforms at one microsecond intervals over one complete integration and readout cycle. A form of run length coding (Zwick 1979) is used to code information over many repetitions of the one microsecond basic clock rate. This technique results in a 12-to-1 reduction in the number of bytes of memory needed to drive a Fairchild CCD202 100x100 element sensor.

The sensors in the Multi-FADS are the Fairchild CCD211 244 by 388 charge coupled device imaging arrays. Initially, the clock generator memory will be programmed to select a 128 by 128 image from the central portion of these arrays. If different sensors or other size images are required, the CCD clock generator ROM can be easily re-programmed.

The central processor unit is a single board computer based on the Intel 8080. An operator controls the system via a control head which allows regulation of frame rate and the selection of frame identification codes stored with each image.

4.2. A High Resolution Image Processing System

Hodgson and McNeill (1982) describe a second generation microcomputer-based image processing system. A block diagram of the system is shown in figure 6. The University of Canterbury Image Processing System described in section 3 is considered a low resolution system and is near the practical limit of the image processing capabilities of an eight bit microprocessor. For higher resolution images, say 256 by 256 or larger, a sixteen bit microprocessor is required. There are two main reasons for this. Firstly, high resolution images require large amounts of memory (64K bytes or more). Most eight bit microprocessors have a sixteen bit address bus which restricts their addressable memory to 64K locations without special memory controllers. Secondly, interactive image processing requires high speed program execution. The sixteen bit processors have throughputs typically five to ten times greater than their eight bit predecessors (Intel 1979). Even with this, studies by Cady (1980) indicate that a practical limit for image processing with sixteen bit microprocessors may extend to only about 400 by 400 element images. Above this size, special dedicated hardware must be constructed to give realistic image processing times.

The system architecture is based on an IEEE 796 standard bus. There are three major subsystems--the camera drive electronics, the central processor unit and non-volatile store, and the display generator.

The imaging array is a GEC Type MA357, giving a resolution of 385 elements horizontally and 576 elements vertically. This array is a frame transfer device (Barbe 1975). A thermoelectric cooling device maintains the temperature at a fixed and low temperature, reducing thermally generated noise. The array is driven by software generated

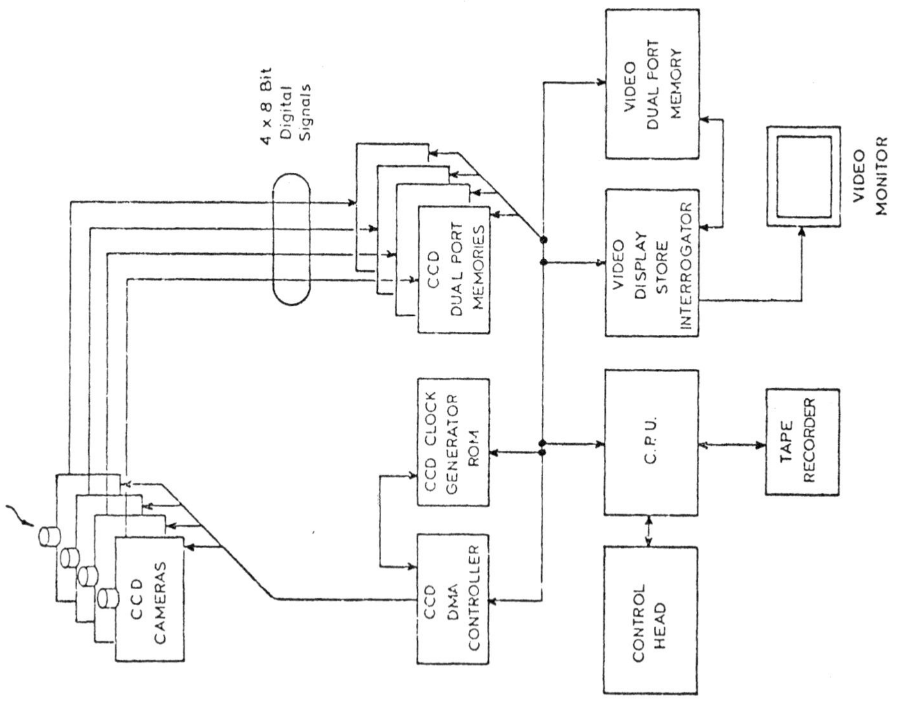

Figure 5. Multispectral imaging system.

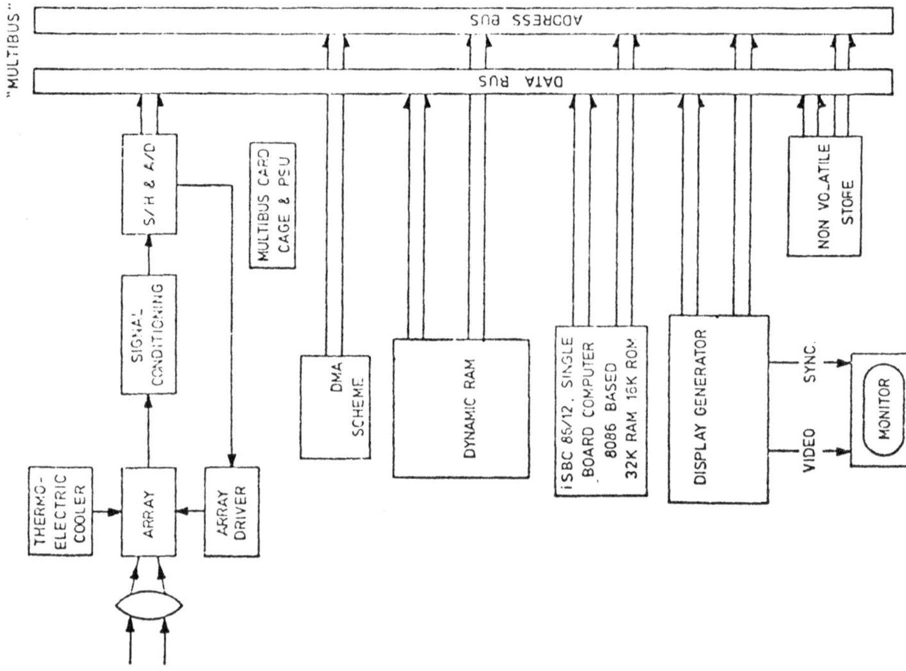

Figure 6. A high resolution system.

clocking signals as described in section 4.1. A sample-and-hold and analog-to-digital converter produce digital data which is transferred to memory using a direct memory access (DMA) scheme.

The central processor unit chosen for this system is the Intel 8086. Its addressing range is one megabyte, allowing several images to be stored in memory simultaneously. The throughput of the 8086 is five to ten times that of the 8080. An additional improvement of five to ten times for arithmetic processing may be achieved by using the 8087 arithmetic co-processor. A 256 by 256 Fourier transform using software floating point routines results in a total run-time of the order of one hour. However, when the 8087 co-processor is used, the transform is produced in 70 seconds, a reduction in time of nearly fifty. A 29 megabyte Winchester disk drive is to be used for mass storage.

4.3. Montana State University Image Processing System

Figure 7 shows an image processing system under development at Montana State University. This system, similar to the HIP system described in section 4.2 differs in that a television camera is used as the input sensor. All other elements of the system are essentially the same.

A television camera was chosen for an input system to test its suitability as a transducer and to compare it characteristics with the charge coupled device camera previously developed. It also allowed commercially available frame grabbers and image frame store and video generation to be used.

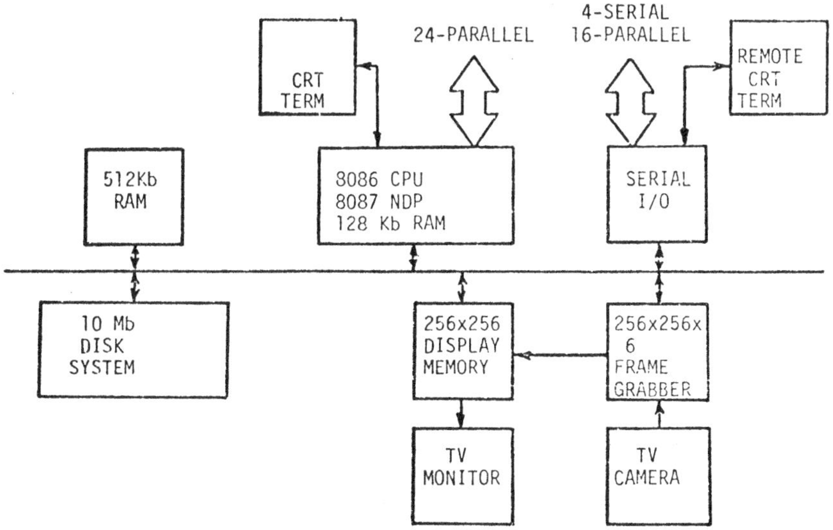

Figure 7. Montana State University Image Processing System

5. Conclusion

A chain of measurements and processes stretching from the object of interest to a representation of that object viewed by an observer has been investigated. Interactive image processing takes place when the observer modifies that representation by invoking a digital processing procedure. To be effective in solving image processing problems, the result of an image processing step should be viewable in a time less than fifteen seconds. Furthermore, the operator should be able to rapidly switch between an image and its processed version to determine if the desired enhancement has been achieved. A microcomputer-based image processing system which has interactive processing capabilities has been described. Although relatively inexpensive, this hardware and software has been very useful. The limitations of a system using an eight bit microprocessor are felt when higher than about 128 by 128 images are required. Two new systems, incorporating sixteen bit microcomputers are under development.

Acknowlegements

Funds for the development of the University of Canterbury Image Processing System were supplied under research contracts with the Physics and Engineering Laboratory, Department of Scientific and Industrial Research, Lower Hutt, New Zealand. Work at Montana State University is supported in part by the MONTS-NSF project ISP-8011449. Some material presented here has been previously published, including Figures 2 and 4 (Cady and Hodgson 1980), Figure 3 (Cady et al. 1981) and figure 5 (Hodgson et al. 1982).

References

1. Amelio, G.F.; Dyck, R.H. (1975) "Fixed pattern noise and cooled photosensor arrays," Proc. Nato Advanced Study Inst. on Solid State Imaging, Lovain-la-Neuve, Belgium, Sept. 1975, pp. 114-122

2. Andrews, H.C. (1974) "Digital image restoration: A survey," Computer, Vol. 7, No. 5, pp. 36-45.

3. Andrews, H.C. (1977) "A potpourri of mathematical techniques applied to image science," Symposium on Current Mathematical Problems in Image Science, Monterey, California, N. Hollywood, California: Western Periodicals, pp. 71-77.

4. Andrews, H.C.; Hunt, B.R. (1977) Digital Image Restoration, Englewood Cliffs, N.J.: Prentice-Hall.

5. Barbe, D.F. (1975) "Image devices using the charge coupled concept," Proc. IEEE. Vol. 63, No. 1, pp. 38-67.

6. Bates, R.H.T.; Cady F.M. (1980) "Towards true imaging by wideband speckle interferometry," Opt. Commun., Vol. 32, No. 3, pp. 365-369.

7. Bentley, J.L. (1979) "An introduction to algorithm design," IEEE Computer. Vol. 12, No. 2, pp.66-78.

8. Booth, J.M.; Schroeder, J.B. (1977) "Design considerations for digital image processing systems," IEEE Computer. Vol. 10, No. 8, pp. 15-20.

9. Cady, F.M. (1980) Applications of Microcomputers in Interactive Image Processing, Ph.D. dissertation, Department of Electrical Engineering, University of Canterbury, Christchurch, N.Z.

10. Cady, F.M.; Bates, R.H.T. (1980) "Speckle processing gives diffraction limited true images from severely aberrated instruments," Optics Letters, Vol. 5, No. 10, pp. 438-440.

11. Cady, F.M.; Hodgson, R.M. (1980) "A microprocessor based interactive image processing system," IEE Proc., Pt. E., Vol. 127, No. 5, pp. 197-202.

12. Cady, F.M.; Hodgson, R.M.; Pairman, D.; Rodgers, M.A.; Atkinson, G.J. (1981) "Interactive image processing software for a microcomputer," IEE J. on Computers and Digital Techniques, IEE Proc., Pt. E., Vol. 128, No. 4, pp. 165-171.

13. Castleman, K.R. (1979) Digital Image Processing, Englewood Cliffs, N.J.: Prentice-Hall.

14. Fairchild (1976) CCD202 Specifications

15. Goodman, T.; Spence, R. (1978) "The effect of system response time on interactive computer aided problem solving," Computer Graphics. Vol. 12, No. , pp. 100-104.

16. Green, W.B. (1977) "Computer image processing - the Viking experience," IEEE Trans. Consumer Electronics. Vol. CE-23, No. 3, pp. 281-289.

17. Hansen, J.V. (1976) "Man-machine communication: an experimental analysis of heuristic problem-solving under on-line and batch processing conditions," IEEE Trans. Systems, Man, and Cybernetics. Vol. SMC-6, No. 11, pp. 746-752.

18. Hodgson, R.M.; Cady, F.M.; Murphy, J.K.N. (1979) "The development of a multispectral scanner to be flown in light aircraft," Proc. of Landsat 79. May 1979, Sydney, Australia.

19. Hodgson, R.M.; Cady, F.M.; Pairman, D. (1981) "A solid-state aircraft sensing system for remote sensing," Photogrammmetric Eng. Remote Sensing. Vol. 47, No. 2, pp. 177-183.

20. Hodgson R.M.; Cady, F.M. (1982) "A system design for a multispectral sensor using 2-D solid-state imaging arrays," IEEE Trans. on Geoscience and Remote Sensing. Vol. GE-20, No. 2, pp. 177-179.

21. Hodgson, R.M.; McNeill, S.J. (1982) "A sixteen bit microcomputer and solid-state array sensor based image capture and processing system," NZIE Annual Conference Proceedings

22. Huang, T.S. (ed.) (1975) Picture Processing and Digital Filtering, Berlin: Springer-Verlag.

23. Intel (1979) The 8086 Family Users Manual. Intel Corporation, October 1979.

24. Mersereau, R.M.; Oppenheim, A.V. (1974) "Digital reconstruction of multidimensional signals from their projections," Proc. IEEE. Vol. 62, No. 10, pp. 1319-1338.

25. Miller, L.H. (1977) "A study in man-machine interaction," Joint Computer Conference.™ Vol. 46, June 13-16, 1977, AFIPS Conference Proc., pp. 409-421.

26. Miller, R.B. (1968) "Response time in man-computer conversational transactions," Joint Computer Conference. Vol. 33, Dec. 9-11, 1968, AFIPS Conference Proc. pp. 267-277.

27. Stockham, T.G. (1972) "Image processing in the context of a visual model," Proc. IEEE, Vol. 60, No. 7, pp. 828-842.

28. Zwick, H.H. "Evaluation results from a pushbroom imager for remote sensing," Canadian J. Remote Sensing. Vol. 5, No. 2, pp. 101-117.

Appendix A. Image Processing Routines

Routine	Process	Size bytes	Execution time - s.
G	Capture data from CCD	10,071	0.16
SAVE,file	Save the picture to the disk	29	20
LOAD,file	Load a picture from the disk	48	6
R	Re-load the last picture	21	6
I,X	Increase picture intensity	26	1
D,X	Decrease picture intensity	31	1
M,X	Mask each pixel with pattern X	23	1
L,X	Plot intensity profile of a line	488	1
EXPND	Linear contrast expansion	166	1
H	Histogram equalization	126	3
E	Exit to system monitor	3	1
F	Bootstrap load disk system	3	7
ON	Turn on the flashing pixel mode	18	1
OFF	Turn flashing pixel off	13	1
SET,XX	Set value of flashing pixel	38	1
GO,XXXX	Unconditional program jump	7	1
BD	Define a batch operation string	82	-
B	Execute batch string	15	-

User Generated Routines

TITLE,text	Write text under the picture	760	1
FIGS,test	Write text onto the picture	700	1
WIPE	Erase text display area	27	1
AVG	Compensate hot spot pixels	67	1
DENSY	Plot intensity distribution	1,100	6
XOR	Exclusive-OR two pictures	23	1
FILTR	2-Dimensional spatial filter	20,500	12
LJUST	Left justify all pixels	67	1
FCNTR	Shift picture so flashing pixel is in center	1,500	3
BCNTR	Find brightest and shift into center	525	3
JOY	Activate joystick	400	-
ADD	Add two pictures	21	1

CHAPTER 7

STATE OBSERVER IMPLEMENTATION VIA MICROCOMPUTERS WITH APPLICATIONS

Naresh K. Sinha
Department of Electrical and Computer Engineering
McMaster University
Hamilton, Canada

ABSTRACT

The microcomputer implementation of a Luenberger-type state observer for dynamic systems is discussed. For the linear case two types of implementations are considered, (a) when a discrete-time model is used, and (b) when the block-pulse function approach is utilized. Since the observations are invariably contaminated with noise, the former approach must include filtering the outputs before applying to the observer. On the other hand, the latter approach was found suitable even without a filter for a small amount of noise. Extension to the more general case of nonlinear systems is also discussed. The Luenberger-type observer can still be used, but the observer gains may have to be adjusted in an adaptive manner. Some results of simulation as well as a brief desscription of the hardware and the software are included.

1. INTRODUCTION

Many control applications require that the states of a dynamic system be available at all instants of time. For example, it is well known that by a suitable design of state-variable feedback, it is possibly to make the system operate in an optimal fashion. Furthermore, for nonlinear systems, it may often be necessary to monitor the values of the state variables in order to track the transition from one approximate linear model to another. This is specially important for systems with regions of bifurcation, leading to catastrophies.

In practice, however, the states are seldom available directly for measurement. Moreover, the measured outputs are often contaminated with noise. For the case of observable linear systems, there are two well-known approaches for estimating the states: one is due to Kalman and Bucy [1961] and the other is due to Luenberger [1964]. Several variations of these methods have been proposed with extensions to nonlinear systems. The Kalman filter was designed to give the optimal

S. G. Tzafestas (ed.), Microprocessors in Signal Processing, Measurement and Control, 121–140.

estimates, in the least-squares sense, in the presence of additive noise, but requires the knowledge of the noise-covariance matrix as well as the initial error covariance matrix. The Kalman gain matrix must then be evaluated by solving a Riccati-type matrix differential equation, although for the stationary case this is reduced to solving an algebraic matrix equation of the Riccati type. The Luenberger observer was originally designed for the deterministic case, i.e. for the case when the output observations are noise-free. In this case, the observer generates a set of states which approach the actual states in an asymptotic manner. The rate at which the error approaches zero can be controlled by a suitable choice of the observer gains. In the presence of noise, the output must be suitably filtered before it is applied to the observer. Both the approaches assume that the dynamic equations of the system are known precisely.

With the availability of inexpensive microcomputers it is natural to investigate their application to the problem of state estimation. Since the Luenberger observer requires much less computation than the Kalman filter, in this chapter we shall discuss the implementation of the observer via microcomputers. A number of modifications will also be considered to take into account the effects of noise and the non-linearities in the system dynamics. Furthermore, although the system itself is usually of analogue nature, the utilization of a digital computer requires using the samples of the input-output data in the observer. Hence the observer has to be designed with an equivalent discrete-time model for the system. Alternatively one may use the continuous-time model with the block-pulse function method. The latter approach appears to have the advantage that the smoothing effect of integration enables the observer to give acceptable estimates in the presence of zero-mean random noise even without a filter.

The chapter is organized in the following way. In the next section the problem of state estimation will be described, along with some practical considerations. This will be followed by the microcomputer implementation of state observers for the case of noise-free observations. Two types of implementation will be discussed, and a subsection on block-pulse functions will be included for completeness. The fourth section describes observers for the realistic case of noisy outputs. This will be followed by a description of schemes for observing the states of a nonlinear system. Finally, some simulation results of the use of microcomputer-based state estimators will be described.

2. THE STATE ESTIMATION PROBLEM: SOME PRACTICAL CONSIDERATIONS

Consider a dynamic system described by the state equation

$$\dot{\underline{x}} = \underline{f}(\underline{x},\underline{u},t) \tag{1}$$

where $\underline{x} \in R^n$ is the state vector, and $\underline{u} \in R^m$ is the input vector. The output of the system is given by

$$\underline{y}(t) = \underline{g}(\underline{x},\underline{u},t) + \underline{v}(t) \tag{2}$$

where $\underline{y} \in R^r$ is the output vector and $\underline{v} \in R^p$ is a zero-mean random noise vector contaminating the output.

Our problem may then be defined as obtaining the best possible estimate of the state $\underline{x}(t)$ from measurements of the input $\underline{u}(t)$ and the output $\underline{y}(t)$.

From the theory of ordinary differential equations, it follows that we can obtain $\underline{x}(t)$ as the unique solution of equation (1) for a given input $\underline{u}(t)$, provided that the initial state $\underline{x}(0)$ is known. However, in practice, $\underline{x}(0)$ is not known. The basic idea behind the Luenberger observer, which was originally proposed for a linear system, is to use a model for the system to generate the estimate of the states for the given input and arbitrarily assumed initial states. The actual output of the system is then compared with the output of the model obtained from the estimated states, and the difference is fed back into the model in such a manner that the estimation error decays asymptotically to zero. This can be understood easily through the following linear model

$$\dot{\underline{x}} = A\underline{x} + B\underline{u} \tag{3}$$

$$\underline{y} = C\underline{x} \tag{4}$$

where it is assumed that the output is noise-free.

The observer is described by the following model

$$\dot{\hat{\underline{x}}} = A\hat{\underline{x}} + B\underline{u} + K[\underline{y}(t) - C\hat{\underline{x}}] \tag{5}$$

Defining the observation error as

$$\tilde{\underline{x}}(t) = \underline{x}(t) - \hat{\underline{x}}(t) \tag{6}$$

It is seen, by subtracting equation (5) from (3), and using (4) and (6), that

$$\dot{\tilde{\underline{x}}} = (A - KC)\tilde{\underline{x}} \tag{7}$$

Thus, the observation error $\tilde{\underline{x}}(t)$ will approach zero asymptotically, if all the eigenvalues of the matrix A-KC have negative real parts. Furthermore, for an observable system, one can select the gain matrix K in such a manner that the eigenvalues of A-KC (called the poles of the observer) can be placed at arbitrarily specified locations. Thus, there is a considerable degree of freedom in the choice of K.

The actual implementation of the observer for a linear system then requires setting up a model of the observer on a computer after the gain matrix K has been selected suitably. The estimate of the states of the system is then obtained from the model. If an analogue computer were to be used for this purpose, the solution of the resulting differential equation would provide the estimated states directly. On the other hand, if a digital computer is used for setting up the model of the observer, we must use the samples of the input, $\underline{u}(t)$ and the output, $\underline{y}(t)$. In this case we must either replace the continuous-time model in equation (3) by an equivalent discrete-time model or use the block-pulse function method while retaining the continuous-time model. With both approaches, it is understood that the sampling interval has been selected carefully, and Shannon's sampling theorem is satisfied. Furthermore, in both cases, the computation for obtaining the estimate of the state must be carried out within a fraction of the sampling interval in order that the procedure may be used on-line.

Another important consideration is that the measurements are almost always contaminated with noise. This will affect the implementation of the observer since the difference between $\underline{x}(t)$ and $\hat{\underline{x}}(t)$ will never be zero. One way to overcome this difficulty is to filter the data before application to the observer. To be effective, this approach requires knowledge of the noise statistics. This will be discussed in Section 4. It may be added that the block-pulse function approach, which is based on approximate integration of the state equations, tends to reduce the effect of noise.

Finally, it must be borne in mind that all physical systems have some nonlinearities. Fortunately, The Luenberger observer can be easily modified to take care of the nonlinearities. The modification will be discussed in section 5.

3. MICROCOMPUTER IMPLEMENTATION OF STATE OBSERVERS: LINEAR NOISE-FREE CASE

The implementation of the state observer with a microcomputer can be done in two ways. In the first approach, after a suitable selection of the sampling interval, a discrete-time model for the system is derived. The use of this model with the observer is very straightforward, as integration is now replaced by matrix multiplication. The only problem is the derivation of a suitable discrete-time model, and for the noise-free case the accuracy of the observer depends upon this model. This will be described in detail in the next subsection.

3.1 State-Observer Using Discrete-Time Approximation

A popular approach for obtaining a discrete-time equivalent of the linear system described by equation (3) is to assume that the input is held constant between the sampling constants. In this case, the discrete-time model is given by

$$\underline{x}(kT + T) = F\underline{x}(kT) + G\underline{u}(kT) \tag{8}$$

where T is the sampling interval, and the matrices F and G are defined as

$$F = e^{AT} = I + AT + \frac{1}{2!}(AT)^2 + \frac{1}{3!}(AT)^3 + \ldots \tag{9}$$

$$\text{and} \quad G = \int_0^T e^{At}\,dt = [U + \frac{1}{2!}(AT) + \frac{1}{3!}(AT)^2 + \ldots]T \tag{10}$$

Since the input to the continuous-time system is seldom held constant between the sampling instants, a much better approximation is to replace equation (8) by

$$\underline{x}(kT+T) = F\underline{x}(kT) + 0.5\,G[\underline{u}(kT) + \underline{u}(kT+T)] \tag{11}$$

The observer is then described by the following difference equation

$$\hat{\underline{x}}(kT+T) = F\hat{\underline{x}}(kT) + 0.5\,G[\underline{u}(kT) + \underline{u}(kT+T)] + K[\underline{y}(kT) - C\hat{\underline{x}}(kT)] \tag{12}$$

Implementation of the observer on a microcomputer is thus obtained simply by carrying out the matrix multiplications and additions indicated on the right-hand side of equation (12) to obtain the estimated state $\hat{\underline{x}}(kT+T)$ for different values of k. To start the algorithm $\hat{\underline{x}}(0)$ may be taken as the null vector, although any arbitrary initial guess will also work. Since the matrices A and B are known, it is only necessary to calculate the matrices F and G initially. With a suitable choice of the sampling interval, the spectral radius of AT is less than 0.5, and the power series given in equation (9) and (10) converge rapidly. Also, the gain matrix, K, must be selected in such a manner that all the eigenvalues of (F - KC) be inside the unit circle.

3.2 State Observer Using Block-Pulse Functions

Although the use of block-pulse functions to linear systems has developed only recently (Sannuty 1977, Shieh et. al. 1978, Palanisamy and Bhattacharya 1981), they appear to be ideally suited for digital processing of continuous-time signals when samples of the latter are available. A brief review of the theory of the block-pulse functions will first be presented in this subsection. This will be followed by application to the state estimation problem.

The jth element of an N-element block-pulse function, over the interval $0 \leq t \leq T_o$, where $T_o = NT$, can be defined as

$$\phi_2(T) \triangleq \begin{cases} 1 & \text{for } (j-1)T < t < jT \\ 0, & \text{otherwise} \end{cases} \tag{13}$$

and is shown in Fig. 1.

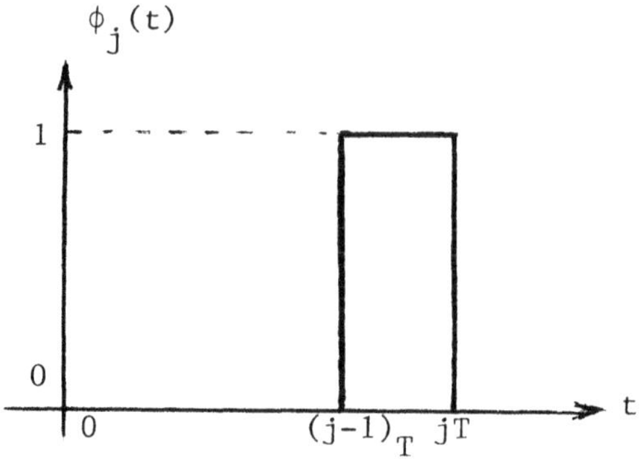

Fig. 1: The jth element $\phi_j(t)$ of a block-pulse function.

The block-pulse function is then defined as the vector

$$\underline{\phi}(t) \triangleq \begin{bmatrix} \phi_1(t) \\ \phi_2(t) \\ \cdot \\ \cdot \\ \cdot \\ \phi_N(t) \end{bmatrix} \tag{14}$$

If we are given a function of time, y(t), which is integrable over the interval $0 < t < T_o$, we can approximate it as

$$\hat{y}(t) = \underline{\phi}'(t)\ \underline{y} \tag{15}$$

where

$$\underline{y} = \begin{bmatrix} y_1 \\ y_2 \\ \cdot \\ \cdot \\ \cdot \\ y_N \end{bmatrix} \tag{16}$$

with the superscript ' representing transposition and

$$y_i \triangleq \text{average value of } y(t) \text{ over the interval } (i-1)T < t < iT \tag{17}$$

It has been shown by Kwong and Chen (1981) that as $N \to \infty$, the block-pulse functions form a complete orthogonal set, so that an arbitrarily small mean-square error in approximation will be obtained by increasing the number of terms in the series. In practice, the sampling interval, T, can be selected on the basis of the frequency spectrum of y(t), following the sampling theorem of Shannon.

Now let us consider the integration of the jth element, $\phi_j(t)$ of the block-pulse function. The integral is shown in Fig. 2.

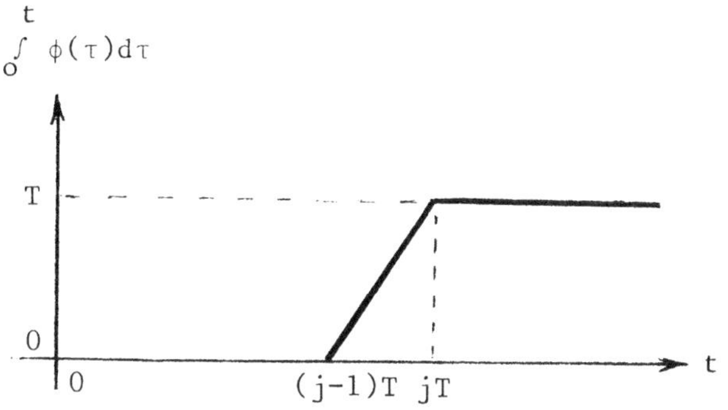

Fig. 2 Integral of $\phi_j(t)$

Using the approximation indicated in equation (15), we get

$$\int_0^T \underline{\phi}(\tau)\, d\tau \simeq TH\underline{\phi}(t) \tag{18}$$

where

$$H = \begin{bmatrix} 0.5 & 1 & 1 & . & . & . & 1 \\ 0 & 0.5 & 1 & . & . & . & 1 \\ 0 & 0 & 0.5 & . & . & . & 1 \\ . & . & . & . & . & . & . \\ . & . & . & . & . & . & . \\ . & . & . & . & . & . & . \\ 0 & 0 & 0 & . & . & . & 0.5 \end{bmatrix} \tag{19}$$

It may be noted that H is an upper triangular matrix of the Toeplitz type. This property is useful for multiple integration, as well as for developing a recursive algorithm.

It follows that the integral of the time function $y(t)$ may be approximated as

$$I_1(t) \triangleq \int_0^t y(\tau)\, d\tau \simeq \underline{I}_1' \, \underline{\phi}(t) \tag{20}$$

where

$$\underline{I}_1' \triangleq \underline{y}' TH = [I_{1,1} \; I_{1,2} \; \cdots \; I_{1,N}] \tag{21}$$

and

$$\left.\begin{aligned} I_{1,1} &= \frac{1}{2} TY_1 \\ I_{1,2} &= TY_1 + \frac{1}{2} TY_2 \\ I_{1,3} &= TY_1 + TY_2 + \frac{1}{2} TY_3 \end{aligned}\right\} \tag{22}$$

etc.

Due to the upper triangular Toeplitz form of H, the following recursive relationship is easily derived

$$I_{1,i} = I_{1,i-1} + \frac{T}{2}(y_i + y_{i-1}) \tag{23}$$

which is seen to be identical to the trapezoidal rule for integration. However, it should be recalled that y_i is the average value of $y(t)$ over the interval $(i-1)T < t < iT$. In case only the samples $y(1T)$ are known, a reasonable approximation is to make

$$y_i = 0.5\, [y(iT-T) + y(iT)] \tag{24}$$

The procedure for integration described above can be directly applied to solving state equations. It is easily shown that the approximate solution of the linear system described by equation (3) is given by

$$\underline{x}(kT+T) \simeq (E - \frac{AT}{2})^{-1} [A\, \underline{I}(k-1) + \frac{T}{2} A\, \underline{x}(kT) + B\, \underline{J}(k) + \underline{x}(0)] \tag{25}$$

where

E is the nxn identity matrix

$$\underline{I}(k) = \underline{I}(k-1) + \frac{T}{2} [\underline{x}(kT) + \underline{x}(kT-T)] \tag{26}$$

$$\underline{J}(k) = \underline{J}(k-1) + \frac{T}{2} [\underline{u}(kT) + \underline{u}(kT-T)] \tag{27}$$

and $\underline{I}(0)$ and $\underline{J}(0)$ are set as the null vectors.

It may be noted that we need the inverse of the matrix $(E - A\frac{T}{2})$. As this is a constant matrix for a given choice of T, it can be calculated and stored for future use.

To implement the state observer using the block-pulse function method, we only have to solve iteratively the state equation for the observer given by (5), which may be rearranged as

$$\dot{\hat{\underline{x}}} = (A-KC)\,\hat{\underline{x}} + B\,\underline{u} + K\,\underline{y}(t) \tag{28}$$

which is similar to equation (3), except that A has been replaced by (A-KC), and we have the additional input K $\underline{y}(t)$. Assuming a zero initial condition for $\hat{\underline{x}}$, we can obtain a recursive solution for (28), similar to that given in equation (25), i.e.,

$$\hat{\underline{x}}(kT+T) \simeq (E-A_o\frac{T}{2})^{-1}\,[A_o\,\hat{\underline{I}}(k-1) + \frac{T}{2}A_o\,\hat{\underline{x}}(kT) + B\,\underline{J}(k) + K\,\underline{P}(k) + \hat{\underline{x}}(0)] \tag{29}$$

where

$$A_o = A - KC$$

$$\hat{\underline{I}}(k) = \underline{I}(k-1) + \frac{T}{2}\,[\hat{\underline{x}}(kT) + \hat{\underline{x}}(kT-T)] \tag{30}$$

$$\underline{P}(k) = \underline{P}(k-1) + \frac{T}{2}\,[\underline{y}(kT) + \underline{y}(kT+T)] \tag{31}$$

and

$\hat{\underline{I}}(0)$ and $\underline{P}(0)$ are null vectors.

The computation indicated in equation (29) is straightforward and can be done on-line if the inverse of $(E - 0.5A_o T)$ is calculated earlier and stored. Furthermore, the integration of the output, as indicated by equation (31), has a smoothing effect on the noise.

4. STATE-OBSERVERS FOR NOISE-CONTAMINATED OUTPUTS

As mentioned earlier, the Luenburger observer does not take into account the effect of noise in the output, which is almost invariably present. In this case, since $\underline{y}(t)$ contains noise, the error between $\underline{x}(t)$ and $\hat{\underline{x}}(t)$ will not approach zero. If information about the statistical properties of the signal as well as noise is available then one may overcome this difficulty by filtering the outputs before they are applied to the observer. This filter may be designed on the basis of the least mean-square error criterion, following the principles of the Wiener filter. It may be noted that the requirement of knowledge of the statistics of the signal and the noise corresponds to the requirement of the knowledge of the covariance matrices in the Kalman filter.

In most practical situations such prior knowledge of the statistical properties of the signal and noise is seldom available. One way out of this difficulty is to use Widrow's adaptive filter utilizing the steepest descent method to seek the optimal solution. The configuration of the filter is shown in Fig. 3, where $\underline{w} \in R^{\ell}$ is the adaptive filter weighting vector, chosen in such a manner as to minimize

$$J = E[\{\underline{y}(k) - \hat{\underline{y}}(k)\}' \{\underline{y}(k) - \hat{y}(k)\}] \text{ for all } k \tag{32}$$

if the output of the filter is

$$\hat{\underline{y}}(k) = \underline{w}(k)' \, Y(k) = Y(k)' \, \underline{w}(k) \tag{33}$$

and

$$Y(k) = [\underline{y}(k-1) \; \underline{y}(k-2) \; \ldots \; \underline{y}(k-\ell-1)]' \tag{34}$$

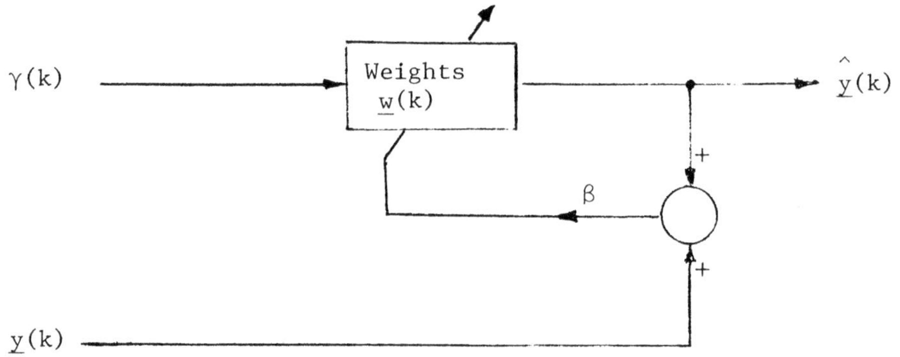

Fig. 3: Configuration of Widrow's Filter

The weighting vector $\underline{w}(k)$ is updated at each iteration according to the relationship

$$\underline{w}(k+1) = \underline{w}(k) + 2c \, Y(k) \, \underline{\beta}(k) \tag{35}$$

where

$$\underline{\beta}(k) = \underline{y}(k) - \hat{\underline{y}} \tag{36}$$

$$c\max > c > 0 \tag{37}$$

and

$$c\max = E[\underline{y}(k)' \, \underline{y}(k)] \tag{38}$$

The implementation of Widrow's filter is shown in Fig. 4.

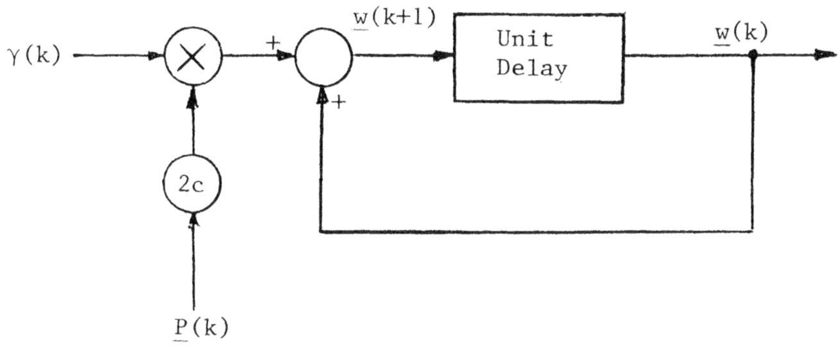

Fig. 4: Implementation of Widrow's Filter

5. OBSERVERS FOR NONLINEAR SYSTEMS

The Kalman filter as well as the Luenberger observer were originally derived for linear systems. The use of the extended Kalman filter for nonlinear systems was subsequently developed. In this case, a linearized version of the nonlinear dynamic model is utilized, with the stipulation that the linearized model is updated after each step using the latest estimates of the states. As a result, the amount of computation required is rather excessive, and the accuracy of the methods may also be affected. Other methods of nonlinear state estimation (second-order filter and invariant imbedding) may give more accurate estimates, but require even more computation (Azim and Sinha 1978). With the present state of art in the microcomputer area, none of these methods can be used for on-line estimation in real time.

A variation of the Luenberger observer can still be used. The nonlinear observer for the system described by equations (1) and (2) may be defined by the differential equation

$$\dot{\hat{\underline{x}}} = \underline{f}(\hat{\underline{x}},\underline{u},t) + K(\underline{y}-\hat{\underline{y}}) \tag{39}$$

where $\hat{\underline{y}}$ is the estimated output obtained from the state estimate $\hat{\underline{x}}$ through the relationship

$$\hat{\underline{y}}(t) = \underline{g}(\hat{\underline{x}},\underline{u},t) \tag{40}$$

In order that the observer may give estimates which converge to their actual value, it is necessary that the state estimation error $\underline{\tilde{x}}$, defined in equation (6), approaches zero as t is increased. Subtracting equation (39) from (1), we have

$$\dot{\underline{\tilde{x}}}(t) = \underline{f}(\underline{x},\underline{u},t) - \underline{f}(\underline{\hat{x}},\underline{u},t) - K[\underline{g}(\underline{x},\underline{u},t) - \underline{g}(\underline{\hat{x}},\underline{u},t)] \tag{41}$$

Hence, the problem is to select the gain matrix, K, in such a manner that equation (41) represents an asymptotically stable system.

It is interesting to note that, unlike the case of linear systems, the right-hand side of equation (41) is not an explicit function of $\underline{\tilde{x}}(t)$. The Taylor series expansions of $\underline{f}(.)$ and $\underline{g}(.)$ about a certain operating point, $\underline{x}_o$, may be used for simplification. Retaining only the first-order terms, we get the following approximation for equation (41)

$$\dot{\underline{\tilde{x}}} \simeq (A-KC)\underline{\tilde{x}} \tag{42}$$

where

$$A \stackrel{\Delta}{=} \left.\frac{\partial \underline{f}}{\partial \underline{x}}\right]_{\underline{x} = \underline{x}_o} \tag{43}$$

and

$$C \stackrel{\Delta}{=} \left.\frac{\partial \underline{g}}{\partial \underline{x}}\right]_{\underline{x} = \underline{x}_o} \tag{44}$$

Hence, one must select the gain matrix in such a manner that the real parts of the eigenvalues of (A-KC) have sufficiently large negative values.

Since equation (42) is an approximation, valid only for small changes in the state, it would appear that one must calculate A and C as the operating point changes, and recalculate K for each case. This would, therefore, require almost the same amount of computation as is required for the extended Kalman filter.

It should be noted, however, that the actual location of the eigenvalues of (A-KC) is not important, as long as it is understood that these eigenvalues have sufficiently large negative real parts. One may, therefore, select the elements of the matrix K in such a manner that this condition for stability is satisfied even if the operating point varies within a large range of values. For very large changes in the operating point, it may be necessary to adjust the values of the elements of K. In many practical problems, it may be possible to divide the state space into a small number of regions so that for each of these, a certain value of the K-matrix will be satisfactory. These values may be conveniently stored in the computer in the form of a look-up table, and used in an adaptive manner.

The block diagram of the observer, including the adapter, as well as the L.M.S. adaptive filter is shown in Fig. 5, where the block called "PLANT" indicates the nonlinear system under consideration.

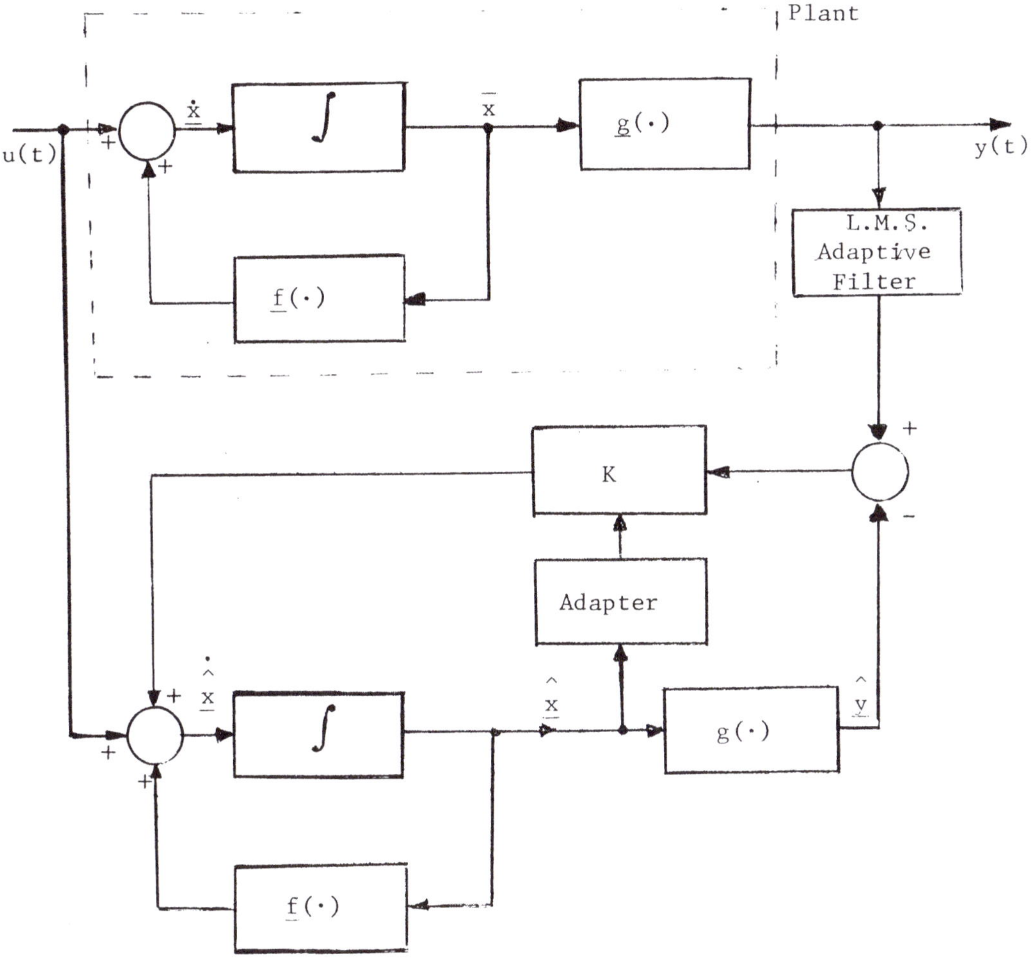

Fig. 5: Adaptive Observer with Filter for Nonlinear System

6. RESULTS OF SIMULATION

Several examples of state observers were simulated for demonstrating the feasibility of the microcomputer-based observer for on-line state estimation. The hardware used for these simulations are described in the appendix.

The first example considered was that of a second-order linear system, described by the following equations

$$\begin{bmatrix} \dot{x}_1 \\ \dot{x}_2 \end{bmatrix} = \begin{bmatrix} 0 & -2 \\ 1 & -3 \end{bmatrix} \begin{bmatrix} x_1 \\ x_2 \end{bmatrix} + \begin{bmatrix} 1 \\ -2 \end{bmatrix} u(t) \tag{45}$$

$$y = \begin{bmatrix} 0 & 1 \end{bmatrix} \begin{bmatrix} x_1 \\ x_2 \end{bmatrix} + v(t) \tag{46}$$

The input applied to the system was given by

$$u(t) = 2 \cos 0.23t - \cos 2.76t - \cos 0.98t \tag{47}$$

The output of the system, in the absence of noise, can be calculated as

$$\begin{aligned} y(t) = c_1 e^{-t} + c_2 e^{-2t} &- 7.758186 \cos(0.23t - 0.283129) \\ &+ 0.971396 \cos(2.76t - 1.562932) \\ &+ 2.641298 \cos(0.98t - 0.909646) \end{aligned} \tag{48}$$

where the constants c_1 and c_2 depend upon the initial conditions.

The linear system as well as the input function were simulated on an analogue computer, and the initial conditions were set such that c_1 and c_2 were zero. A sampling internal of 0.05 seconds was selected, and both the versions of observer described in section 3 were implemented for two cases, (a) noise-free output, and (b) the output contaminated with white noise so that the noise-to-signal ratio was 10 percent.

The equivalent discrete-time model for the system can be easily calculated in the form given by equation (11) with

$$F = \begin{bmatrix} 0.9976 & -0.0928 \\ 0.0464 & 0.8584 \end{bmatrix} \tag{49}$$

and

$$G = \begin{bmatrix} 0.0547 \\ -0.0916 \end{bmatrix} \tag{50}$$

The eigenvalues of F are located at 0.9512 and 0.9048. In order to place both the eigenvalues of the observer at 0.5 in the Z-plane, the gain matrix was obtained as

$$K = \begin{bmatrix} 5.246 & 0.8561 \end{bmatrix} \tag{51}$$

Setting the initial condition

$$\hat{x}(u) = \begin{bmatrix} 0 \\ 0 \end{bmatrix} \tag{52}$$

it was found that for the noise-free case the states of the observer were almost indistinguishable from those of the original system within 15 sampling intervals. In the presence of noise and without the filter, however, the performance of the observer was quite unsatisfactory since, even after 40 sampling intervals, there was considerable difference between the states of the observer and those of the system. With the filter added, the performance was considerably improved, and the difference between the states was reduced to about 1% after 25 sampling intervals.

Next, the observer using block-pulse functions was simulated. In this case the gain matrix was selected as

$$K = [-34 \quad -9] \tag{53}$$

placing the poles of the observer at -6 in the s-plane. The observer was tested with zero initial condition as before. As expected, this observer worked quite well in the presence of noise even without a filter. After about 25 sampling intervals, the error was reduced to about 2%. With the filter the error was less than 0.5% after 25 sampling intervals.

Next, a second-order nonlinear system was simulated. This system is described by the following equations

$$\left.\begin{aligned} \dot{x}_1 &= x_2 \\ \dot{x}_2 &= -3x_1 = -3x_1 - 2x_1^2 - x_2 + u \end{aligned}\right\} \tag{54}$$

and

$$y = x_1$$

Linearizing the system about $x_1 = 0$ and $x_2 = 0$, a second-order observer was designed to have eigenvalues at -3 and -4. It is given by the following equations

$$\left.\begin{aligned}\dot{\hat{x}}_1 &= \hat{x}_2 - 6y \\ \dot{\hat{x}}_2 &= -3\hat{x}_1 - 2\hat{x}_1^2 = \hat{x}_2 + u - 3y\end{aligned}\right\} \quad (55)$$

The nonlinear system was simulated on the analogue computer and a step input was used. The estimates were obtained for the noise-free case as well as with 10% noise added to the output. The results are shown in Fig. 6. It will be seen that the nonlinear observer with the filter works quite well even in the presence of noise.

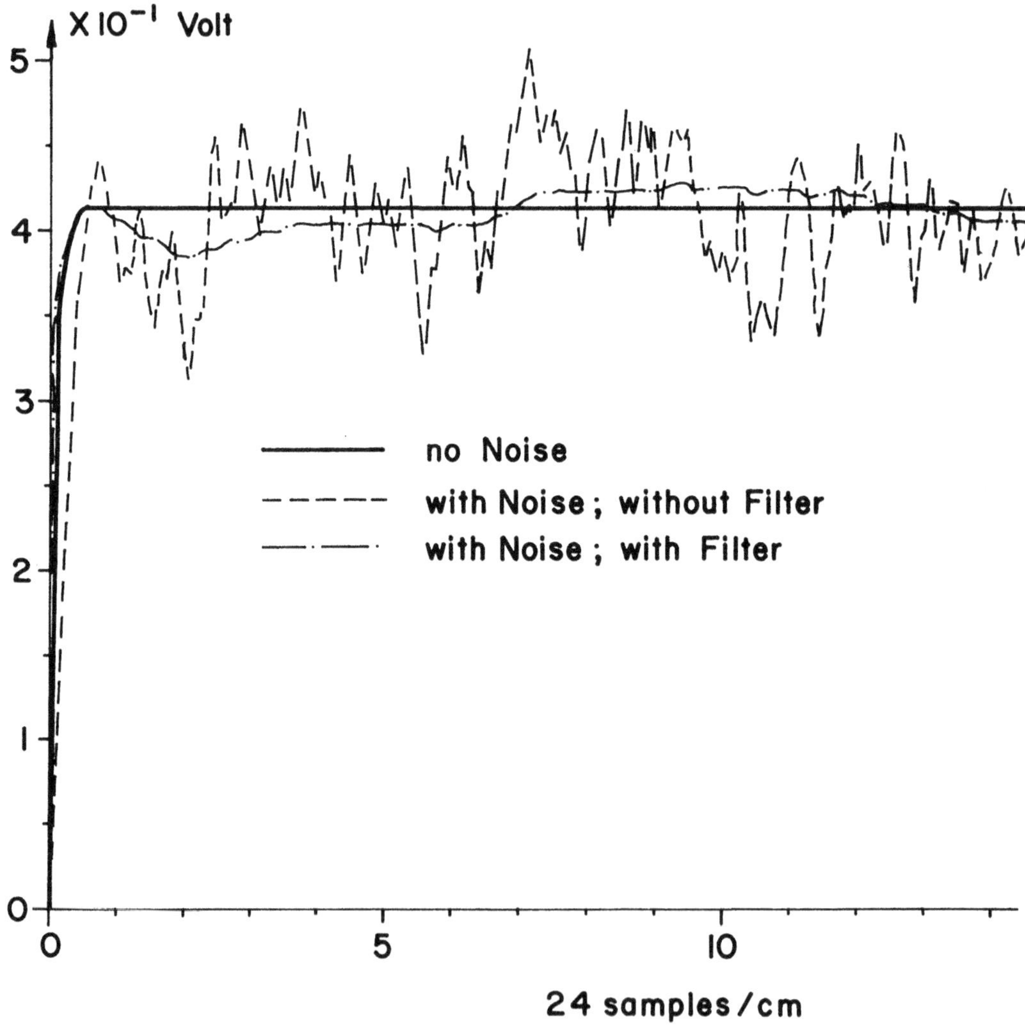

Fig. 6: Estimates for the Nonlinear Case

The estimation scheme shown in Fig. 5 was also tried for estimating the orbital states of a communications satellite from ground-based measurements of its range, azimuth and elevation. The dynamic equations of the motion of the satellite are nonlinear and of the sixth order. Measurements are made every ten seconds, which is sufficient for solving the observer equation on a microcomputer. The results of simulation (Li, 1980) again indicate that the observer states track the actual system states very closely. However, stability of the algorithm could be guaranteed only if the initial error is small.

7. CONCLUSIONS

We have discussed two methods for implementing the Luenberger observer for a linear system via a microcomputer. When the discrete-time model of the system is used for the observer, it is normally necessary to filter the system output before application to the observer since some noise is always present. On the other hand, with the block-pulse function implementation, a filter is not necessary for low noise-level. In both the cases one must select the microprocessor carefully. For on-line operation, it is necessary that the computations for the observer be completed within one sampling interval. The choice of the sampling interval is also important, and depends upon the system time constants. As a general rule of thumb the sampling interval T must be selected so that $0.5 \leq wT \leq 0.2$, where w is the magnitude of the system pole farthest from the origin of the s-plane.

As most physical systems have some nonlinearities, extension of the observer to the nonlinear case has also been discussed. It is seen that this is fairly straightforward with a microcomputer, but in some cases one would need an adaptive observer for altering the observer gains as the system states change. One can store in the computer a look-up table whereby this adaptation can be carried out conveniently. As the adaptation is based on the estimated states, the scheme may turn out to be unstable if the initial estimation error is large. This requires further work.

8. REFERENCES

Azim, S.A. and Sinha, N.K., "Comparison of different methods of nonlinear state estimation with applications to a satellite", 1978, Proc. 16th Annual Allerton Conf. on Communication, Control and computing (Monticello, Ill.), pp. 202-212.

Kalman, R.E. and Bucy, R.S., "New results in filtering and prediction", 1961, Trans. ASME J. Basic Eng., sec. D, pp. 95-108.

Kwong, C.P. and Chen, C.F., "The convergence properties of block-pulse series", 1981, Int. J. Systems Science, vol. 12, pp. 745-751.

Li, M.H., "Nonlinear observer for the orbital states of an earth satellite", 1980, M. Eng. Trans., McMaster University.

Luenberger, D., "Observing the status of a linear system", 1966, IEEE Trans. Mil. Electronics, vol. MIL-8, pp. 74-80.

Palanisamy, K.R. and Bhattacharya, D.K., "System identification via block-pulse functions", 1981, Int. J. Systems Science, vol. 12, pp. 643-647.

Sannuti, P., "Analysis and Synthesis of dynamic systems via block-pulse functions", 1977, Proc. I.E.E., vol. 124, pp. 569-571.

Shieh, L.S., Yeung, C.K. and McInnis, B.C., "Solution of state-space equations via block-pulse functions", 1978, Int. J. Control, vol. 28, pp. 383-392.

Sinha, N.K., Law, S.S.Y. and Li, M.H., "Microcomputer-based adaptive observer for nonlinear systems", 1981, IEEE Trans. on Industrial Electronics and Control Instrumentation, vol. IECI-28, pp. 136-140.

APPENDIX: DETAILS OF HARDWARE AND SOFTWARE

The hardware used for these studies consist of three different types of equipment. The first is the TR-20 analog computer. The integrators, the summing amplifiers, the potentiometers, and the multipliers are used for simulating the system.

The second type of major equipment is the INTEL MDS 800 microcomputer together with MDS-016 16K RAM, increasing the memory size to 32,000 words, and an MDS-DDS dual density disk system which allows half-million-byte data capcity. the CPU module uses INTEL's powerful NMOS 8-bit 8080 microprocessor. Its 2-µs cycle time, 78 instructions, unlimited subroutine nesting, vectored interrupt, and DMA capabilities are fully utilized in the system. Bus control logic resolves bus conflicts between the CPU module and other modules capable of acquiring control of the bus. The CPU module interfaces with a 6-line address bus a bidirection 8-line data bus. The 8080 status signals are decoded and utilized for memory and input/output operations. An 8-level nested interrupt priority system, complete with an interrupt priority push-down stack, resolves contention for 8080 interrupt servicing. The front panel control module provides system initialization, priority arbitration, and real time clock functions through a 1-ms interrupt request generator, which can be disabled under program control. The system is also equipped with a general-purpose input-output module, a high-speed math board, a cathode-ray terminal, and a high-speed line printer. Loading and recording or programs can be done easily through the dual floppy disks.

The analogue and digital computers are coupled together through the third type of equipment, the interface panel. Mounted on this panel are the sampling devices and the A/D and D/A conversion boards. Six channels of A/D conversion with 12 binary digits for signals not exceeding 10 V and two channels of D/A conversion are available. Control circuits to coordinate the sampling process are also mounted on this panel.

Control of the analogue computer is performed through a manual mechanical switch. However, the program provides a feature for coordinating the on-off switching of the analog computer and the starting of the identification algorithm. This is done by making the computer repeatedly check the data buffer of one channel at the beginning. If the analog computer is off, this data buffer is zero. As soon as a certain threshold value is detected in this data buffer, it is understood by the program that the analog computer has been switched on and it will proceed with the rest of the program. The nonzero threshold is necessary due to the presence of a small amount of noise in the system. Experience showed that a level of 75 mV was adequate.

The INTEL MDS-800 microcomputer was used to process the input and output from the analogue computer after analogue-to-digital conversion. The observer equations were implemented on the microcomputer with the SBC-310 Mathematical Board from INTEL to speed up the arithmetical operations as compared to a complete software realization. The programs were written in the 8080 assembly language with hardware interrupt capability. The flow chart of the program is shown in Fig. 7.

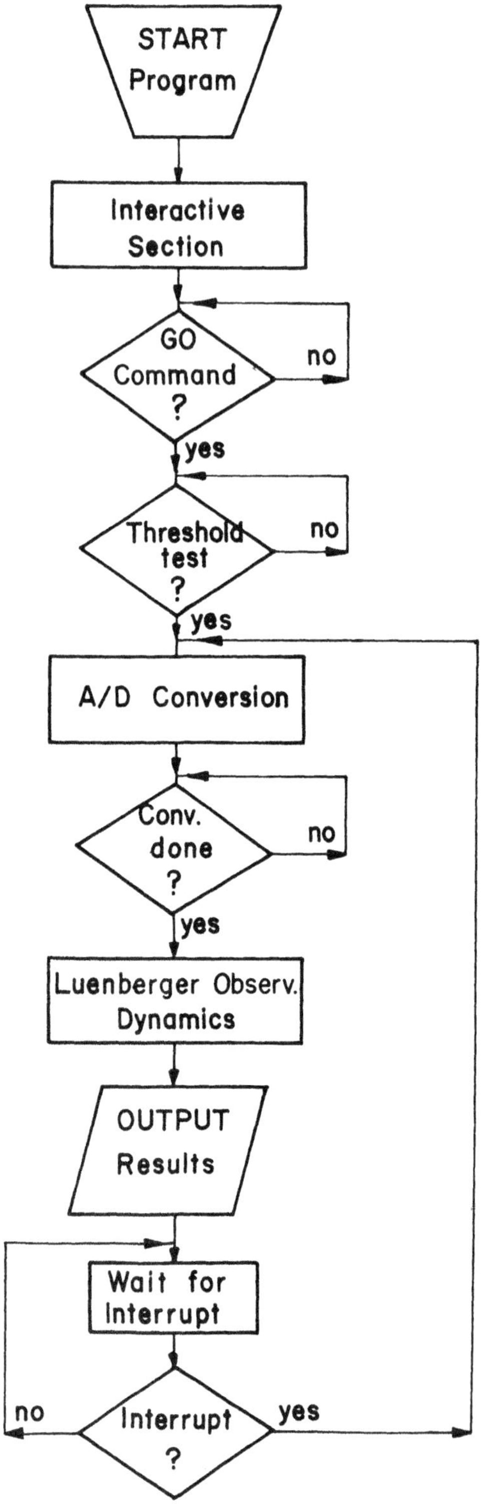

Fig. 7: Flow Chart

CHAPTER 8

STATE ESTIMATION USING MICROPROCESSORS FOR PROCESS SUPERVISION AND CONTROL

Andrea Holmberg and Jussi Orava
Systems Theory Laboratory
Helsinki University of Technology
02150-Espoo 15, Finland

ABSTRACT: Microprocessor-based state estimation in connection with process supervision and control is treated in this article. The state-of-the-art is briefly analysed on the basis of a literature survey, which confirms that very few on-line applications exist. The most commonly used state estimation algorithms, i.e. the Luenberger type observer, the Kalman-Bucy filter, the extended Kalman filter, as well as a nonlinear extension especially suitable for on-line computation, are reviewed with some comments concerning their applicibility to microprocessor-based process control. Applications concerning on-line state estimation of the activated sludge waste water treatment process and a pH control process are introduced.

1. INTRODUCTION

A question always appearing at some stage of the planning and implementation of process control systems is: Are the variables required for control measurable on-line? Unfortunately, the answer is often no. Sometimes no measurement devices or sensors exist for monitoring the relevant variables. Very frequently, off-line analyses are required, which add a time-delay to the measurements. In some cases the relevant control variables are obtained indirectly by utilizing their more-or-less complex relationships with measurable variables. The increased number of process computer installations, however, improves the possibilities of using mathematical estimation algorithms for determining unmeasurable state variables. In many situations, knowledge of unmeasurable variables is of importance not only for control purposes but also for process monitoring and for detection of malfunctions. Such information is also particularly valuable in process industry applications in which routine analyses are made only 1-2 times a day or even less frequently. Continuously updating state estimates, even if not exact, reveal the trends of process behaviour and can thus contribute in detecting severe process disturbances at an early stage.

The literature concerning state estimation algorithms is rich. Starting from the Kalman-Bucy filter (1960, 1961) and the Luenberger observ-

S. G. Tzafestas (ed.), Microprocessors in Signal Processing, Measurement and Control, 141–158.

er (1964) for linear systems, numerous variations of these algorithms, including extensions for nonlinear systems, have been published. Efficient utilization of these estimation methods require thorough knowledge of the model parameters and of the noise affecting the system. These comparatively stringent requirements are reflected in practice by the very low number of realized state estimation applications in process industry. In fields in which the process models are very accurately known, such as space technology, servomechanics and electrical engineering, state estimation methods have, however, been successfully applied.

Estimation algorithms generally require much computation time and computer storage. The crucial point when considering mini- and microcomputer implementations is therefore whether the requirements of storage capacity and computation time can be met.

1. STATE-OF-THE-ART

In order to obtain an overall view of the field of microprocessor-based state estimation a literature search was carried out in March 1982 using the ESA Information Retrieval Service. The results are summarized below.

Search	Key Words	Number of publications
1	State estimation $\cap$ (Microprocessor $\cup$ Microcomputer)	9
2	Supervision $\cap$ (Microprocessor $\cup$ Microcomputer)	26
3	State estimation $\cap$ Process control	82

Search 1 gave only 9 references, of which two treated estimation algorithms and 5 concerned electric power plants. None of the papers treated process industry applications.

Search 2, in which 'state estimation' was replaced by 'supervision' yielded 26 references, most of which treated electrical applications such as transmission network control. A large number of the articles were general in nature and discussed trends and general aspects of microprocessor applications for supervision purposes rather than concrete applications.

In Search 3 the key words 'state estimation' and 'process control' were used in order to provide some insight into the background of microprocessor-based state estimation. Of the 82 references obtained in this search, a rough classification based on the abstracts of the papers showed that there was a marked distribution between different fields of process industry. Twenty-six of the references concerned applications in chemical or related industry, to which 5 papers concerning biochemical applications could also be added. Other branches such as nuclear reactors, power plants, metallurgical industry and the paper industry were represented by less than 3 papers each. In addition there were about 20 theoretical papers, mainly treating algorithms. The majority of the applications were further theoretical in the sense that they studied simulated processes.

An observation concerning the methods used in the process control studies is that the Kalman filter and its modifications was used in more than 20 cases and Luenberger type state observers in less than 10 papers.

Process-oriented approaches in which the states were reconstructed by simulation and parameter estimation were used in about 20 papers, while the remainder treated various types of nonlinear filter algorithms. The great number of papers concerning state estimation of chemical processes is not incidental but shows that the need for estimation or reconstruction of nonmeasurable state variables is great in this field. Chemical process variables can often not be measured on-line because sensors and analyzers do not exist. The need for replacing such analyses with mathematical algorithms is thus obvious.

3. STATE ESTIMATION ALGORITHMS

It is generally considered that the development of state estimation algorithms began with the pioneering work of Kalman and Bucy in the beginning of the nineteen-sixties after the introduction of modern control theory, which utilizes the state, and not only the process output, for feedback control.

There is some ambiguity concerning the terminology associated with state estimation. In this article we use the terms *state estimation* and *state reconstruction* as synonyms meaning generally a mathematical method for determination of nonmeasurable state variables. There are various *state estimation algorithms* or *state estimators*. Depending on which algorithm is used the state estimator may be an *(Luenberger) observer*, a *Kalman-Bucy filter*, an *extended Kalman filter (EKF)* or some type of nonlinear filter.

The estimation algorithms may be derived for systems given in continuous-time or discrete-time form. In order to improve readability, only continuous forms are given in this article. Equations of this form can be implemented straightforwardly using digital computers after choosing a suitable integration method for the differential equations in question.

A common state space model of a nonlinear continuous process can be expressed by the equations

$$\dot{x}(t) = f(x(t), u(t), t) + w(t) \tag{1}$$

$$y(t) = g(x(t), u(t), t) + v(t) \tag{2}$$

in which x is the n-dimensional state, u the given m-dimensional input, and y the r-dimensional output or measurement while w and v are additive process and measurement noises. The corresponding linear time-variant system model is given by

$$\dot{x}(t) = A(t)x(t) + B(t)u(t) + w(t) \tag{3}$$

$$y(t) = C(t)x(t) + D(t)u(t) + v(t) \tag{4}$$

in which A(t), B(t), C(t) and D(t) are time-varying matrices. The state estimation task is to determine the state x from the known measurement y.

The reconstructed state denoted by $\hat{x}$ can then be used for process supervision and feedback control (Fig. 1)

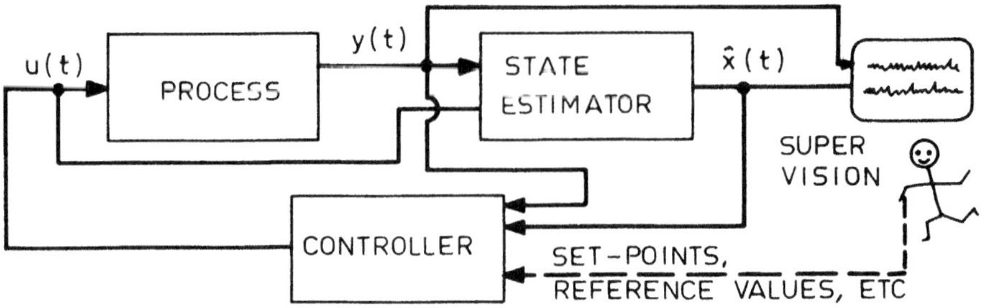

Figure 1. State estimation for process supervision and control.

According to a common formulation the state estimator of the system (1)-(2) attains the form

$$\dot{\hat{x}}(t) = f(\hat{x}(t), u(t), t) + K(t)[y(t) - g(\hat{x}(t), u(t), t)] \tag{5}$$

in which K(t) is the estimator gain. The behaviour and the convergence properties of the estimator (5) essentially depend on the choice of gain matrix K(t). The distinction between different estimation algorithms is based on the performance criterion or the heuristic approach used for determining the gain K(t).

In the following the most commonly used estimators, the Luenberger type state observer, the Kalman-Bucy filter and the extended Kalman filter are described. Further a new special algorithm (Halme 1980) is presented to serve as a representative of the class of nonlinear filters, which are especially suitable for implementations via microcomputers.

3.1 Luenberger type observers

The Luenberger state observer (1964, 1971) was originally derived for the deterministic linear time-invariant system, which can be obtained as a special case of eqs. (3) - (4) with zero noise i.e. $w(t) = v(t) = 0$ and time-invariant matrices A, B, C and D, which must be completely known a priori. The state estimator (5) then takes the form

$$\dot{\hat{x}}(t) = A\hat{x}(t) + Bu(t) + K[y(t) - C\hat{x}(t) - Du(t)] \tag{6}$$

in which the gain matrix K is constant. The stability of this observer depends on the location of the characteristic values of the matrix A-KC, i.e. of the observer poles. A stable observer has the property that the reconstruction error $x(t) - \hat{x}(t) \to 0$. In many practical situations the Luenberger observer, however, also works when applied to stochastic systems. Due to disturbances the convergence is of course not complete but by suitable data filtering satisfactory results have been obtained (Sinha, 1981). The Luenberger observer has also been extended to nonlinear systems (Thau, 1973).

The computations required when applying the Luenberger observer are simple, as they consist only of integrating the system equation (6). The time-invariant matrix K can be predetermined. This estimator can thus easily be implemented using microcomputers. Difficulties may appear if the process dynamics is very fast or stiff-like or contains nonlinearities which require very small integration steps or a complicated integration method demanding much computation time. Applications in which Luenberger observers have been implemented using microcomputers have been reported by Ciccharella et al. (1980) and Sinha et al. (1981).

3.2 Kalman-Bucy filter

Real processes are always affected by disturbances. In the models (1) - (4) this is expressed by additive noise. In the following it is assumed that the state excitation noise w(t) and the measurement noise v(t) are uncorrelated zero-mean white Gaussian processes with

$$E\{w(t)w^T(\tau)\} = R_w(t)\,\delta(t-\tau)$$
$$E\{v(t)v^T(\tau)\} = R_v(t)\,\delta(t-\tau) \qquad (7)$$

in which $R_w(t)$ and $R_v(t)$ are positive semidefinite covariance matrices and $\delta(t-\tau)$ the Dirac delta function, and E denotes the expectation value operator.

In the Luenberger observer the constant estimator gain can be arbitrarily chosen if only the observer poles are located in the left half-plane. In the Kalman-Bucy filter the gain matrix K(t) is optimally chosen in the sense of minimizing the variance of the estimation error $x - \hat{x}$. For the derivation, see e.g. (Kalman and Bucy 1960, 1961, Jazwinski 1970, Bryson-Ho 1969, Kwakernaak 1972).

The optimum (minimum variance) estimator (Kalman-Bucy filter) for the linear continuous system (3) - (4) with (7) consists of the following equations:

$$\dot{\hat{x}}(t) = A(t)\hat{x}(t) + B(t)u(t) + K(t)\,[y(t) - C(t)\hat{x}(t) - D(t)u(t)] \qquad (8)$$

$$K(t) = Q(t)C^T(t)R_v^{-1}(t) \qquad (9)$$

in which the covariance matrix Q(t) is the unique solution of the matrix Riccati equation

$$\dot{Q}(t) = A(t)Q(t) + Q(t)A^T(t) - Q(t)C^T(t)R_v^{-1}(t)C(t)Q(t) + R_w(t) \qquad (10)$$

with the initial conditions

$$\hat{x}(t_0) = E\{x(t_0)\} = x_0$$
$$Q(t_0) = E\{(x(t_0) - x_0)(x(t_0) - x_0)^T\}, \quad \text{positive semidefinite} \tag{11}$$

Determination of the gain matrix K(t) from (9) - (11) is independent of the measurements y(t) and u(t). Thus the covariance and gain matrices can be computed beforehand and stored if the parameters of the system and the measurement equations are known. In such a case the matrix Riccati equation (10) need not be solved in real time, which considerably decreases the required on-line computation time. This is particularly convenient for time-invariant systems for which the solution of the Riccati equation (10) converges to a unique constant value Q.

In ordinary on-line computation the demands of both computer memory and computation time are heavily dependent on the number of state variables n in the process model. Mendel (1971) carried out a comprehensive study concerning the computation time and storage requirements of Kalman filters. Roughly speaking, data storage varies with n^2 and filter computation time with n^3 (Wilson, 1980).

3.3 Extended Kalman filter (EKF)

The extended Kalman filter (EKF) (Jazwinski, 1970) is frequently used for estimating the states of nonlinear systems. The EKF for the continuous nonlinear system (1) - (2), (7) consists of estimation via

$$\dot{\hat{x}}(t) = f(\hat{x}(t), u(t), t) + K(t)[y(t) - g(\hat{x}(t), u(t), t)] \tag{12}$$

$$K(t) = Q(t)G^T(\hat{x}, u, t)R_v^{-1}(t) \tag{13}$$

$$\dot{Q}(t) = F(\hat{x},u,t)Q(t) + Q(t)F^T(\hat{x},u,t) - Q(t)G^T(\hat{x},u,t)R_v^{-1}(t)G(\hat{x},u,t)Q(t) + R_w(t) \tag{14}$$

in which

$$F(\hat{x}, u, t) = \frac{\partial}{\partial x} f(x, u(t), t)\Big]_{x = \hat{x}(t)} \tag{15}$$

$$G(\hat{x}, u, t) = \frac{\partial}{\partial x} g(x, u(t), t)\Big]_{x = \hat{x}(t)} \tag{16}$$

Note that for the EKF Q(t) and K(t) cannot be precalculated but must be determined on-line since they are coupled to the current estimate $\hat{x}(t)$ through the linearization procedure (15) - (16).

3.4 Nonlinear filters

In addition to the extended Kalman filter a great number of nonlinear filters have been developed in recent years (e.g. Control and Dynamic Systems 1976). The following nonlinear filter algorithm, developed by Halme (1980), is presented here because it is very convenient for microprocessor implementations as it does not require on-line determination of Q(t) in (14).

Consider the system defined by (1) - (2) and (7). Suppose that the function f and g are expressed in the form

$$f(x(t), u(t), t) = Ax(t) + f_1(x(t), u(t), t) \tag{17}$$

$$g(x(t), u(t), t) = Cx(t) + g_1(x(t), u(t), t) \tag{18}$$

in which A and C are constant time-invariant matrices. The nonlinear filter consists of the following algorithm (Halme, 1980)

$$\dot{\hat{x}}(t) = f(\hat{x}(t), u(t), t) + K(t)[y(t) - g(\hat{x}(t), u(t), t)] \tag{19}$$

$$K(t) = QG^T(\hat{x}, u, t) R_v^{-1} \tag{20}$$

in which G is defined by (16) and Q is the positive definite steady state solution of the matrix Riccati equation

$$\dot{Q}(t) = AQ(t) + Q(t)A^T - Q(t)C^T R_v^{-1} CQ(t) + R_w \tag{21}$$

in which R_w and R_v are constant matrices.

The filtering equation (19) differs from the extended Kalman filter in that on-line calculation of Q is not required. The main difference lies in the fact that the time-invariant matrices A and C can be determined in advance, e.g., by linearizing the model with respect to state at a fixed nominal value instead of evaluating the matrices from the estimated $\hat{x}$ at each time instant. When computing the gain K(t) the instantaneous value of the estimated state $\hat{x}(t)$ is, however, used in a similar manner as in the EKF. This filter is very suitable for on-line use and has proved to work well in many applications (Halme and Selkäinaho, 1981).

3.5 Comparison of estimation algorithms

The calculations required for determining Q and K when using the different estimation algorithms can be summarized as follows

System model \ Estimator	Luenberger observer		Kalman filter		EKF		Halme's filter	
	Q	K	Q(t)	K(t)	Q(t)	K(t)	Q	K(t)
Linear time-invariant	1)	off-line	off-line $Q(t) \to Q$	off-line $K(t) \to K$	2)	2)	2)	2)
Linear time-variant	1)	off-line	off-line	off-line	2)	2)	off-line	on-line
Nonlinear	1)	off-line	not suitable		on-line	on-line	off-line	on-line

1) not required, 2) coincides with that of Kalman filter.

4. APPLICATIONS

In the previous section a short survey of some frequently used state estimation algorithms was given. These methods, however, have not yet found very widespread use in applications in process industry. The main reason for this is that the methods usually require rather exact a priori knowledge of the process models, of the model parameters and of noise characteristics. These and other difficulties will be illustrated in connection with the following examples concerning state estimation using microcomputers.

4.1 State estimation of the activated sludge waste water treatment process

4.1.1 Process description. The activated sludge process (Fig. 2) is the most commonly used process for biological waste water treatment.

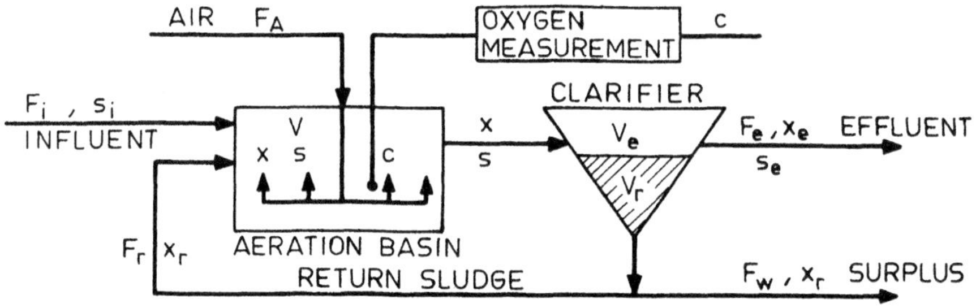

Figure 2. The activated sludge process.

Waste water with a concentration s_i of polluting organic substrates, measured by its biochemical oxygen demand (BOD_7), is fed to an aeration basin (volume V) at a flow rate F_i. BOD removal takes place in the aeration basin where microorganisms (concentration x) grow with consumption of organic substrate and oxygen. The oxygen needed for growth is brought to the process in the form of air at a flow rate F_A. From the aeration tank the solution is brought to the secondary clarifier where it is settled. One part of the settled sludge (concentration x_r) is recycled to the aeration basin at a flow rate F_r while the surplus sludge is removed from the process at a flow rate F_w. The purified effluent water has a BOD_7 concentration s_e and the biomass concentration x_e.

4.1.2 Process model. The dynamic behaviour of the activated sludge process can be described by the following model based on mass balances (Holmberg 1981, 1982).

$$\dot{x} = \mu(s)x - \frac{F}{V}x + \frac{F_r}{V}x_r \tag{22}$$

$$\dot{s} = -\frac{1}{Y}\mu(s)x - \frac{F}{V}s + \frac{F_i}{V}s_i \tag{23}$$

$$\dot{x}_e = \frac{F_e}{V_e}[k_x F(t-\Delta_e)x(t-\Delta_e) - x_e] \tag{24}$$

$$\dot{x}_r = \frac{F_r + F_w}{V_r}\left[\frac{1 - k_x F_e(t-\Delta_r)}{F_r(t-\Delta_r) + F_w(t-\Delta_r)} F(t-\Delta_r)x(t-\Delta_r) - x_r\right] \tag{25}$$

$$\dot{c} = k_L a(F_A)(c_s - c) - (a\mu(s)x + bx) \tag{26}$$

$$s_e = s + kx_e \tag{27}$$

in which $\mu(s)$ is the specific growth rate of sludge, Y is the yield constant for conversion of substrate into active sludge, $k_L a(F_A)$ is the airflowdependent oxygen transfer rate coefficient, c is the concentration of dissolved oxygen, c_s is the saturation value of dissolved oxygen, a and b are parameters related to oxygen consumption, k_x is a separation parameter, k is the BOD_7-equivalent of the effluent sludge and Δ_e and Δ_r are timedelays of the settler. Influents and effluents with negligible influence have been omitted from the mass balance equations (22) - (26). The other variables are defined in Fig. 2. The time arguments have been omitted in (22) - (27) except those where the delays must be considered.

4.1.3 State estimation problem. Many of the variables of eqs. (22)-(27) are of great interest for supervising and controlling the plant. For example, the BOD_7 value of the effluent s_e is a measure of the overall purification result. Water authorities normally confirm the maximum permitted values of x_e and s_e for each plant. Supervision of these variables is thus important. Further knowledge of the biomass concentration of the aeration basin x and the recycle x_r are required for controlling the sludge age and sludge load.

The greatest problem with utilizing the above-mentioned variables for supervision and control is, however, the lack of on-line information. No sensors exist for on-line determination of these variables. They require laboratory analyses the duration of which are 7 days for the BOD, and 3-4 hours for the sludge concentrations. At most plants these analyses are carried out only the required 2-3 times per month as they are rather expensive. It is thus obvious that mathematical estimation methods are required to enable e.g. continuous monitoring of the purification efficiency, and also for implementing possible control schemes.

The process model, eqs. (22) - (27) consists of 5 state variables, only one of which, the dissolved oxygen concentration c, can be directly on-line measured. By denoting the state vector by z and the measurement by y and by using the general form described by eq. (5) yields

$$z = [x \ s \ x_e \ x_r \ c]^T, \quad y = [0 \ 0 \ 0 \ 0 \ 1]\, z$$
$$\dot{\hat{z}}(t) = f(\hat{z}(t), u(t), t) + K(t)[c(t) - \hat{c}(t)] \tag{28}$$

The choice of estimation gain for the nonlinear system (28) is, however, not a straightforward task. In addition to the unmeasurable states the BOD_7 of the influent, i.e. s_i, is also unknown. Although s_i

often follows a diurnal rhythm, changes in the load cannot be predicted as they depend on e.g. randomly appearing industrial effluents. Further, the process model contains many parameters which, although they are separately updateable, constitute a remarkable source of uncertainty. Characterization of the process noise would also be a very difficult task.

4.1.4 State estimation algorithm. Because of the difficulties discussed in the preceding paragraph, the state estimation algorithm presented in the following has been developed in a process-oriented way. From (26) it can be concluded that the dissolved oxygen c is connected with the other state variables only indirectly through a term describing the oxygen uptake rate

$$OUR = a\mu(s)x + bx \tag{29}$$

The dissolved oxygen dynamics can thus be separated from the rest of the system and can be used specifically for determination of OUR. By substituting the OUR value from (29) into eqs. (22) - (23) a simple state estimation algorithm can be obtained. The estimation procedure may be summarized by the following two steps:

STEP 1. Determine OUR from (30) (obtained by substituting (29) into (26))

$$\dot{c} = k_L a(F_A)(c_s - c) - OUR \tag{30}$$

using recursive least squares. This has been treated in detail by Holmberg (1981).

STEP 2. Substitute

$$\mu(s)x = \frac{OUR - bx}{a} \tag{31}$$

obtained from (29), into eqs. (22) and (23). Then the following simple estimator can be obtained

$$\dot{\hat{x}} = -\left(\frac{F}{V} + \frac{b}{a}\right)\hat{x} + \frac{F_r}{V}\hat{x}_r + \frac{OUR}{a} \tag{32}$$

$$\dot{\hat{s}} = -\frac{1}{Ya}(OUR - b\hat{x}) - \frac{F}{V}\hat{s} + \frac{F_i}{V}s_i . \tag{33}$$

$\hat{x}_r$ and $\hat{x}_e$ can be obtained from (24) and (25) by substituting $\hat{x}$. In practice when s_i is unknown, eq. (33) has been simplified using some biologically motivated assumptions (Holmberg, 1982). An example of the use of this algorithm for on-line estimation carried out at a plant in Suomenoja, Finland is shown in Fig. 3.

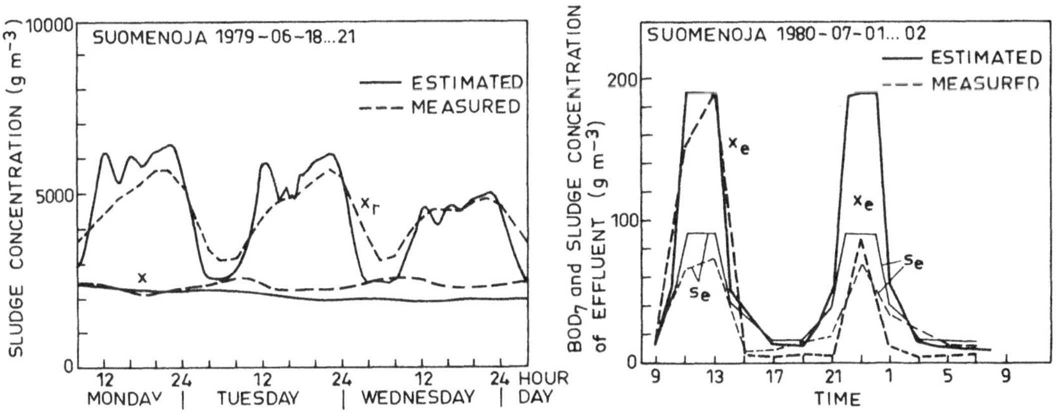

Figure 3. Example of on-line estimation compared to measurements.

4.1.5 Microprocessor realization. The estimator described in this article was implemented and tested using a microprocessor at the research plant in Suomenoja, Finland (Holmberg, 1982). This plant has the capacity to treat 15 m^3/h waste water, which corresponds to the municipal load from 1000 - 1500 inhabitants. The volume of the aeration basin is 90 m^3 and that of the secondary clarifier 15 m^3.

All computing, estimation and data storing tasks were performed using the UL-300 process control station (Fig. 4) of the plant. The system consists of an Intel SBC-80/20 microprocessor and is equipped with 15 kBytes EPROM memory for the operating system and 3 x 16 kBytes RAM memory for the application program and necessary devices for interface and input/output activities.

The microprocessor implementation has 5 analog inputs (water temperature, dissolved oxygen, air flow, liquid flow F_e and sludge blanket height of secondary clarifier) and 3 analog outputs or control signals to the process (liquid flows F_i and F_r and air compressor effect). The

Figure 4. UL-300 process control station.

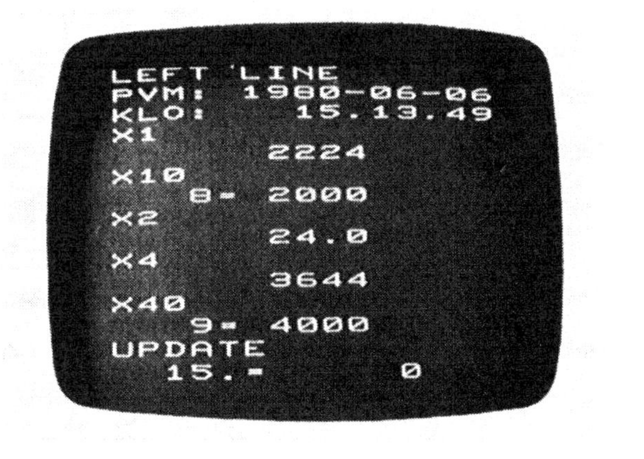

Figure 5. Microprocessor display for estimates of x (x_1), x_e (x_2), and x_r (x_4).

analog signals were read at 1 second intervals. Instantaneous information concerning the state of the process was obtained using different display configurations, one of which is demonstrated in Fig. 5.

The program developed for supervising, controlling and estimating the activated sludge process is presented in Fig. 6 in the form of a simplified block diagram. The program includes the following subroutines: Digital data filtering, digital PI-control, recursive least squares estimation, integration of differential equations and routines for calculating averages and for storing results.

The most time-consuming of the blocks is the recursive least squares estimation used for estimating OUR from (30). The estimation time when using the UL-300 system was approximately

$$7 \times n^2 + 9.3 \times n + 5.8 \text{ ms} \tag{34}$$

in which n is the number of parameters to be estimated. This shows that estimation of 10 parameters would require about 800 ms while 2 parameters can be estimated in about 60 ms. Thus up to 10 parameters can theoretically be estimated during the 1 second cycle.

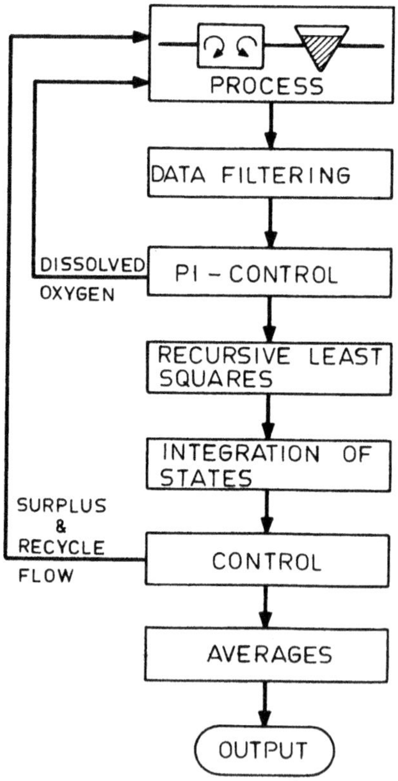

Figure 6. Simplified block scheme of the microprocessor program.

4.2 An estimator for chemical concentrations of pH-processes in through-flow reactors

4.2.1 Process description. Controlling the pH-value is an integral part of modern waste water treatment which usually takes place in stirred tank reactors with a controlling reagent flow (Fig. 7). The pH-value of the liquid in the reactor is to be set to a reference value depending on the purpose of the process; e.g., the coagulation, or neutralization of the effluent into receiving waters.

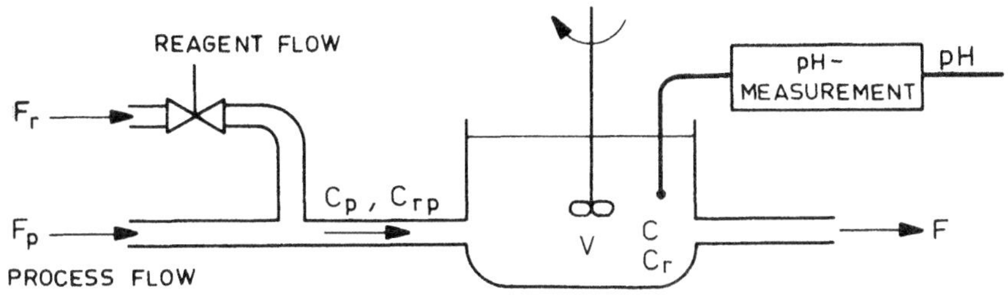

Figure 7. Reactor for pH control.

4.2.2 Process model. By utilizing material balance, the conditions for reaction equilibrium and electroneutrality, the dynamic equations

$$\dot{C} = q(C_p - C), \qquad q = F/V \tag{35}$$

$$\dot{C}_r = q(C_{rp} - C_r) \tag{36}$$

and the static equation

$$C^+ - \frac{K_w}{C^+} - H(C^+)C - H_r(C^+)C_r = 0 \tag{37}$$

are obtained (Jutila and Orava,1981). The pH of a dilute water solution is by definition

$$pH = -\log(C^+/(\text{mole/litre})) \tag{38}$$

In (35) - (38) C^+ is the molar concentration of hydrogen ions and C is a vector of molar concentrations,

$$C = [C_A - C_B, C_{\alpha 1}, C_{\beta 1}, C_{\alpha 2}, C_{\beta 2}, \dots]^T \tag{39}$$

in which C_A = total concentration of strong acids, C_B = total concentration of strong bases, $C_{\alpha i}$ = total concentration of weak i-valued acid α_i , $C_{\beta i}$ = total concentration of weak i-valued base β_i, C_r = total concentration of the reagent (acid or base, strong or weak). The coefficient matrices in (37) are given by

$$H(C^+) = [1, \frac{K_{a11}}{C^+ + K_{a11}}, -\frac{K_{b11}C^+}{K_{b11}C^+ + K_w}, \frac{K_{a21}C^+ + 2K_{a21}K_{a22}}{C^{+2} + K_{a21}C^+ + K_{a21}K_{a22}}$$

$$-\frac{K_{b21}K_wC^+ + 2K_{b21}K_{b22}C^{+2}}{K_{b21}K_{b22}C^{+2} + K_{b21}K_wC^+ + K_w^2}, \cdots] \qquad (40)$$

$$H_r(C^+) = \begin{cases} 1, -1, \text{for strong acid, base, respectively} \\ \dfrac{K_{ra11}}{C^+ + K_{ra11}} \quad \text{for weak 1 - valued acid} \\ \text{etc, cf. (40)} \end{cases} \qquad (41)$$

in which K_w = ion product of water, K_{aij} = dissociation constants of weak acids, K_{bij} = dissociation constants of weak bases.

4.2.3 State estimation problem. The pH value can be measured with good accuracy using electric sensors. Thus equations (37) - (38) can be considered to form the output equations of the system. As the system model (35) - (38) is nonlinear with respect to the measured pH but linear with respect to the concentration C, it was essential to develop concentration control algorithms (controlling C) instead of conventional pH control algorithms. Unfortunately, there are no reliable and cheap on-line sensors for measuring the components of C. In addition, even the identity of component species is often unknown or varying in practical waste water applications. Consequently, only the pH of liquids can be measured, and the concentrations must be estimated.

4.2.4 State estimation algorithm. The measurable output, i.e. the pH value of the reactor is related to the C^+ concentration by (38). An estimate for this concentration can therefore be obtained by

$$C^+ = 10^{-pH} \quad \text{(mole/litre)}. \qquad (42)$$

As a transformed output the quantity

$$y = C^+ - \frac{K_w}{C^+} \qquad (43)$$

is chosen. The state reperesentation of the system for the concentrations to be estimated is then given by

$$\dot{C} = q(C_p - C) + qw \qquad (44)$$

$$y = H(C^+)C + H_r(C^+)C_r + v \qquad (45)$$

Note that (45) is apparently linear with respect to the state C and the variable C_r although the coefficient matrices H and H_r depend on these variables in a complex manner (cf. (37)-(41)), from which C^+ cannot be explicitely obtained. By applying (42) the matrices H and H_r can be regarded as explicitly given functions of time and C_r is considered as an auxiliary quantity given by (36). The linear time-variant Kalman-Bucy filter can thus be applied. The following estimation algorithm can then be obtained.

$$\dot{\hat{C}}_r = q(C_{rp} - \hat{C}_r), \qquad \hat{C}_r(0) = C_{r0} \tag{46}$$

$$\dot{\hat{C}} = q(C_p - \hat{C}) + K(t)[y - H(C^+)\hat{C} - H_r(C^+)\hat{C}_r], \quad \hat{C}(0) = C_0 \tag{47}$$

$$K(t) = Q(t)H^T(C^+(t))R_v^{-1} \tag{48}$$

$$\dot{Q}(t) = -2qQ(t) - Q(t)H^T(C^+(t))R_v^{-1}H(C^+(t))Q(t) + q^2R_w, \quad Q(0) = Q_0 \tag{49}$$

R_v and R_w are the covariances of the noises v and w. The estimation procedure is illustrated in Fig. 8. Further details of the practical applications of this filter are given in (Jutila and Orava,1981, Nihtilä and Jutila,1982).

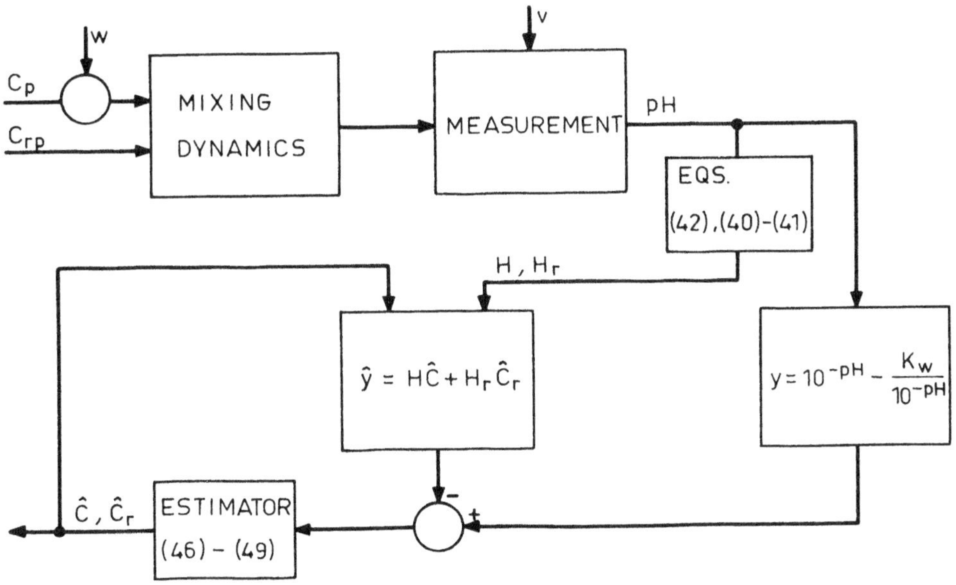

Figure 8. Scheme of state estimator for pH process.

The functioning of the algorithm described above is demonstrated in Fig. 9 with the aid of a simulated example. The process water considered in this example consists of four weak protolytes, the concentration of one of which is considered to change stepwise to another value and after some time back to the original value. The process was simulated using the Runge-Kutta integration method and an integration step of 0.02 min. The Riccati equation was integrated by the Euler method and the integration step was 0.002 min. For further details such as parameter values see (Jutila and Orava, 1981).

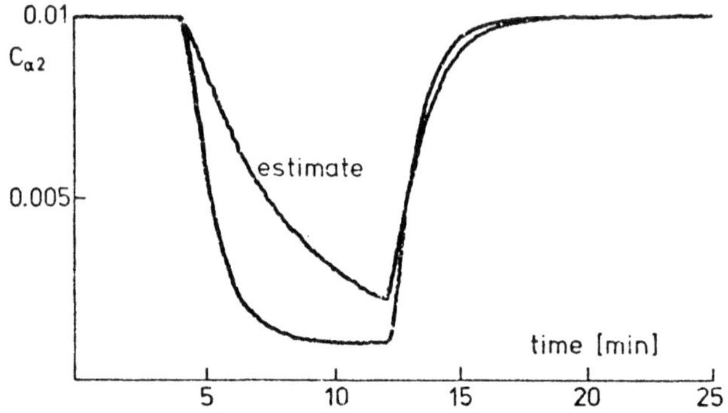

Figure 9. Simulated real and estimated concentrations of a weak acid after a step change.

4.2.5 <u>Suitability for microprocessor implementation</u>. The estimation algorithm described above requires much computation capacity as the dimension increases. The solution of the matrix Riccati equation (49) is particularly time-consuming because of the small integration step. By determining the steady-state solution Q of the Riccati equation with constant H and using this solution for computing the filter gain K. the computing time can be reduced considerably. The algorithms have been tested on real processes using a MINC microcomputer equipped with a LSI microprocessor. The estimation and control algorithms required about 10 kBytes of memory and estimation of 4 concentrations was performed in 2.5 seconds. The control interval was 3-12 seconds. This was the critical number of estimable variables. Only in certain cases there has been success in estimating 5 variables.

5. CONCLUSIONS

State estimation of nonmeasurable variables was considered with special emphasis on the possibilities of implementing algorithms using microprocessors. It was found that there are not as yet many installations in process industry. The particular problems concerning microprocessors are related to storage requirements and computation time. In many practical cases, however, the algorithms can be simplified using robust approximations so that auxiliary equations of the algorithms such as the Riccati equation need not be solved on-line.

As an increasing number of industrial processes in the near future will be equipped with different kinds of computers, the opportunities for implementing many kinds of advanced algorithms, for control as well as for estimation at reasonable cost, will expand remarkably. In this respect the introduction of new cheap microprocessor technology has certainly had a positive influence on the practical utilization of well-known estimation algorithms and on the development of new algorithms, and this influence will increase in the future. By using microcomputers, connected to industrial processes, it is possible to test theoretically developed estimation algorithms in practice, whereas previously they have usually been tested only in simulated process models.

As was seen from the practical examples, state estimation in practice is usually not merely a straightforward application of known algorithms. Rather, it can be considered as an art in which the specifications of the process under consideration must be thoroughly considered. There must be a sound compromise between simplifications and assumptions on the one hand and accuracy on the other. In this way the optimal solution with respect to minimum storage space and computing time can also best be achieved.

REFERENCES

Bryson A.E. and Ho Y.-C.: 1969, *Applied optimal control*, Ginn and company.

Ciccharella G., De Carli A. and Verrocchi P.: 1980, *IFAC Conf. on Application of microprocessors in devices for instrumentation and automatic control*, London, England, Nov. 18-20, 1980. Instr. Meas & Control, London, England, pp. 137-147.

Control and Dynamic Systems. Advances in theory and application. Vol. 12, 1976, ed. C.T. Leondes.

Halme A.: 1980, In O.L.R. Jacobs, N.H.A. Davis, M.A.H. Dempster, C.J. Harris and P.C. Parks (ed.) *Analysis and Optimization of Stochastic Systems*. Academic Press, New York, pp. 373 - 385.

Halme A. and Selkäinaho J.: 1981. *20 th IEEE Conf. on Decision and Control*, San Diego, California, Dec. 16 - 18, 1981, pp. 71-73.

Holmberg A.: 1981, Int. J. Systems Sci. 12, pp. 703-718.

Holmberg A.: 1982, Water Research, 16, pp. 1233 - 1246.

Holmberg A. and Ranta J.: 1982, Automatica 18, pp.181 - 193.

Jazwinski A.: 1970, *Stochastic processes and filtering theory*, Academic Press, New York and London.

Jutila P. and Orava P.J.: 1981, Int. J. Systems Sci, 12, pp. 855-875.

Kalman R.E.: 1960, Trans ASME, J. Basic Eng., Vol. 82 D, pp. 35-45.

Kalman R.E. and Bucy R.S.: 1961, Trans. ASME J. Basic Eng., Vol. 183 D, pp. 95-108.

Katz P.: 1981, *Digital Control Using Microprocessors*, Prentice Hall.

Kwakernaak H. and Sivan R.: 1972, *Linear Optimal Control Systems*, Wiley - Interscience.

Luenberger D.G.: 1964, IEEE Trans. Mil. Electron., Vol. MIL-8, pp. 74-80.

Luenberger D.G.: 1971, IEEE Trans. Aut. Control, Vol. AC-16, No. 6, pp. 596-602.

Mendel J.M.: 1971, IEEE Trans. Aut. Control, Vol. AC-16, No. 6, pp. 748-758.

Nihtilä M. and Jutila P.: 1982, In preprints of *Symposium on Application of Multivariable Systems Theory* 26.-28. October, 1982, Plymouth, UK.

Sinha N.K., Law S.S.Y. and Li M.H.: 1981, IEEE Trans. Ind. Electr. Contr. Instr. Vol. IECI - 28, No. 2, pp. 136-140.

Thau F.E.: 1973, Int. J. Control, 17, pp. 471-479.

Wilson T.: 1980, Eurochem 80, 24-26 June, Bingham, England.

Part III

MEASUREMENT

CHAPTER 9

MICROPROCESSOR-BASED MEASUREMENT SYSTEMS

Andrzej Sowiński
Industrial Institute of Electronics
Warsaw, Poland

1. INTRODUCTION

A microprocessor is a Large Scale Integration (LSI) component that implements most of the functions of a processor on a single silicon chip. It consists of two units: the arithmetic and logic unit (ALU) and the control unit (CU). The arithmetic and logic unit performs arithmetic and logical operations on the data and the control unit fetches, decodes and executes the successive instructions of a program stored in the memory.

The microprocessor could be used as the basis for control of a microprocessor-based system. Integration of the memory unit and the input/output unit with the microprocessor then forms a microprocessor system which allows the introduction of programming to the system as well as the coupling of the system with external devices and/or systems. Most of the microprocessor systems in current use can be programmed as digital equipment as shown in Fig. 1.

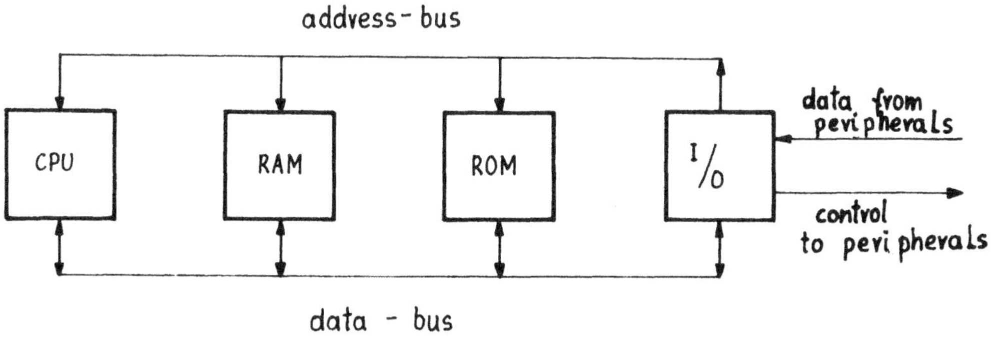

Fig. 1. Programmable digital equipment.

S. G. Tzafestas (ed.), Microprocessors in Signal Processing, Measurement and Control, 161–188.

The vast variety of microprocessors calls for the need to establish basic features which characterize microprocessor systems. These are as follows:

- length of a word
- memory capacity
- interruption systems
- stack arrangement
- instruction set
- addressing technique

The technical and economic advantages of introducing microprocessor systems are:

- system miniaturization and reduced physical volume
- reduction of number of system components and the resulting reliability improvement
- reduced power consumption
- increased functional flexibility of a system
- decreased system cost
- reduced system design time
- increased system versatility
- simplified design and easy modification

In the application of a microprocessor in a measurement system, the system designer must trade off two factors in order to achieve the best solution:the desire to increase the versatility features of the system and the level of complexity, and the additional cost.

The functions of a microprocessor in a measurement equipment can be divided in the following categories:

(1) Automated zeroing and calibration.
(2) Automated choice of measurement range.
(3) Processing of measurement results.
(4) Introduction of measurement results in appropriate form.

Automated zeroing and calibration. Between measurement cycles, the system input is disconnected and the standard of measured quantity turned on, and then the level accepted as zero is measured. The difference of these results is the correction signal. In this way, one can obtain the accuracy of measurement which is most close to the accuracy of own standard.

Automated choice of measurement range. Procedure of this function is carried out with the aid of logical analysis of the results, which gives the best accuracy.

Processing of measurements results: It is possible to operate on both single-measurement results as well as on the whole set of measurement data precompiled in memory. The majority of equipment of some parameter quantities is capable to carry out the measurement of data mean and dispersion, and to schedule the data set. The application of appropriate processing algorithms allows, in many cases, the direct measurement of such quantities, which cannot be obtained with classical methods.

It is also easy to implement the algorithm of deviation control of subsequent value measured from nominal value, together with the signalling of the time instant when the border given is overcome. This application is typical for testing goods and the inspection of technological processes.

Introduction of measurement results in an appropriate form. It concerns, in particular, multi-parameter measurements, when the results need to be introduced in various physical units and in various forms, in the form of alfa-numeric print, projection on displays and in the form of tables and diagrams, what requires a variety of intermediate processing sometimes in different codes. Procedures applied here serve simultaneously the control of external equipment which cooperates with a given measurement system.

Additional advantages of the application of microprocessors in measurement systems should also be mentioned. These are as follows:

- wide possibilities of processing, treatment and reduction of measurement data
- possibility of implementing more complex operation algorithms without increasing the complexity of control circuit
- interface signals can be easily obtained. This ensures the mutual connection of equipment and system, as well as with external equipment via the mutual data bus
- flexibility of system organization and simplicity in initiating changes. Modification of operation algorithm can be carried out solely by a change of a program, without the necessity to change-over the the separate blocks.

Some of the functions mentioned above could be carried out simultaneously. That is why multi-processor circuits are being used more and more often in systems. The multi-processor contains, apart from the main processor, additional auxiliary microprocessors or dedicated LSI circuits which carry out separate functions. The application of dedicated circuit of 8279 type for constant control of digital indicators and the reading of a key-board may serve here as an example.

2. STRUCTURE AND ORGANIZATION OF MICROPROCESSOR-BASED SYSTEMS

The design of microprocessor-based systems needs a slightly different outlook as opposed to designing equipment and systems without a microprocessor. The following require, among others, special attention:

- integral organization of the system with all of its components
- organization of cooperation of the operator and the system
- simultaneous design of hardware and software
- autotesting of microprocessor circuits before and during work
- system testing

Because of their numerous functions in measurement equipment, microprocessors become central blocks in the organizational structure of measurement systems. Fig. 2 shows a typical structure of a measurement system with a central processor that carries out the functions of system work control and data processing. The measurement circuit contacts the microprocessor through the control data registers (programming) and the output data registers which involve information about the measurement results, as well as about the state of the measurement circuit.

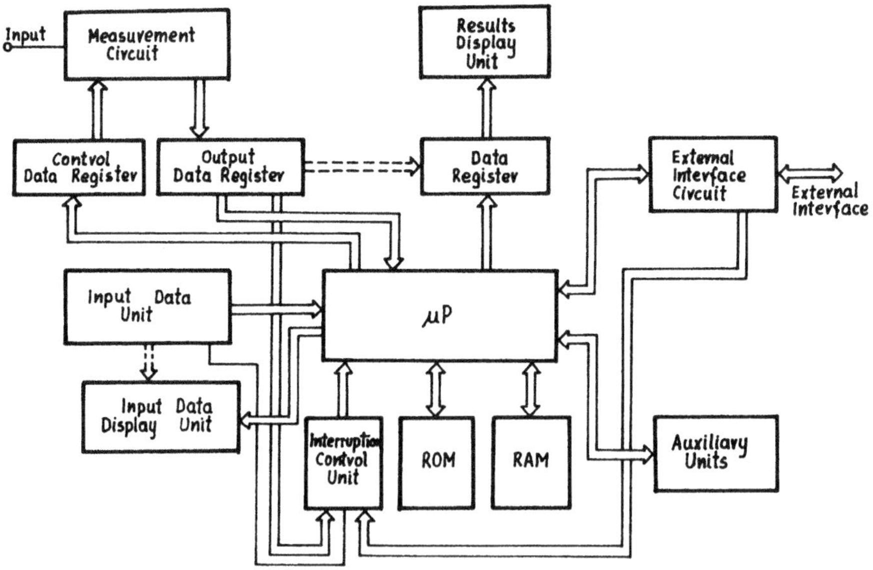

Fig. 2. Equipment with a central microprocessor.

The presentation of measurement results can be made in alfanumerical form (data display, print), as well as in a diagram

form on an oscilloscope screen. In the first case, information about the measurement results can be taken directly from the output register of a measurement circuit;this is usually the case of simple equipment in which the processing of measurement results is unnecessary.
The control of equipment work is achieved with the aid of data storage. The standardized solution of this set is a functional-numeric key-board that allows to fulfill the requirements in a sufficient way (easiness of service even in the case of a hardly competent operator) by storing data controlling element which simplifies the contact of the operator with the measurement equipment. In the simplest form, it may be a reproduction field on which the values of parameters are shown; those introduced with the alfa-numerical key-board. The operator evaluates visually whether he has not made any error when putting a parameter;an affirmative evaluation is the basis for the data storage in the microprocessor memory. Another way to test the solution is to charge the microprocessor with it,which, of course, can test input values only from the logical and formal points of view. Values classified as correct may be either displayed on the reproduction field or printed.
Communication of measurement equipment with the surroundings (at the level of cooperation within the system) is ensured by the external interface circuit. Presently, the interface of series-bit transmission of information, which satisfies either IEC or IEEE recommendations, is most commonly applied. The rule is that any microprocessor-based measurement equipment performs permanent programs which are brought out automatically, if necessary, according to the established algorithms of equipment functioning, or with the aid of a functional key-board. These programs are stored in permanent memories of a microprocessor. The execution of mathematical and logical operations also needs the operational memory in equipment. The microprocessor in the circuit shown in Fig.2, works as a multi-task system, while the majority of tasks concern work in real time (work control, original data processing) and only some of them do not have distinct time restrictions. It places determined requirements ont the software, and especially onto its control part;in many cases, it appears impossible to fulfill all the requirements. This is the reason why in more complicated equipment, there is a tendency to introduce microprocessor hierarchical structures, in which dedicated processors execute separate tasks, under the supervision of a control microprocessor.

Fig.3 shows the general form of measurement equipment of such a structure. Dedicated microprocessors find their application in original fast processing of measurement data in real time, and then in display control of measurement results, as well as in the external interface circuits.

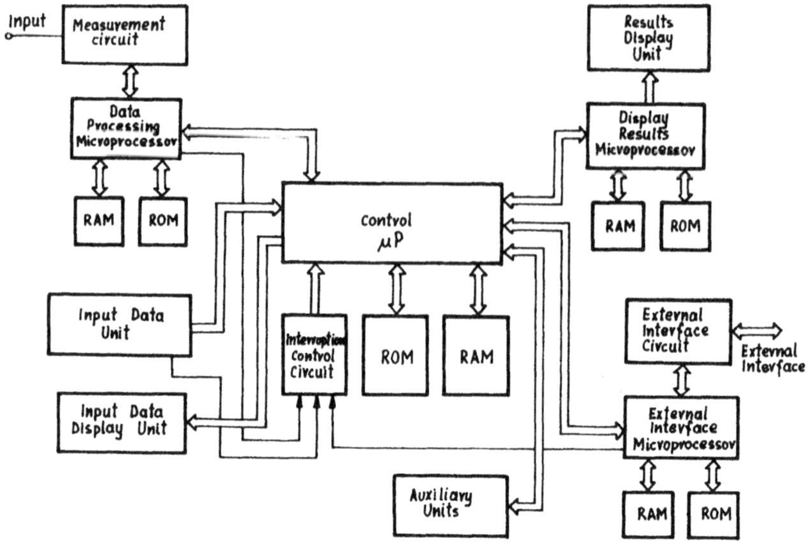

Fig. 3. Equipment with dedicated and control microprocessors.

Thus here we deal with multi-processor systems;there are some possibilities that just a couple of microprocessors may share the work. It may be (Fig.4):

- -a system with one common bus;this system is switched over in time, but synchronous work should not appear here,
- -a system with many buses, which allow for synchronous work,
- -a crossbar system, with matrix of switch-over, which is switched-over in space,
- -a multi-bus system with multi-port memories, on every input,
- -an asymmetric system, which is non-homogeneous;it has a microprocessor regarded as the main one, and other microprocessors regarded as peripheric,
- -an array system, where microprocessors (except from the main one) are situated in the form of an array matrix.

The internal organization of a measurement system with a microprocessor is determined, above all, by the establishment of information flow that passes through the main system units. These characteristics determine the necessary length of a word of basic information, frequency of transmission, as well as time links with external information (transmission in real time or with acceptable delay). Data obtained in this way constitute a sufficient basis for determining the block stru-

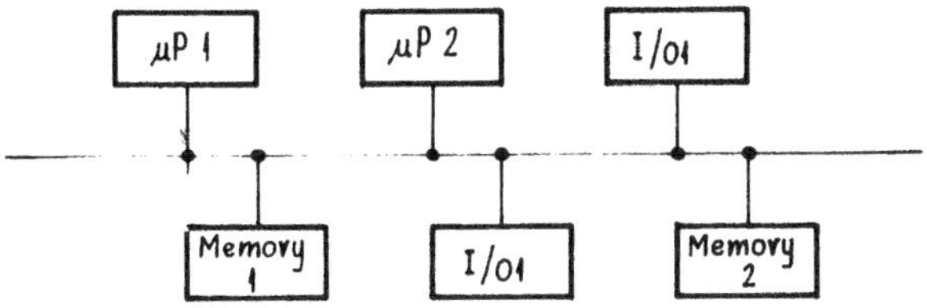

Fig.4a. Common-bus system.

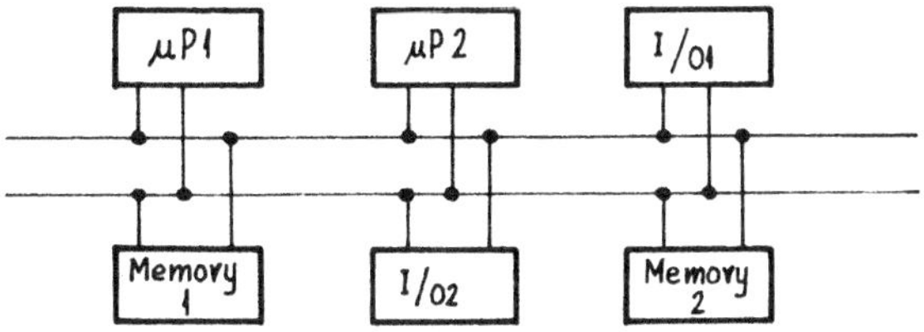

Fig.4b. Multi-bus system.

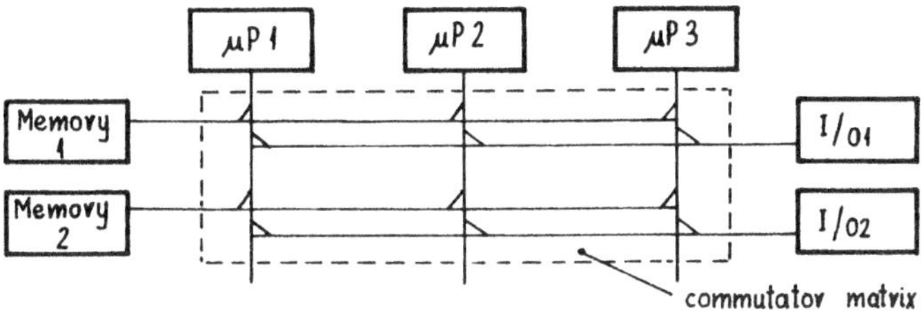

Fig.4c. Cross-bar system

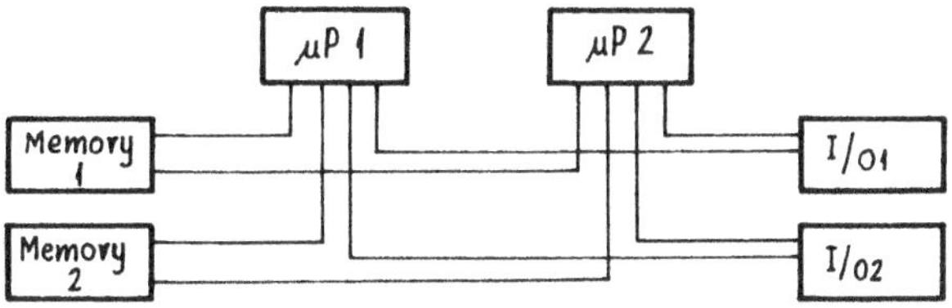

Fig. 4d. Multi-bus system with multi-input memories.

cture of a measurement system and the organization of information from inside this system.

Block structure of a system with one central microprocessor is useful only when the number of tasks carried out by a microprocessor is not too large, and the speeds of information processing are not too high. The evaluation of whether a given microprocessor system, to be applied in the equipment designed, is going to be efficient in time, is not simple. This can be done with the analysis of quantity, type and time of execution of the most time-consuming basic operations which appear in the algorithms of system work, and taking

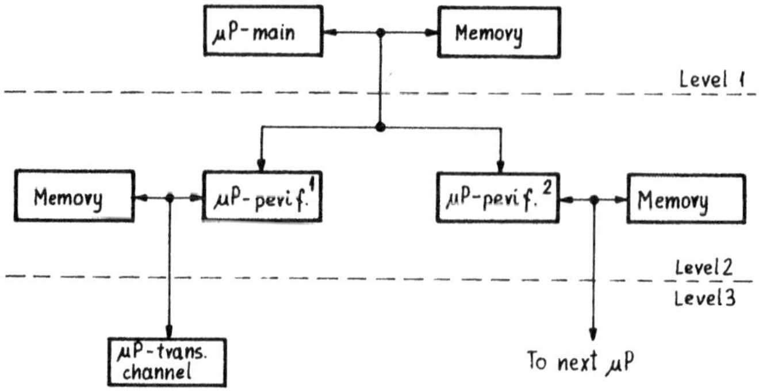

Fig. 4e. Asymmetric system.

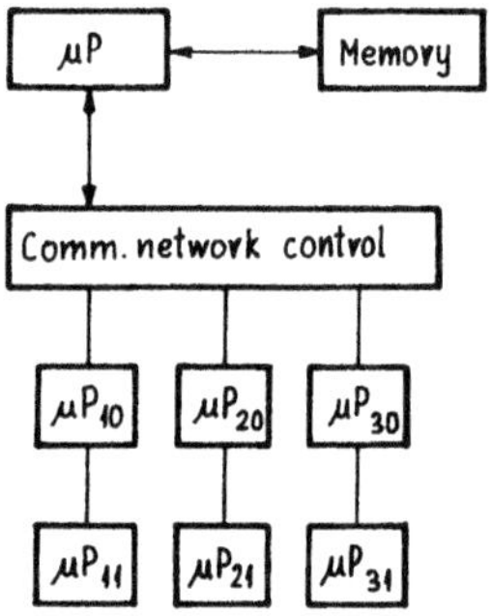

Fig. 4f. Array system.

into account their total with a security factor of about 30-50 per cent.

Tasks carried out in real time usually need a cyclic allowance of microprocessor time. In general, two different ways of processor time allowance in one work cycle are possible:

- allowance of permanent time periods for the completion of each task,
- time-sharing according to the task priority principle.

The first way is applied very often because of the circuit and program simplicity, yet it is a little economic in time because of the security margins on the connections of adjoining tasks. The second way, especially applied in connection with the creation of temporary data sequence, is mainly applied in situations where some tasks need more rarely time allowance of a processor than the remaining ones, and this makes it possible to utilize this time better. In this case, the microprocessor is more sophisticated and developed.

In cases where analysis indicates that time borders of the capabilities of a given microprocessor are reached, or when the number of various tasks to be performed by a microprocessor is large, the acceptance of a double-or multi-processor structure is better, and sometimes the only possible solution.

The charge of a dedicated microprocessor with a single task allows the selection of its structure and technical parameters in an optimum way;its software is autonomous and hence it is simpler and easier to be designed. The command microprocessor software is substantially simplified. The utilization of dedicated microprocessors that cooperate with a measurement block or with an interface block, also simplifies considerably the solution of galvanic separation of the blocks mentioned;it is often indispensable because of the work conditions of analogue circuits in an equipment. The structure preferred is the multi-processor, as may be generally indicated the various the reasons mentioned above.

The communication between the internal blocks of a measurement system is generally carried out with the aid of internal interface. The choice of the internal interface structure depends, to a great extent, on the character of information flow exchanged between blocks, Fig.5 shows three basic interface structures which are suitable for application in a measurement equipment.

The simplest solution is to connect separate blocks of a system directly to the internal interface of a microprocessor (Fig. 5a). Separate blocks are addressed in the same way as memory cells;active blocks which transmit information announce the service requirement through the break circuit. The advantage of such a way of connection is the simplicity of block addressing, and the disadvantage is the necessity to participate in each basic data transmission between blocks by the microprocessor. This is the reason why the structure described is suitable for application in those systems in which information transmission between system blocks occurs not so often, or there is substantial time reserve of the microprocessor work. Generally, the rate of information exchange between separate blocks of a system is not the same; disagreements of such a structure may be decreased by using additional direct connections between blocks which usually cooperate with themselves. (Fig.5b). This solution, however, complicates, to some extent, the hardware of the whole circuit, and it is suitable for the simpler structures.

Very fast data transmission between blocks can be obtained

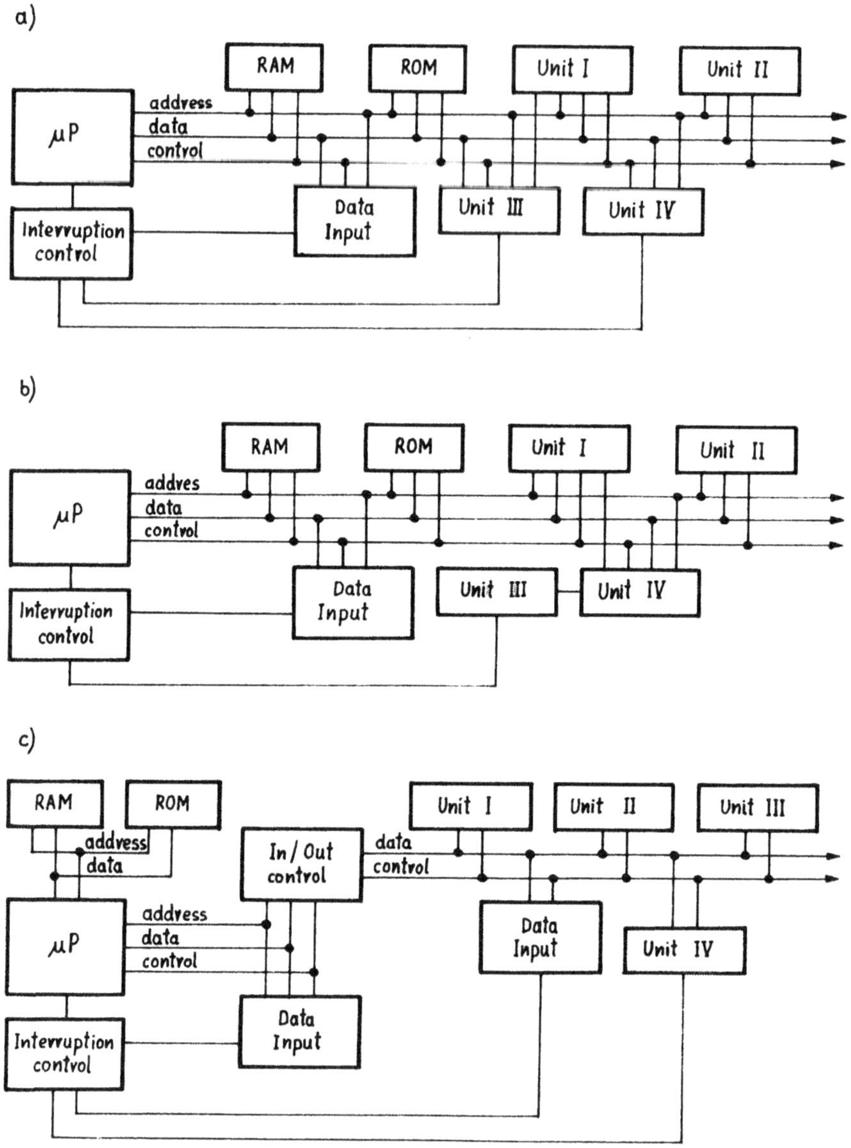

Fig.5. Structures of data transmission.

by using the channel of direct access to memory. In case where there is an intensive communication between blocks in a measurement system, the most effective way is to separate the microprocessor interface from the interface to which fast blocks of a system, cooperating with themselves, are adjusted (Fig. 5c). Then a microprocessor just controls the cooperation of blocks, while not participating in the data transmission itself. Such a structure may also be satisfactorily applied in multi-processor circuits.

The placement of a key-board on the side of a microprocessor internal interface is logical. External interface may be much simpler; only the command lines and data bus are necessary.

3. MICROPROCESSORS IN MEASUREMENT SYSTEMS

3.1. Measurement functions

The basic tasks performed by a measurement system are:

- The measurement operations.
- The processing of measurement results.
- Communication whith an operator or external equipment.

The measurement process decides about the basic parameter of a measurement, i.e. its accuracy. The accuracy should be determined by the stability of parameters of a measurement system as functions of time and conditions of the surroundings, such as temperature, pressure, etc. as well as by the effect of interferences and noises.

The utilization of a microprocessor in a measurement process allows the decrease of requirements about the components quality, from which the measurement circuit is constructed. The introduction of periodic and automated processes of zeroing and calibration into the equipment operation algorithm, allows the programmed compensation of changes in parameters of measurement components, as well as the elimination of influence of these changes on measurement accuracy. Automated zeroing depends on the periodical testing of circuit responses to the zero input signal. Autocalibration is carried out by the sequence of measurements for different values of reference signal on the equipment output, and for different equipment scopes. Responses are stored in the equipment memory, and they serve to correct the results before they are transmitted to the operator.

In some cases, automated calibration is carried out over the full range of input signal changes. Reference signals change in a linear way (in time), while the table of pattern responses is used for every output state of a converter.

A digital voltemeter 7115 of Synstron-Donner Corp. could be the typical example of a microprocessor applied for autocalibration. The process of automated calibration of voltmeter of constant voltage is carried out every 100 or 1000 measurements, and after the equipment is turned on. Autocalibration cycle consists of 16 measurements, the results of which are stored in the form of tables. Numerical values placed in tables, determine the dislocation of zero and the intensification of every sub-range of voltmeter. The measurement result is calculated by a microprocessor, using the following expression:

$$W_c = \{\pm(|W_m| - 200)X\}Y$$

where W_m is the measured and non-corrected value, W_c is the result after the correction, and X, Y are the values of a given function determined from the results of the autocalibration test.

Correction of results of variable current measurements is more complex, while the direct and variable voltage intensification is taken into account, as well as direct current zero shift (DC offset), and RMS converter zero shift. Possible nonlinearity of converter characterstic of RMS value is corrected by the transformation of appropriate polynomials. Application of the method of auto-corelation has reduced the number of elements that influence the size of measured error from 33 to five in a critical way. A microprocessor also increases the equipment accuracy by the elimination of disturbances.

The next example (Fig.6) describes the measurement method of network disturbances, compared to a classical circuit method that utilizes an integrator. The circuit of an integrator needs gate impulses of stable parameters, and precise integration capacitor, while $U_{DC} = U_{WE}dt$. The utilization of a microprocessor makes it possible to compensate the network disturbances by making two measurements displayed on a half of disturbance course period, and to calculate an average value from two intermediate results.

In measurement circuits, many functions carried out in a classical way with quite complicated analogue circuits, can be carried out with the aid of a microprocessor. These may be comparator networks (Fig.7), analogue memory circuits, or resistance networks.

In the thermometer Doric's Digitrand 200 which has thermocoupled detectors, the resistive linear networks are replaced by the appropriate measurement program that corrects the nonlinearity of thermo-couples' features.

Another example of substantial reduction of the number of measurement circuits is an admittance measurement bridge {eg type 2230} (Fig.8). The bridge equipped with a microprocessor always measures, both parallel capacity and conductance. The row inductance L_s, the row capacity C_s, and the Q factor are obtained as a result of calculations. In relation to conventional solutions, the above solution allows one to avoid the five different configurations of a bridge.

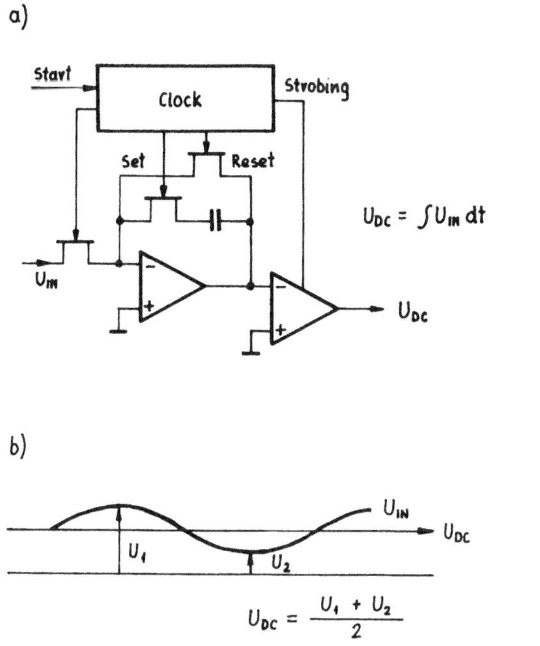

Fig.6. (a) Circuit of mains' noise elimination.
(b) Method of noise elimination by microprocessor.

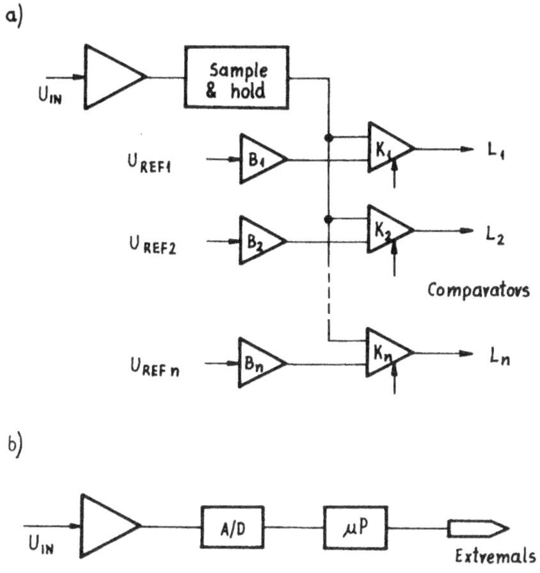

Fig.7. (a) Analog multilevel comparator.
(b) Microprocessor as comparator function.

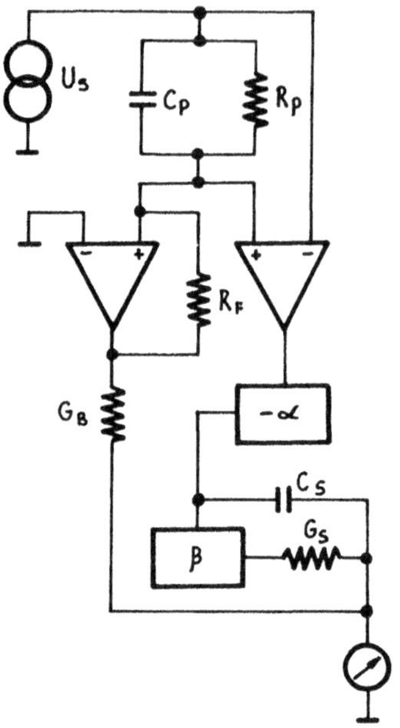

$$C_P = \alpha C_S G_F R_L$$

$$D = \frac{\beta G_S}{\omega C_S}$$

Fig.8. Admittance measurement bridge.

3.2. Operator-system communication.

The operator of a complex system usually meets problems of the following type:
- -is the system efficient after being engaged?
- -in what order should its jobs be programmed?
- -are the introduced data correct?
- -is the work of the system run properly?
- -what are the settings at an instantof measurement?
- -in what physical units is represented the result obtained?

In "traditional" solutions, the increase of the number of program parameters increases the number of required hand-wheels, switches, controllers, etc., resulting in more com-licated front equipment boards, while the costs grow very fast. The application of microprocessors allows, generally, the replacement of handwheels and switches by digitally fun-ctioning key-boards.

In this case the simplification of service results mainly, from the possibility of applying the procedures that carry out in automatic way (without any intervention of the opera-tor) sequences that program a bigger number of equipment blocks connected in a complex way. In account of the fact that there is a large number of settings which are stored simultaneously, the number of program and tested parameters grows up.

An additional and important advantage is the possibility of testing the correctness of operator instructions in a programmed way. Thanks to this, wrong data may be eliminated, such as these that establish forbidden measurement conditions or which lead to conflicts, resulting from the wrong selection of sets of independently programmed parameters. An example of a typical error of this kind may be the drive to establish impulse parameters of width exceeding the length of impulse repetition cycle. Data introduced from a key-board are displayed on digital or alfa-numeric indicator units, while each one of the settings, introduced before, may be tested at any moment.

In the case where a microprocessor controls communication in the direction of the operator, then the capabilities of measurement equipment, in relation to traditional solutions, also increase substantially. Measurement results which are obtained through a multiple transformation of a set of indirect results, can be presented to an operator in appropriate sizes and technical units. For example, an oscilloscope 1722A of Hewlett Packard or 2001 of Norland Instruments equipped with a microprocessor, makes automatic calculation of impulses, of time, increase of areas of pin impulses, of distances between them, of effective values of the course, of average values n of measurement points, of proportional relations of amplitudes, etc.

The communication between the various equipments is a very important problem that influences the automatization of measurements, and the possibility of easy design of systems. Now, both microprocessors and programmed implementation of interface functions allow a substantial circuit simplification. Instead of individual sets of I/Os for each measurement block, there may be used central circuits of interface, for bigger number of sets, which connect separate equipment registers by appropriate buses.

The programmed implementation of the transmission process is, generally, 20 to 50 times slower when compared with direct communication. Thus it can be applied only when the speed of transmission between units of a system plays a secondary part.

3.3. System interface.

System interface, also called external interface, is usually designed according to IEC recomendation for series-byte transmission of information. This recommendation determines all interface functions together with the required procedures, and it does not restrict the way of implementation.

The design of interface circuits using integrated components of small and medium integration scale is troublesome enough, for it needs 40 to 50 circuits. So there is a tendency to utilize LSI elements for this purpose. There are three possible ways of interface circuits construction using LSI elements:

- the application of dedicated integrated circuits which carry out basic interface functions,
- the application of programmed logical networks with memories,
- the implementation of interfaces with the utilization of a microprocessor.

Dedicated LSI circuits which cooperate with transmitter-receiver interface lines, allow the implementation of basic interface functions in a simple way that guarantees maximum transmission speed of interface signals.

The application of programmed logical networks with memories is the solution that makes it possible to carry out interface in an economic way, being adapted to the structure of a given equipment with the application of a small number of integrated circuits. Here, the implementation of separate interface functions is applied, based on the work of a synchronous automaton of the Mealy type.

The execution of interface functions with the aid of a microprocessor is a natural solution whenever a microprocessor cooperates with the measurement part of the system or when communication with the external interface is achieved with the storage or data processing in the wider sense. The employment of a microprocessor in the second case is generally more economic than the design of the necessary circuits based on elements of small integration scale.

Specific time requirements imposed in the implementation of SH-transmission test, and AH-reception test functions cannot be practically met by typical microprocessors in a programmed way, and hence their circuit implementation is used. All other interface functions can be easily carried out in a programmed way, although it is usually impossible to obtain, even in this way, maximum transmission speeds as required by IEC recommendation.

The implementation of external interface with a microprocessor requires the application of interrupt circuits in order to receive signals ATN, REN, IFC and EOI (when applying PP functions) to be used as initiating signals. Signals are interpreted with the aid of appropriate sub-programs which are called out by the interrupt circuit.

System testing. The complication of equipments and systems based on microprocessors raises some difficulties at their mass production and at their service, because a substantial number of errors or damages, difficult to be foreseen, may occur. This is why the designer of equipment should find adequate simplifications which allow one to carry out simple tests of operation on separate blocks and components,and to detect easily the conditions of a possible damage.

The testing of a microprocessor- based system needs appropriate, generally complex enough, tests to be designed. Test programs may be designed in such a way so as the system alfa-numeric indicator also serves as an error display, which displays up to a maximum number of damaged memory modules. This is useful especially for the service test. The alfa-numeric indicator itself should also be tested using decoder and control circuits. Simple tests based on the display of separate segments of indices and on appropriately selected combinations of signs may serve this purpose, the former being put-on manually or with the aid of a microprocessor. The testing of functional and numeric key-boards is also relatively easy; it is carried out by displaying the code numbers of separate keys one by one.

Despite some increase in price, the addition of the above diagnostic facilities, is very economic for both the producer and the user, allowing to substantial simplification of maintenance and repair operations.

The testing of measurement systems with microprocessors involves the examination of the microprocessor, and of the functions fulfilled by it. When stating that these functions are carried out correctly, one may assume with say 85 per cent guarrantee that the whole system works correctly. This percentage becomes higher when the system works with dedicated microprocessors.

The methods of microprocessor testing in a system are presented below. The irregularity of microprocessors internal structure makes it impossible to test them with repeatable procedures which are applied for memory circuit tests. Sequences of instructions, data and responses of microprocessors have a character of non-repeatable courses, and they cannot be obtained with the aid of micro-programmed tests generators. There are three well-known methods of microprocessor testing:

- testing in an utilitarian system,
- comparison with a pattern circuit,
- method of stored answer.

Testing in an *utilitarian system* is the simplest method of microprocessor examination. The test is based on the placing of a circuit, examined into a microprocessor system previously tested, and observing the program work results. An affirmative result is the evidence that the circuit can be successfully applied to such a system, when the program with which it was tested is used. The testing in a given utilitarian system does not embrace the whole set of parameters and working conditions of a microprocessor. That is why an affirmative result does not ensure that the circuit tested fulfills all the requirements posed by the designer. Circuits that were tested and gave an affirmative result in one system, may not work, at all, in another system which utilizes different address space, and has different control-test signals, or imposes different dynamic requirements. The testing in a utilitarian system does not allow the information on digital values of separate parameters of the circuit tested to be obtained. It simply permits to get through potentially faulty circuits of parameters which are on the border of correct operation of the system. The method of testing in a utilitarian system does not require any special test equipment (this is actually its basic advantage), and hence it finds application by those users who utilize a small number or microprocessors.

In the method with a *pattern circuit*, the testing depends on the comparison of an answer of the microprocessor under test with the answers of a pattern circuit that is controlled by the same sequence of signals as the circuit tested. The basic disadvantage of this method is the necessity to have a circuit, which can be assumed in any case as a pattern one. One simplification is the buffering of pattern microprocessor signals, which act, in all respects, with the advance of one clock cycle in relation to the circuit tested. It lets the pattern microprocessor off, and avoids the necessity to have pattern and dynamic parameters, while reducing the restrictions upon the functional correctness of the operation and the required frequency of the clock generator. The testing is implemented during the implementation by the examined and pattern circuits, of a special test program which includes sequence of orders and data stored in memory of a pattern microprocessor system connected to the tester. The test program is usually executed repeatedly during the test, at various levels of supply and control voltage, as well as at various delays of circuits stimulating and comparing answers of the microprocessor under test. The method of comparison with a pattern circuit is used for the examination of microprocessors and other LSI circuits of irregular structures, with the aid of relatively cheap testers employed by both producers and users. Among the equipment based on the application of this method, are the Megatest Q-8000 and Adar MT-17 testers.

The method of *stored answer* depends on the control of the circuit examined by signals taken from fast buffer memory that include , apart from stimulating signals, pattern answers of the microprocessor under test. When using this method in testers, in order to decrease the capacity of buffer memory needed to store a long sequence of some scroes-bit words (including instructions, data, and pattern answers) there are special sequence processors that allow one to make loops, sub-programs and stacks. Sequence processors allow the decrease of the capacity of the buffer memory, necessary to test the microprocessor, three to five times.

The application of this method requires to prepare the data and pattern answers of the microprocessor examined before the appropriate series of orders. Sequences of orders and data are usually prepared in a symbolic language, based on the logical scheme and functional parameters of the microprocessor under examination. The translation of a program from a symbolic language to a machine one, as well as the generation of pattern answers, is usually carried out in a computer way, with the aid of a special program called the *tests generator*. The tests generators, known, are not universal, and they can be used only in relation to a concrete microprocessor type. The method of stored answer is the only method applied to test microprocessors with the aid of universal test systems, used by producers of LSI circuits for characterizing tests.

Microprocessor testers, produced industrially, utilize two methods:the method of comparison with pattern circuits and the method of stored memory. As it has been mentioned before, testers based on a rule of comparison with pattern circuits are instruments which are relatively small and several times cheaper than systems which utilize the method of stored answer. The mostly known tester which utilizes pattern circuits, is the model Q-8000 of Megatest. In this tester there are changeable modules, namely those which create the so-called "natural environment"of pattern circuit (Fig.9).

Modules assigned for the testing of LSI circuits of irregular structures, such as microprocessors, are equipped with pattern microprocessor systems, in the memory of which the test program, composed of adequate sequences of instruction and data, is stored. The module constitutes the buffer memory which delays the signals taken from pattern system for one clock period. This makes the results of comparison to be independent of dynamic features of this system (within one cycle). The time control block included in the module introduces program time delays of the command signals of the circuit under examination, and determines the instants of comparison of the examined and pattern circuits. This allows to make quantitative evaluation of dynamic parameters. Microprocessor functional

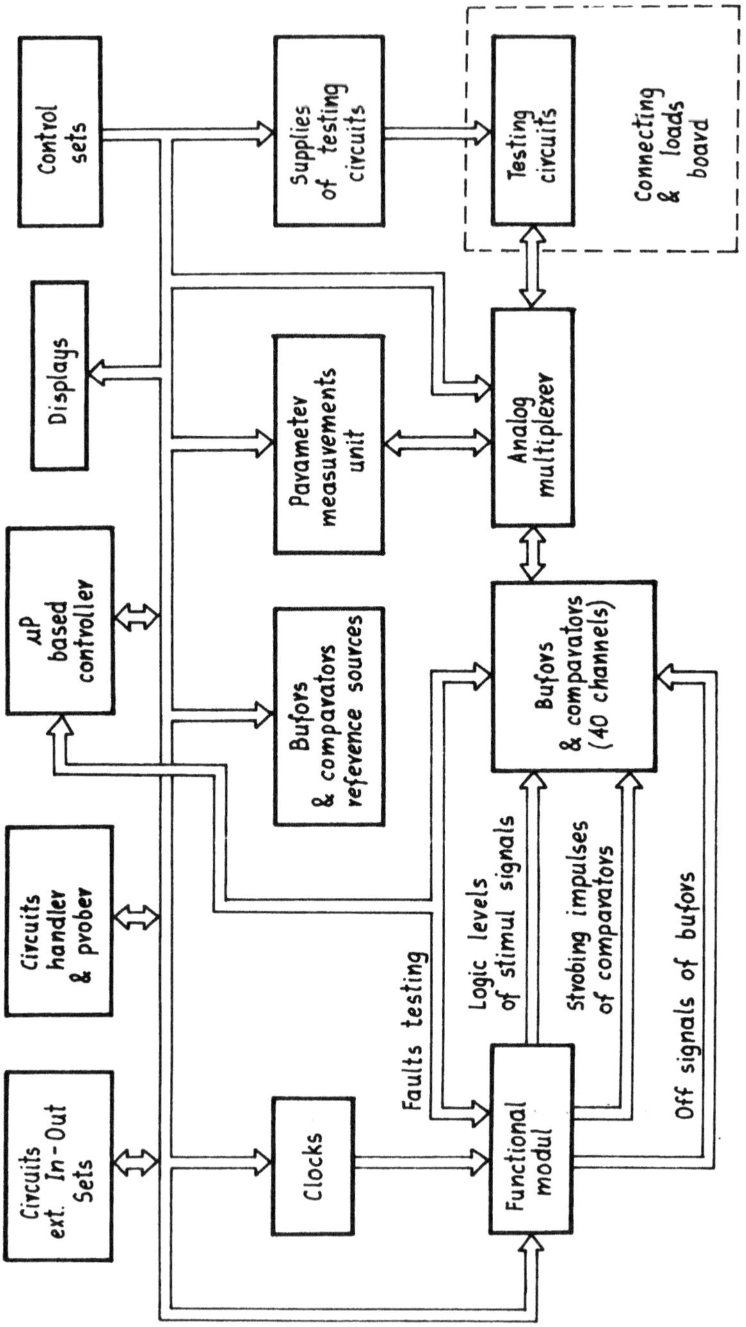

Fig. 9. Tester Megatest Q-8000.

modules of the tester Q-8000 may be utilized both for the testing of microprocessors as well as of other LSI circuits. The test programs of these circuits are written in the microprocessor language which is included in the functional module. In this case, a microprocessor is the source of command signals for both pattern and examined circuits (Fig.10). This is especially convenient in the case when testing circuits are assigned for the application in microprocessor systems.

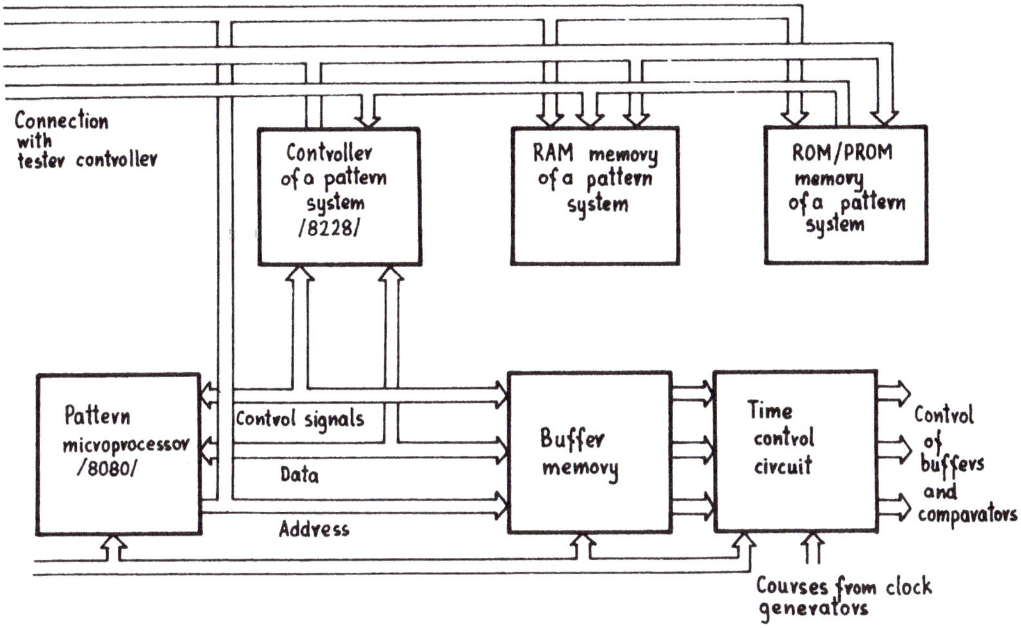

Fig.10. Functional module of a tester Q-8000 to test a microprocessor (8080).

The settings of separate blocks of a Q-8000 tester are regulated in a programmed way, with the aid of a tester, working with the application of a 8080 microprocessor cooperating with a program memory of capacity 16 k bits. The test program of separate circuits may be introduced into the controller memory with the aid of a key-board, or from magnetic or paper tape. A special higher level language allows the easy introduction of data, setting on tests and modification of parameters.

In the category of universal systems that utilize the rule of answers, the mostly known devices include testers of Sentry type, having been produced by Fairchild for a number of years.

The basic disadvantage of these systems is the excessive complexity of hardware which, in turn, causes big investment costs. This is the reason why their usefulness for the users of microprocessors, as well as for the designers of microprocessor based systems, is quite small. For these people, the universal and simplified systems are more interesting.

The most modern solution of a simplified universal system is the model Sentinel of Fairchild. This system maintains the basic structure of Sentry universal system, and it is capable to carry out programs of testing LSI circuits designed on these systems (Fig. 11):

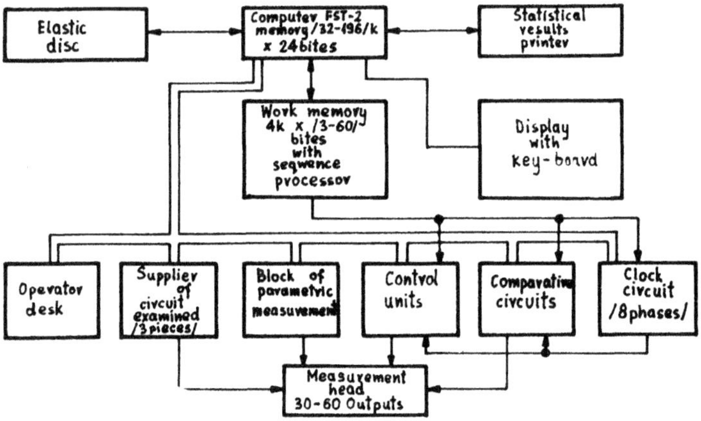

Fig. 11. Block diagram of Sentinel system.

Taking into account the unique technical and economic advantages of these new structures, one can assume that they will overcome the whole sphere of production examination of microprocessors in the nearest future. Producers of electronic equipment who use microprocessors in their instruments, are the most keen to do examinations on such equipment by employing identical programs with the producers of circuits utilized. Such an activity makes it possible to avoid any misunderstanding occurring between suppliers and consumers due to the different interpretation of measurement results. The lack of possibility to buy small testers, as well as to apply big systems from the producers side caused that, until recently, also some users of microprocessors were employing such systems. Taking into account the relatively small number of circuits examined, as well as the substantial costs of purchase and maintenance of big systems, such a solution is far from the optimum.

3.4. System development

Not only the design, but also the process of putting into motion and examination the systems are carried out, to a great extent, in a heuristic way. Algorithms of putting into motion and examination newly designed systems may be formulated quite generally in practice. Too many details of these algorithms would lead to the disadvantage of unexpected and often "malicious" errors, which cannot be avoided either on the design or the execution stages. Because of the above reasons, the methodology of putting into motion and examination a computer system concentrates, in a natural way, on the means of work rather than on the algorithms of action. It concerns the initiation of motion and examination of both software and equipment, as well as their functional "connections", with special attention being given to real time work.

In the case of microcomputer systems, one should have more specialized means of putting into motion and examination, than in the case of a big computer system. Regarding the big computers, observe the following.

- In general, they are being designed in bigger research and development laboratories.
- They are usually equipped with hardware-software diagnostic means: debugging, programs, check-tests, as well as appropriate consoles equipped with convenient switches, displays, etc., and not only with basic software such as: operation system, assembler, compilers, etc.
- They have relatively large operation memories of random access.

On the other hand, in the microprocessor systems, especially the smaller ones (which are the most typical) the situation is reverse.

Nowadays, three principal groups of initiating and examination means of microcomputer systems exist:

- *Cross-software* (i.e. dedicated assemblers, compilers and simulators) written in a higher level language (Intel, for example, provides them in Fortran language) and assigned for other bigger and universal equipment. An advantage of such software is its comfort of application, under the condition that one has an appropriate computer such as IBM 370, ODRA 1305, or a minicomputer such as PDP-11 or MERA 400, for which there are, among others, dedicated assemblers of a microprocessor, say a 8080 . A disadvantage of dedicated software is the need to put into motion the utilitarian programs beyond an equipment "of destination". This is often done under artificial conditions and with a speed that is considerably lower than that required for real time work.

- *Logic state analysis* that serves, above all, to solve hardware problems;starting from the location of faulty (no go) integrated circuits, and finishing with the discovery of time conflicts and disturbances difficult to be noticed. These are multi-channel instruments (typically 16 lines) which record, in a synchronous or asynchronous way, two-state approximated electrical motions, and display these courses in a way convenient for the user. Several ways of displaying may be followed;from displaying courses as functions of time, via maps of memory, up to displaying rows of zeros and ones, or texts in a symbolic language. Especially, courses on address buses or data buses, or in I/O ports of a microprocessor may be recorded.
- *Software development systems* which serve, not only to put software into motion (starting from the edition of source program texts and their translation) but also to gradual integration of this software with anequipment "of destination" of a system based -on the so-called principle of *in-circuit-emulation*. An advantage of development systems is that it is possible to test the work of the whole system, and to correct the software under conditions approaching very closely the real ones (also including matters of time) by just including a program in the RAM. The introduction of this program into the EPROM memory is done only at the end state of the work.

The above equipment (e.g. assemblers as well as dedicated and resident compilers, which constitute the equipment of development systems) is partly doubled in a sense. On the other hand they partly supplement each other,as it can be noticed in the case of developed systems and analysers of logic states. The former ones are manufactured mainly "from the point of view" of software, but the latter are designed following the "hardware (equipment)"approach.

Above all, systems are being offered by manufacturers of microprocessors, as a "hinterland"for their products. Nevertheless, there appeared already interesting development systems based on some of the most common microcomputers of various manufacturers, and designed by "indepedent" companies; the system 8002 of Tektronix may serve here as an example. In both cases, one changes the mostly used optional paquets-adapters or discs for a specific microcomputer type.

A typical development system is equipped with hardware-software means of emulation and diagnostics, it has a RAM memory of relatively large capacity, as well as a disc memory. An example of the design phases of a utilitarian system based on the employment of a standard development system is shown in Fig. 12. A typical example is the system MDS INTEL which serves as a "hinterland"for the family of 8080.

3.5. Hardware-software relation and tradeoff

Until now, the designers of measurement equipment of microprocessor-based systems have been interested only in hardware. In parallel with the development of microprocessor technology, they were forced to find a common language with the software educated people, or to learn it by themselves. This has been, and still is, very difficult, since it appears that it has to be decided at the same time what is to be automated in equipment and to what degree, in order for its price and characteristics to be accepted by the user's market. Most designers have the opinion that one should follow an attitude similar to that followed after the transistor had appeared. This means that one should consider a microprocessor as one more element of components in the sub-assembly arsenal, serving the development and engineering of electronic equipment. This attitude seems too simplified, since the microprocessor is the component which does what the program elaborated by us wants it to do. In general, most of the designers agree with the fact that in equipment, the technology of microprocessors will be utilized more and more, to make the service easier and to eliminate the errors made by the operator. Speaking in a different way, microprocessors can help in the control of measurement equipment and in the analysis of data collecting, but they cannot in fact improve its "measurement parameters" .

The users on the other hand have slightly different opinion and look at the matter from the point of view of the evaluation of the quality of the objects being examined.

One of the first questions of the user is whether this equipment is to be fully automated or controlled manually. The users state that manual control with still remain for a long time. Most of the measurement devices have access to the keyboard and to the display control existing in the systems. This creates the possibility of varied utilization of the key-board for routinized software. The monitor should read all the functions of automated measurement, including generated reports, anticipated directions of an error, and the set of "historical" data. The movable "test stands" which are controlled from the central processor, are more and more common. These stands are easily exchangeable in their configuration, thus making easy the introduction of new possibilities. Multifunctionality is extremely important for the general ap-

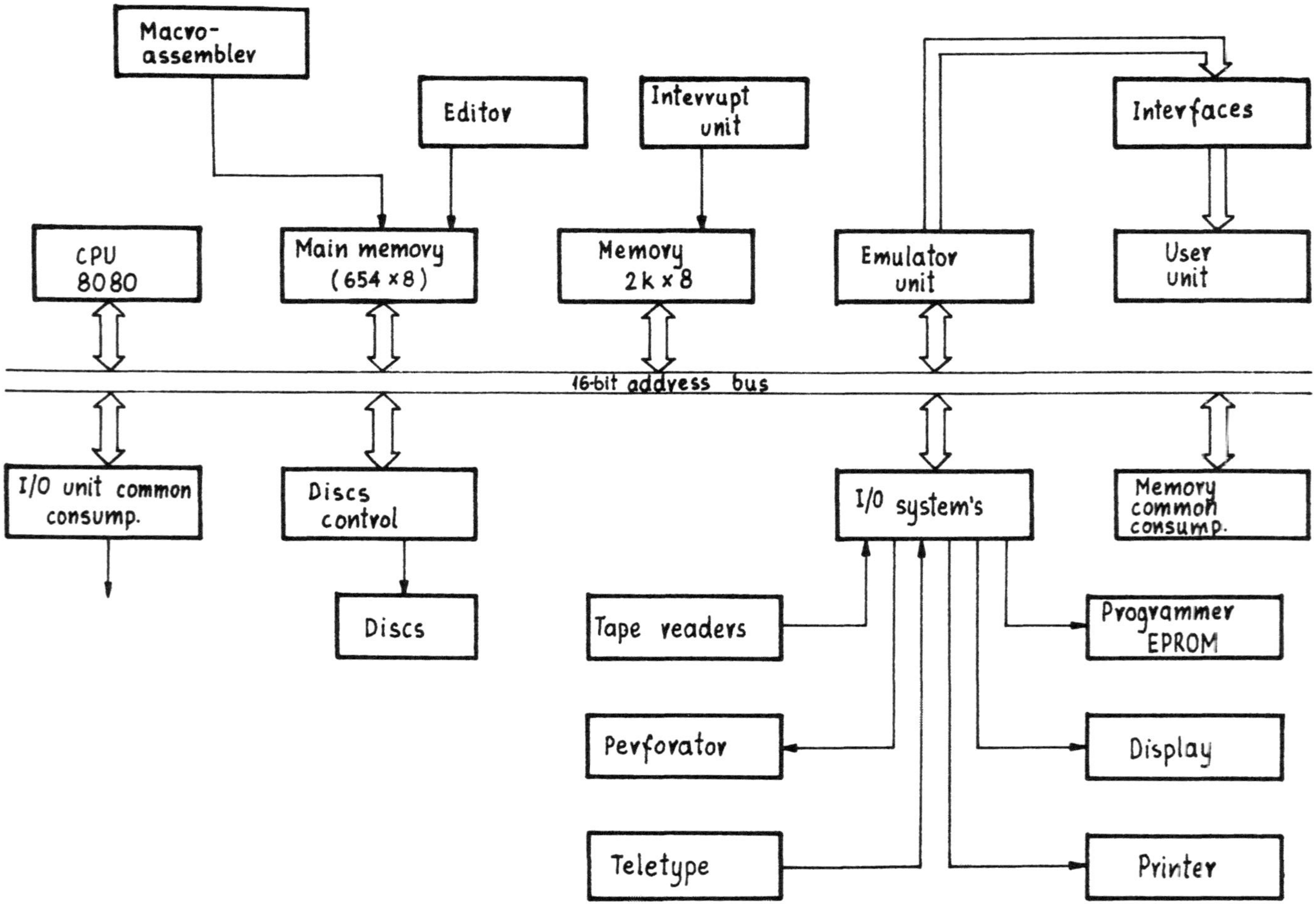

Fig. 12. Structure of a standard development system.

plication, because the procedures of dedicated tests, such as low frequency, high frequency, parametric measurements etc, concern the equipment with single operator's interfaces and microprocessor controllers. There is still an important distinction between the laboratory and production applications. The multifunctional application is usual in laboratories, and the dedicated one in production. Nevertheless, one can assume that the high cost of multifunctional equipment influences the specialized application, and often draws the attention of small scale "not so wealthy" users.

On the other hand, producers of equipment must see the growth of flexibility of applications, and thus they must make more effective use of equipment at every stage of the production and exploitation of the product. The equipment which does repeatable work, with checked control, will use software as the programming mean, and the equipment of non-repeatable, and independent-of-check-control work, will use the elements controlled manually. The number of these elements, however, should be reduced to a minimum, and it should be adapted to the work of poorly qualified staff. As more qualified personnel, is engaged one can use a sub-panel with an appropriate control, or a connecting block, that makes it possible to connect a key-board (or an appropriate input-output circuit) to programming, and to obtain full capability of an equipment.

4. THE FUTURE

Although in the seventies a fast development and expansion of microelectronics of general application was achieved, this development does not keep up with the requirements of modern measurement systems. The digital LSI circuits of low power intake are urgently necessary, and particulary those which are simultaneously very fast. Also new types of analogue LSI circuits, and LSI circuits for the control of LCD and LED, compact supply sources, etc., as well as dedicated LSI circuits (e.g. for fast Fourier transformation) are needed. There is also a lack of A/D and D/A converters, and micro-peripheral equipment suitable to work with microprocessors. On the other hand, despite these lacks, the development of the hardware and measurement systems is nowadays very much ahead of the development of software of these systems. The utilisation of standard interfaces allows a considerable speed-up of the design of system hardware, but it also requires to create an adequate software. The control problem in systems with microcomputers in among the most difficult to be programmed. It requires, e.g. approximately many working months for a program of 1000 instructions. This results from the limitation of time for program implementation as well as from store capacity limitation depending on the microcomputer characteristics.

As it is known, software can be divided into that prepared by the computer producer, and that being elaborated by the user of a given system. Minicomputers (more rarely microcomputers) are equipped by the producer with software in the form of operation systems. These systems allow the multi-programmed work to be done with the consideration of priority tasks, automated diagnostics of computer elements, etc., and offer the possibility to program usable tasks in high level languages. The user can employ a typical software elaborated for various tasks, e.g. data processing, and can always create his software which is determined from the specific tasks to be performed, and from structural configuration and properties of the system concerned.

A separate problem concerns the systems with distributed units (there is an analogy with microwave circuits having distributed parameters). The installation of A/D converters and microprocessors directly near the measurement sensors makes it possible to avoid the transmission of analogous measurement signals at a distance. In such systems, it is reasonable to use the distributed data cases, which minimizes the loading of connection channels. Such data are the measurement results and the pattern data. In distributed systems, it is essential to distribute microprocessors and memories, as well as their mutual cooperation, in a rational way. Of course, the complexity of software grows here.

Curently, two simply opposed trends are in development in the leading laboratories of the world. Must we introduce the relatively cheaper VLSI circuits in the place of expensive software, or are we to develop the latter? Clearly, the opinions on this do not coincide. Also one can notice the supremacy of those following the increasing hardware, which is somehow going to replace software. This opinion is shared by such companies as Fairchild, Texas Instruments, and National Semiconductor. The idea behind this trend is summarized as follows. The measurement system develops for a considerable number of functional blocks, which being designed as VLSI circuits, can be cheaper and cheaper in the mass production. Thanks to the large number of functions carried out by these blocks, one can diminish considerably the number of instructions required to make the program at hand. Taking into consideration the high costs of programming and the time necessary for it, any decrease of these costs is worthy. In other words, the total production cost of hardware is lower than that of software for a given system, and moreover, the hardware elements can be utilized in various systems, where a separate software should be elaborated for each one of them.

Some new control strategies must be found in connection with this. It is hoped that this barrier can be easily overcome with the present measurement and computing knowledge.

CHAPTER 10

MICROCOMPUTERS IN NUCLEAR PLANT DATA MEASUREMENT

Hans Toffer
UNC Nuclear Industries
P.O. Box 490
Richland, Washington
United States of America

Computers from large mainframes to thumbnail sized microprocessors have had a dramatic impact on modern technology. This fact is particularly true in the nuclear industry where the need for safety and reliability are critical. Although the development of nuclear power began before the widespread availability of electronic computers, it is obvious that the sophisticated designs and control techniques now available would not be possible without high speed computers.

This chapter will focus on the uses of microcomputers for acquisition and display of nuclear plant data. Rather than attempt to cover the broad range of applications in the Nuclear Industry, this chapter will instead discuss the uses of microprocessors at the Hanford N Reactor, a reactor operated by UNC Nuclear Industries for the United States Department of Energy. Three applications will be discussed in detail, they are, 1) The acquisition and analysis of flux data using an intelligent terminal, 2) The use of a network of microcomputers to monitor vibration levels on rotating machinery, and 3) The use of a digital image processor to display nuclear reactor control parameters.

INTRODUCTION

Most modern technological advances are closely correlated with the development of computers for numerical processing and systems control. Although development of nuclear power in the 1940's started without the use of electronic computers, the need for automated numerical processors to perform the required sophisticated design calculations required for power reactor development became obvious. Since the 1950's the evolution of nuclear power has been strongly supported by the emergence and advances of computer technology. The use of sophisticated computers in the development and design of advanced reactor concepts such as breeders or fusion devices and the operation of power reactors has become invaluable.

S. G. Tzafestas (ed.), Microprocessors in Signal Processing, Measurement and Control, 189–204.

In addition to the use of computers for design, safety evaluations and economic studies, electronic digital processors and controllers in various forms have become an integral part of an operating reactor. Control system responses, component monitoring, online safety evaluations, system performance displays and automatic data acquisition have been assigned to mini and microcomputers and microprocessors. Such efforts have reduced human interactions and have significantly increased plant reliability, safety and plant performance. As an example, the Canadian Bruce Nuclear Generating Station, one of the largest in the world, extensively uses computer technology for control, data acquisition, and data display. This plant has a gross lifetime capacity factor of 80%, one of the best in the world (Dumka, 1981). A dual computer system is employed to supply most plant control functions to regulate reactor thermal power and boiler pressure, to provide load following capabilities and to govern moderator control adjustment.

It is difficult to cover computer or microprocessor application to reactors in general terms or even consider the application to one particular reactor type such as a Pressurized Water Reactor, a Boiling Water Reactor, a High Temperature Gas-Cooled Reactor, a Heavy Water Pressure Tube Reactor or a Light Water Cooled Graphite Moderated Reactor.

Instead this chapter focuses on some specific novel uses of microprocessors and minicomputers at the Hanford N Reactor, a reactor operated by UNC Nuclear Industries for the United States Department of Energy. The N Reactor was designed and built in the late 1950s and early 1960s. Computer applications for process control were minimal, the devices that were used were vacuum tube based, small, and required extensive maintenance. The applications considered illustrate the introduction of current computer systems such as parallel processors, microprocessors and minicomputers for nuclear plant data acquisition, processing and display. The use of such computer systems represents an upgrading effort, an introduction of current technology to an old facility to improve plant operating performance and safety. The particular examples are sufficiently general to be of use to other nuclear plants or industrial applications.

A DESCRIPTION OF THE HANFORD N REACTOR

The Hanford N Reactor is a 4000 MW(th), graphite-moderated, pressurized water reactor. It is called a dual-purpose reactor since the waste heat developed in producing nuclear products for the United States Department of Energy is used to generate up to 860 MW electrical power for use throughout the Pacific Northwest region of the continental United States. The reactor core is a graphite cuboid approximately 10 m square at the face and 12 m long. A total of 1003 horizontal Zircaloy-2 pressure tubes penetrate the graphite stack. A fuel charge for each tube is 10.6 m long and is made up of 16-19 fuel elements. A fuel element consists of two concentric low enriched uranium metal cylindrical shells clad with Zircaloy-2. The fuel is cooled on all surfaces by pressurized water.

Perpendicular to the pressure tubes are 86 horizontal control rods used for shaping the power distribution and for shutdown control of the reactor. The control rods enter from both the left and right side of the reactor. Vertical to the pressure tubes are 107 channels which can be filled with neutron absorbing balls to provide backup shutdown capabilities.

Most N Reactor instrumentation is outside the core except for some thermocouples and flux wires. Neutron flux mapping in N Reactor is accomplished using nine flexible carbon steel wires. These flux wires, called the Traveling Wire Flux Mapping (TWFM) system are periodically inserted parallel to the pressure tubes in nine distinct regions of the core to obtain axial neutron flux distributions. Such distributions are essential to determine axial power peaking and monitoring neutron absorber distributions in the core. An expanded view of the reactor core displaying the major reactor components is shown in Figure 1.

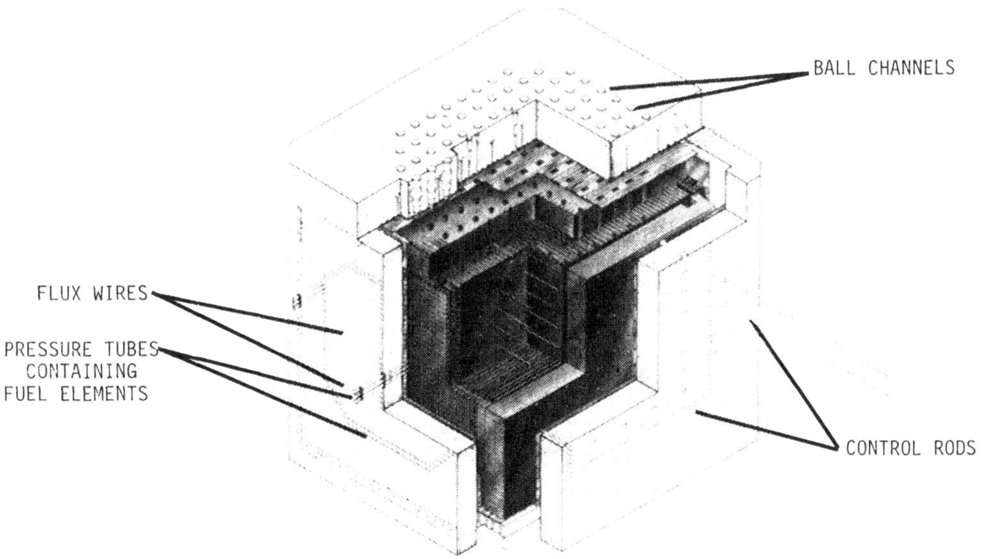

Figure 1. N Reactor Major Components.

The N Reactor coolant flow, and coolant temperatures are monitored continuously for each of the 1003 pressure tubes. From this information, individual tube powers are calculated. Reactor power data in conjunction with control rod positions need to be displayed in a manner as to be readily surveyed and assimilated by control room personnel.

N Reactor has a large number of rotating machines: turbine driven primary pumps associated with the six steam generator cells, emergency cooling system pumps and diesel engines, high pressure injection pumps, secondary cooling system pumps, an auxiliary turbine generator station, and other numerous motors and blowers. The equipment is dispersed over a much larger site than in the compact designs of Light Water Reactor plants. The rotating equipment at N Reactor is aging and requires surveillance and frequent maintenance.

Although N Reactor was built in the beginning of the computer age, significant retrofitting and upgrading with solid state and micro chip technology at the reactor plant has occurred during the last eight years. An integrated reactor information system based on microprocessors is in place to monitor extensive arrays of reactor parameters. Data General minicomputers are being used to perform operations support and safety calculations. However, the reactor is still manually operated requiring rapid means for displaying pertinent reactor data to the operators.

DISCUSSION OF EXAMPLES

The three examples discussed in this chapter illustrate specific applications of microprocessors and minicomputers to the acqusition, processing and display of reactor plant data. The first example shows how a commercially available microcomputer, a Tektronix 4051 intelligent terminal, can be expanded into a sophisticated data processing and evaluation station with extensive computational capabilities. This development was in support of the N Reactor TWFM system analysis.

The second example deals with the use of microprocessors and a minicomputer linked together in a data communications network. The microprocessors function as semi-autonomous data acquisition stations that are linked to a control computer. The system is being implemented to provide continuous routine vibration and signature monitoring of rotating machinery at the N Reactor.

The third example illustrates the use of a digital image processing system to display large amounts of data. Such efforts are important to improve man machine interactions by providing the reactor operators with readily recognizable representation of core operating conditions.

MICROPROCESSOR BASED EVALUATION AND ANALYSIS SYSTEM

A special purpose, relatively inexpensive, data processing station was developed around the Motorola 6800 microprocessor, contained in a Tektronix 4051 Intelligent Terminal (Peterson et. al., 1980) This system is used to evaluate N Reactor TWFM data. As a result of this effort rapid and more reliable computations of axial peaking factors are possible and systematic evaluation methods are in place for monitoring short term and long term flux perturbations in the core.

The axial TWFM wire is a flexible carbon steel wire that is exposed to the core neutron flux for approximately 30 minutes. Impurities such as Mn-54 in the steel are activated. The wire is pulled from the core past a sodium iodide (Na-I) crystal detector at a rate of 7.6 cm/sec. The output from the detector is sensed by a picoammeter and recorded on a strip chart recorder. The induced activity along a TWFM wire is directly proportional to the axial neutron flux distribution in the neighborhood of the wire. The use of a lead collimator in front of the NaI crystal assures a spatial resolution in the activation traces of approximately 1.0 cm.

Prior to the development of the microcomputer based analysis system, the information derived from the TWFM traces was used only to determine axial peaking factors employing laborious error prone manual methods. The new evaluation and analysis system permits rapid peak to average flux value calculations and the extraction of additional information about reactor core conditions from the TWFM data. Figure 2 is a typical strip chart recording of a TWFM wire activation profile.

The computer system shown in Figure 3 is an enhanced microprocessor based calculator having some of the capabilities of a minicomputer.

The interactive system is programmed in BASIC which is easy to use for the required signal processing. The slow BASIC system is supplemented by machine language programmed "ROM" (read only memory) modules. The ROM modules provide increases in computational speed approaching those of a minicomputer.

The computer system includes an analog to digital converter with 12 bit resolution and a maximum sampling rate of 27,000 samples per second. The A/D converter digitizes the signal sensed by the picoammeter. Two signal processing ROM modules contain routines for integration, differentiation, array manipulations. The ROM modules also include a fast Fourier transform routine and other signal processing functions. A separate mathematical function processor ROM is used to speed up calculations.

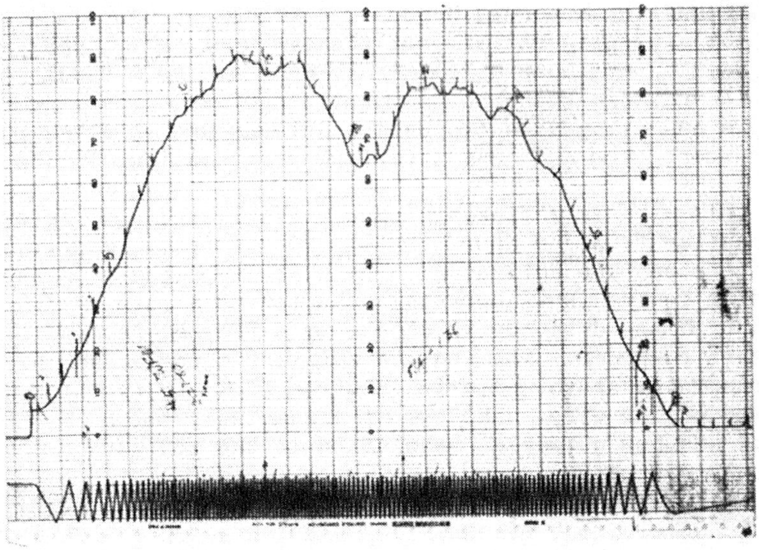

Figure 2. TWFM Wire Activation Profile.

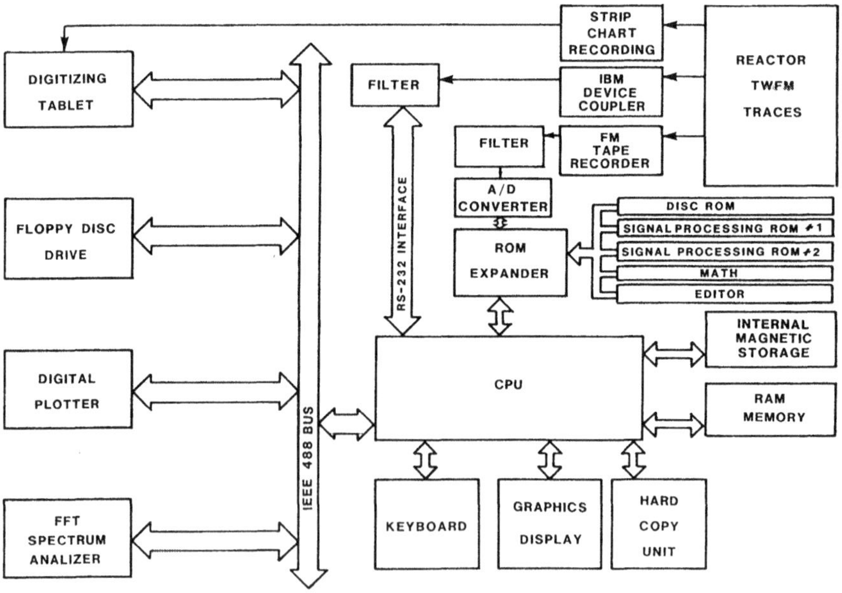

Figure 3. Block Diagram of TWFM System.

The computer has high resolution graphics capabilities. Detailed plotting of digital data or mathematical functions are possible. The resolution of the storage display is sufficient to display any feature observed in the analysis of the flux wire mapping system. The graphics system is entirely controllable by the operating system so that flux distribution data can be displayed in any desirable form. The system contains two interfaces, a serial interface used for communications with mainframe computers and bit serial devices, and a parallel interface conforming to the IEEE-488 standard for use with peripheral instruments and a hard copy device.

The hard copy unit and the digital plotter are used for data display and production of permanent records. An internal magnetic tape drive and an external flexible disk system greatly increase the storage capabilities of the computer.

Special software had to be developed for processing and evaluating the TWFM trace data. Since the computer contains only 26K bytes of memory, the entire software system cannot reside in memory but must be stored on disc. Extensive file manipulation between the disc and computer is time consuming. Therefore, use is made of the computer's twenty user definable keys. A separate controller program is first called into memory which defines the functions of each key. Pressing a particular key will bring into memory the routine to perform a required function. This technique always reserves as much memory as is necessary for data storage.

The TWFM data that are evaluated by the system are obtained from previously recorded strip charts using a computerized digitizing tablet, current data recorded on a high quality FM data tape recorder, or directly digitized traces stored on discs.

Prior to digitizing the raw data, any high frequency noise is filtered. This operation eliminates any possible aliasing problems. Following this step the analog signals are converted to digital information by means of the A/D converters. Typically there are 500 to 600 data points in a TWFM trace. The digitized data are indexed and stored on floppy discs.

When a TWFM trace is processed the data are centered, background corrections are made, the axial peaking factor is calculated, correlations with control rod patterns are established, the data are searched for unusually steep flux gradients, and a frequency spectrum of the data is computed. The results of the computations are viewed on a high resolution display terminal, hard copy prints are made of any interesting intermediate results, and final figures are generated with the multicolored digital plotter. A typical result is shown in Figure 4.

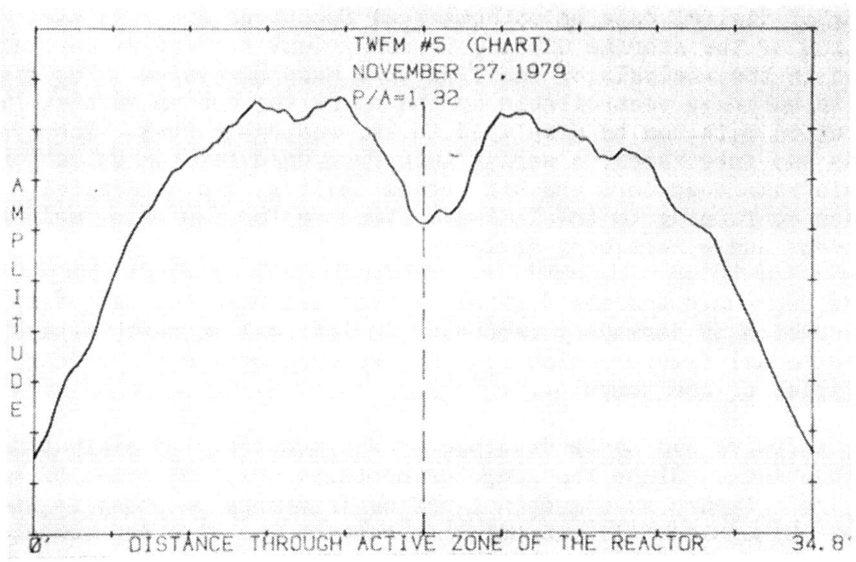

Figure 4. Processed TWFM Trace.

The analysis system has been successfully applied to rapidly provide more consistent axial peaking factors, parameters essential to the safe and efficient operation of the reactor. In addition, frequency domain analysis has been applied to the data to identify periodic structures such as ball channels, control rod locations, and fuel element positions.

The extensive TWFM data base is being used to identify and qualify the regions of influence associated with control rods, ball channels and other neutron absorbing materials in the core. The possibility of monitoring structural changes in a graphite core associated with prolonged neutron exposure is being investigated using the TWFM trace analysis system.

The microprocessor based data evaluation and analysis system has proved to be very useful in the immediate evaluation of TWFM traces. Analysis of the TWFM data base has identified local flux depressions and long term trends in core conditions.

LINKED MICROPROCESSOR - MINICOMPUTER DATA COMMUNICATIONS NETWORK

Vibration monitoring and signature analysis of rotating machinery can be used to determine the mechanical integrity of plant equipment by analyzing the unique characteristic vibrations of machinery during operation. The technique is based on the observation that machinery degradation and mechanical breakdown is preceded by changes in characteristic rotational vibrations. Recording the vibration data using accelerometers, velocity meters and proximeters when the machine is in good operating condition will provide a baseline against which changes in characteristic vibrations can be compared and predictions about machine degradation or anticipated failure modes can be made. Industry experience indicates that a month to several months advance warning can be provided for failure modes given that proper data are available.

Use of vibration meters, portable analysis equipment and periodic data acquisition has been successfully employed in many industrial application to improve preventive maintenance procedures, to avoid costly tear down inspection and to identify the cause of equipment failure. However, to establish a predictive rotating machinery vibration analysis and signature identification program requires continuous consistent data availability with rapid access to an interactive computer system with sophisticated analysis capabilities.

At the Hanford N Reactor, the development of an automated vibration monitoring and signature analysis system entails the installation of semi-autonomous data collection stations at specific concentrations of rotating machinery (Harrington et. al., 1982). The stations are linked to a large minicomputer using a computer network approach. The data collection stations are microprocessor based and include components to digitize the vibration data and perform a cursory check of the digitized vibration level. The minicomputer provides control of the data collection stations, stores the vibration data for long term trend and signature analysis, provides the computational power for complex data analysis, and communicates with the operator's console.

Each data collection station consists of a multiplexer, analog-to-digital converter (ADC), microcomputer with random access memory (RAM), communications interface, and digital annunciator. A block diagram of the major components of the data collection station is shown in Figure 5. The multiplexer operates under control of the microprocessor to switch each sensor to the ADC. Programs in the microcomputer compute the spectrum of the digitized vibration data using the fast Fourier transform algorithm. The computed vibration level at each frequency is compared with preset limits stored in RAM. If the vibration level exceeds the preset limit, alarms are initiated locally and a remote alarm is sent to the minicomputer through the communications link.

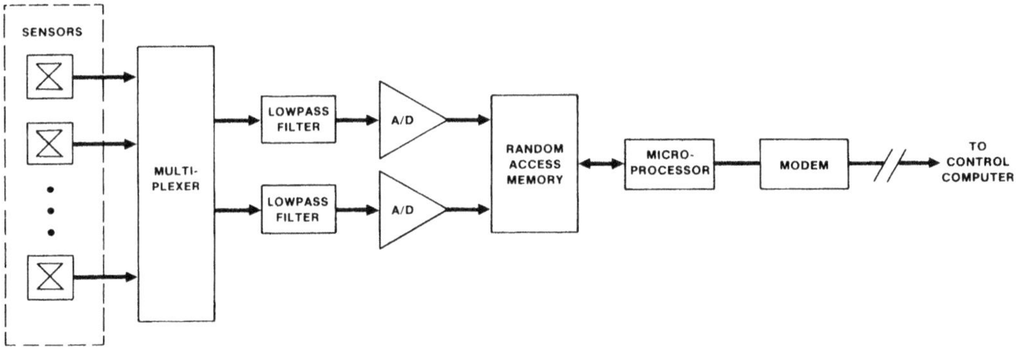

Figure 5. Block Diagram of Data Collection Station.

The minicomputer controlling the operation of the data collection stations consists of a high speed processor, large main memory, disk storage, magnetic tape transport, communications interface, printer and plotter. The primary activities of this computer are; to direct operation of the stations, record vibration data periodically for trend anaysis, and communicate with the operator's console. This minicomputer also has the computational capability to perform complex time series and spectral analysis on the vibration data. A block diagram of the main features of the minicomputer is shown in Figure 6. From the control computer, the operator can download programs into the stations, determine the functional status of each station, and instruct the station to collect data from a particular sensor.

Data communications between the control computer and the data collection stations are accomplished using a computer network approach. This approach has several advantages for this application 1) the software permits downloading programs and data to the satellite stations, 2) store and forward operations, 3) transparent message rerouting between the control computer and a "down" station, 4) messages accounting which guarantees that each message is definitely received and that it is received only by the station to which it is directed, and 5) remote operator control and program development. The network communications hardware and software supports the High Level Data Link Control (HDLC) protocol, CRC-16 error checking, buffer management, and direct memory transfers.

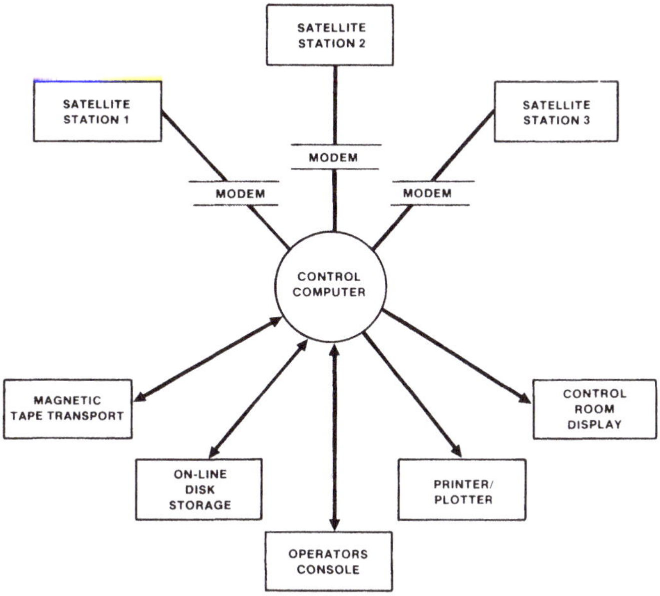

Figure 6. Block Diagram of the Minicomputer System.

In the network the control computer and each station is assigned a unique node identification number. Remote operator commands and user programs reference the number of the node with which they wish to communicate. For example, if the control computer is linked to the first of three stations that are linked in series, and if the control computer is assigned node number 1 and the three satellite stations are assign node numbers 2 through 4 then the control computer is connected to nodes 3 and 4 through node 2. For the computer to communicate with either node 3 or 4, messages must be routed through node 2. This capability of the network software is called store and forward. Nodal addressing with store and forward capability offers important advantages. Hardwired or modem links can be shared reducing initial costs for hardwired connections or operating costs for data communications. If redundancy is required in the network, a ring configuration can be used. In this design multiple paths connect the control computer to each of the stations. In a ring network, dynamic rerouting of messages can bypass a "down" station. This capability increases the overall reliability of the entire network.

The hardware interfaces which implement the network protocol are themselves microprocessor based. These microprocessors are programmed to manage the routine communications processing freeing the station for vibration monitoring. Under control of on-board firmware, the microprocessor converts commands into actions, such as establishing the communications link and loading/unloading data from the onboard buffers to the station. The interface also performs protocol generation and interpretation, error checking, and error recovery by retransmission, all without the attention of the station.

The unique features of this vibration monitoring system are the use of microcomputer based data acquisition staions and the computer network approach to communications between the control computer and the stations. The data collection stations replace the vibration meters and manual data acquisition method. The control computer permits more sophisticated analysis techniques to be applied to the vibration data, and the logging of vibration data for long term trend and signature analysis. The network linking the data collection stations to the central computer permits the vibration information from all of the machinery on the plant to be analyzed, displayed, and stored at a central location. Implementation of this system should significantly enhance plant availability and assure safe operation of the Hanford N Reactor.

HIGH SPEED DATA PROCESSING FOR CONTROL ROOM DISPLAYS

New techniques for information assimilation, organization, processing, and display are being implemented in nuclear reactor control rooms. These techniques provide improved guidance to control room personnel during normal reactor operation and permit rapid diagnosis of problems and quick action response during abnormal occurrences. Such techniques are especially important to manually operate and control a nuclear reactor.

Advanced microprocessor technology in the form of a digital image processor can be effectively used for control room data processing and display. At the Hanford N Reactor a digital image processor, a Comtal Vision One/20, is being used to display reactor core data in a manner that is readily surveyed and assimilated by control room personnel (Cook et. al., 1982). The utility of this system is illustrated with N Reactor power map processing and display.

The Comtal Vision One/20 is a complete image processing system with built-in interactive processing and control capabilities. The system, as configured, has a spatial display resolution of 512x512 picture elements or pixels and can generate over 16 million different colors or 256 gray levels. Figure 7 is a blocked diagram of the Vision One/20 showing the main components of the system. The Digital Equipment Corporation LSI-11 microprocessor controls the operation of the image processor. The LSI-11 processor utilizes a combination of random access and read only memory to provide interactive control and processing. The Vision One/20 incorporates both hardware and software developed specifically to display, manipulate, and process images. The hardware design of the Vision One/20 incorporates pipeline processors in the form of function memories which can be loaded with arbitrary values under interactive control. The function memory can be used to perform realtime histogram modification, a technique used to change the brightness or contrast of display image. Graphics overlays for displaying informational text directly on an image are also provided.

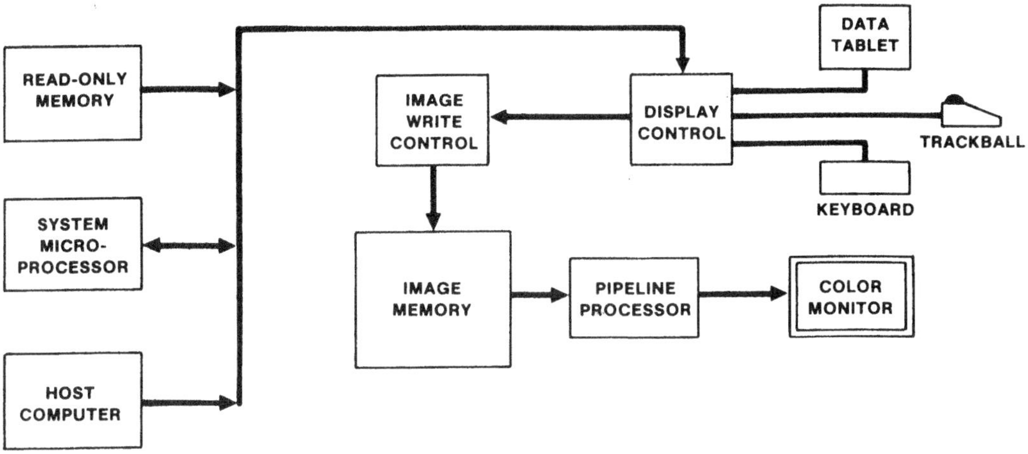

Figure 7. Block Diagram of the Image Processing System.

When a power map is initially transferred to the Vision One/20, it appears on the 19" color monitor as a small square about 2cm on a side. The zoom capability of the Vision One/20 is required to display the enlarged image. Pixel replication used in the zooming causes the individual process tubes to take on a blockly appearance when enlarged. As shown in Figure 8 this makes it very easy to determine the power level of each process tube. In Figure 8, the power in each process tube is associated with a particular gray level. The gray scale bar at the bottom of the Figure gives the correspondence between gray level and power. The dark gray shades correspond to very low tube powers while the tubes with the lighter gray shades have the highest power. Once the power map has been enlarged, several simple processing techniques can be used to enhance the presentation of the information. Two techniques that have proven especially useful are level slicing and graphics overlays. Level slicing is a technique that can be used to identify all tubes that fall within a given range of power levels. In this case, all of the pixels associated with tubes in a particular power range are visible while all remaining pixels are dark. This enhancement technique is used to emphasize particular regions of the power map that are of interest. The technique can be particularly useful when examining the power profile to determine optimum control rod movement.

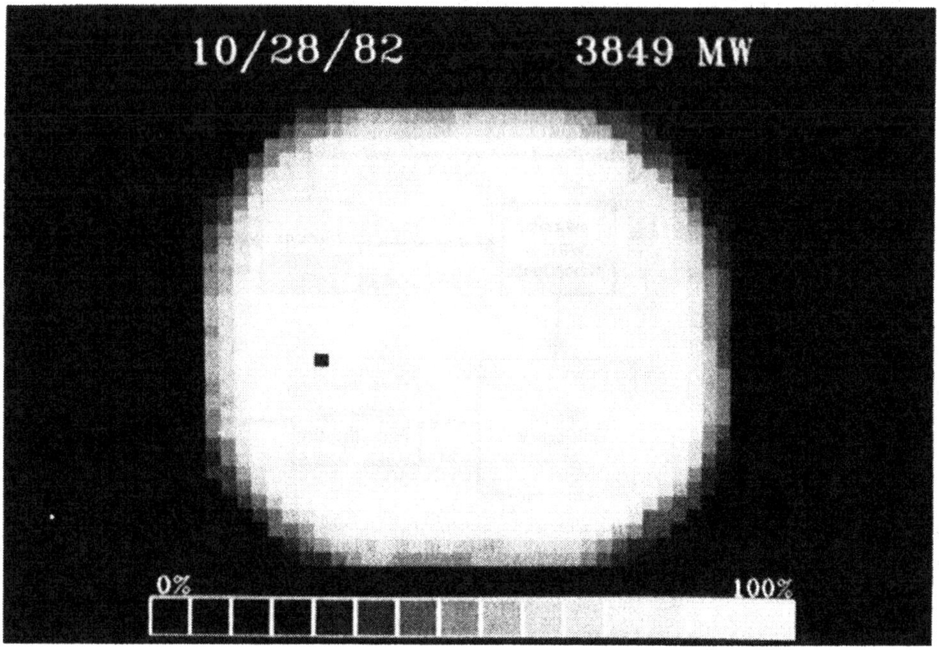

Figure 8. Power Map Enlargement Using Pixel Replication.

The graphics overlay capability superimposes informational text and figures on the image. This capability is useful for displaying control rod positions and for labeling the power map with information such as maximum tube power, date, and time. Graphics overlays are used in Figure 9, to show the positions of the control rods at the time that the power map data were gathered. The numbers used to label the control rods are coded to provide the operator with location information for each control rod bank. The position of the tip of the control rod within the reactor core is indicated by the arrow head. The orientation of the arrow indicates the side of the reactor from which the control rod entered. The rod identification numbers indicate the location of the rods along the length of the reactor core. The date and time and the maximum tube power for this particular set of data are also displayed on the power map.

An image processing technique that has also proven useful is differencing. In this technique, small changes in reactor power can be observed. Figure 10 is a difference power map obtained by subtracting the individual tube powers obtained at two different power levels. The gray scale in this display corresponds to changes in power. Dark grays and blacks represent the least amount of change in power level while the lighter grays and white show the maximum changes. The control rod positions for the two power levels are superimposed on the map. The

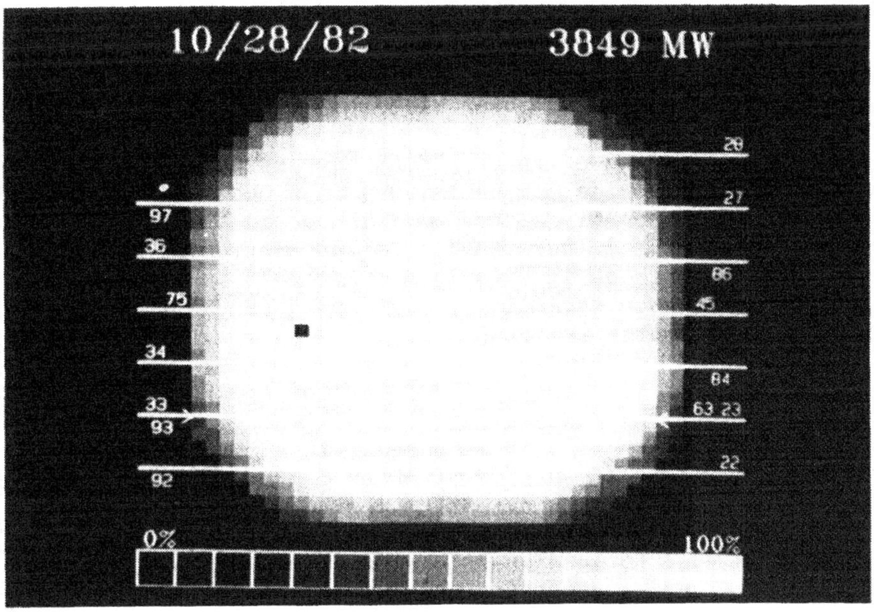

Figure 9. Power Map With Graphics Overlays.

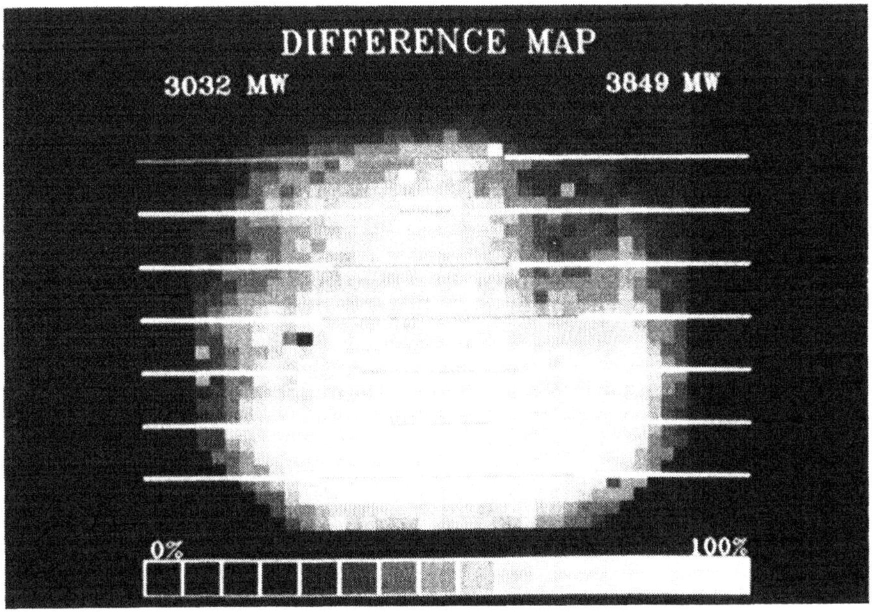

Figure 10. Difference Map.

difference power map displays very clearly the changes in local power associated with rod movement. Such difference maps can be generated very rapidly with the image processor. They provide control room personnel with useful insight to control rod effectiveness and rod movement to achieve balanced power distribution.

Digital image processing of reactor data represents a distinct improvement in man-machine communications in a nuclear reactor control room. Normal and abnormal reactor conditions can be monitored and evaluated by control room personnel. The effectiveness of the displays and processing techniques is chiefly attributable to the use of advanced microprocessor technology, high speed random access memory, and specialized computer architecture.

These three examples illustrate microprocessor applications to data acquisition, processing and display at a nuclear plant. The applications are concerned with very specific activities at the Hanford N Reactor omitting the discussion of other projects and programs using microprocessers in radiation detection monitoring and tracking. Yet, the basic concepts described in this chapter such as: expanding a small microprocessor to an elaborate data processing station, the use of microprocessor based satellite stations for plant data acquisitions, and the implementation of special purpose microprocessor systems for rapid informative data displays could find many applications in chemical processing plants, in manufacturing, and in the medical field.

REFERENCES

Cook, S. A., T. P. Harrington, H. Toffer, "Digital Image Processing Improves Man-Machine Communication at a Nuclear Reactor," IAEA-SM-265 Proceedings of International Symposium on Nuclear Power Plant Control and Instrumentation, October 1982.

Dumka, B. R., "Comparison of Ontario Hydro's Performance With World Power Reactors - 1981," NGD-12 (1981), April 1982.

Harrington, T. P., S. P. Roblyer, H. Toffer, "Vibration Monitoring Using a Computer Network Approach," UNI-SA-106, UNC Nuclear Industries, Richland, Washington, September 1, 1982.

Peterson, D.G., H. Toffer, S. P. Sapp, "Microprocessor Based Flux Wire Evaluation and and Analysis System," UNI-TR-5, UC-80, January 1980.

Part IV

CONTROLLERS

CHAPTER 11

MICROPROCESSOR-BASED INDUSTRIAL CONTROLLERS

Gautam A. Tendulkar
Eskilstuna Str. 17
8520 Erlangen, FRG

This chapter describes the application of microprocessor-based industrial controllers in hierarchical and stand-alone configurations. Programmable systems have considerable advantages over hard-wired systems, such as the ease of configuration and modification with user-friendly graphic programming languages. After describing the architectures of standard microprocessors, three special architectures for programmable controllers are analyzed. The development of a "universal" programmable controller for both logic and fast closed-loop control is technologically possible and an innovative architecture for such a controller is described in detail. Trends in architectures for industrial controllers, the influence of software and the convergence of microcomputer and programmable controller architectures are discussed.

1. THE APPLICATION OF INDUSTRIAL CONTROLLERS

Industrial controllers are found not only in the largest processes and plants such as oil refineries, steel factories and power stations but also in smaller installations and machines. Industrial controllers in distributed configurations can be linked together by a process bus in large scale automation systems. They can also be used as stand-alone embedded systems for controlling machinery, robots, machine tools and drives. Tasks of modern industrial controllers include not only logic and closed-loop control but also monitoring, signalling, measurement, data acquisition and data concentration. Increasing emphasis is placed today, by the users of industrial controllers, on sophisticated diagnostics for trouble shooting faults and for graphic display of plant variables.

Before the advent of digital technology, industrial controllers were based mainly on mechanical, hydraulic or hard-wired electrical devices. Examples of such controllers range from flywheel governors to combinations of electromechanical relays, timers and cam-type sequencers. Extensive engineering and manual adjustments were necessary

S. G. Tzafestas (ed.), Microprocessors in Signal Processing, Measurement and Control, 207–230.

for configuration, starting up or if modifications had to be carried out after installation. Such systems were complex in design but limited in capability. Controllers based on "soft-wired" computers offer considerable advantages over "hard-wired" systems. Computer-based systems could be produced in large quantities and easily modified through programming as required. The introduction of the digital computer to industrial control in the fifties initially only had an impact on the largest systems; for instance chemical plants, where the high cost of the mainframe was a small proportion of the total cost of the plant. The available hardware used slow drum-type memories and was massive in size compared to what is possible with present day microelectronics. In spite of this, mainframes were used as centralized systems for optimization, logging, alarm signalling and plant modelling. Since all signals had to be connected to the computer over large distances, interface design, start-up and maintenance were difficult. Although the operating speeds and performance of mainframes have increased significantly since the early days, their application in real-time control tasks is still limited by the high installation cost, which compels them to be used for the control of many simultaneous tasks. This restricts their application in the real-time control level in industrial control systems where response must be fast. Furthermore the real-time operating systems required for handling multiple concurrent and cooperating tasks in the face of interrupts and alarms are very complex. Operating system overheads required for task scheduling and data synchronization add significantly to the reaction time. Abrupt system degradation is unavoidable in the case of failure of the single mainframe. This is unacceptable at the machine-oriented control level, as dangerous operating conditions and severe damage can result. For this reason mainframes are used primarily in the highest levels of the industrial control hierarchy; for planning, optimization, scheduling and logistics, mainly in batch processing modes.

The availability of powerful minicomputers and the low cost and flexibility inherent in microprocessor-based controllers has led to their increasingly widespread use in real-time logic and closed-loop industrial control applications. A microprocessor-based controller can be used in widely different applications through suitable programming. The availability of high level application-oriented user programming languages in addition to machine level assembler languages is essential for rational and efficient programming. General purpose high level languages such as FORTRAN, PASCAL, BASIC and ADA can be used in microprocessor-based industrial controllers. However these languages and their variants are not at present optimal in terms of real-time response or memory utilization. This may gradually change with decreasing memory costs and the availability of faster microprocessors with special on-chip hardware support for high level language execution.

An important factor in industrial controllers is the requirement that non-specialized operating personnel should be able to configure, debug and modify the user programs. For this, a problem-oriented high

Ladder Diagram

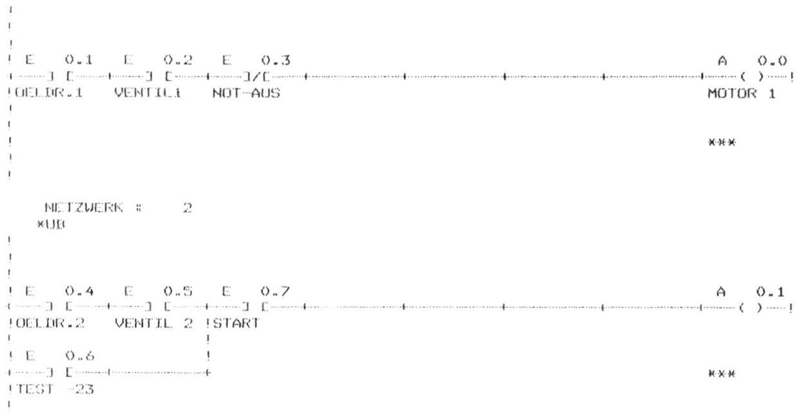

Logic Gate Diagram

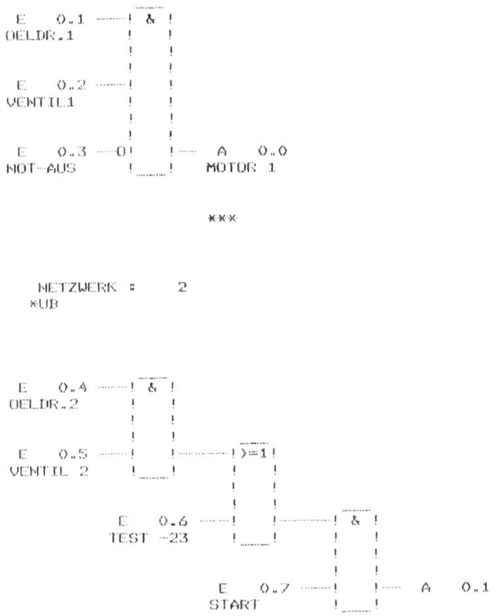

Control List

```
U    E     0.1    OELDR.1
U    E     0.2    VENTIL1
UN   E     0.3    NOT-AUS
=    A     0.0    MOTOR 1
BLD        255

U(
U    E     0.4    OELDR.2
U    E     0.5    VENTIL 2
O    E     0.6    TEST -23
)
U    E     0.7    START
=    A     0.1
BLD        255
```

Fig.1 High Level Languages for Programmable Controllers

level language which can be readily understood in terms familiar to the user is required. Graphic ladder diagram and logic gate type programming languages are found in programmable controllers in addition to the control list (see Fig.1). Initially these graphic languages were designed to emulate the logic elements such as relays and contactors used in the wiring diagrams of hard-wired controllers. However high performance programmable controllers offer more sophisticated software functions such as drum timers, sequential controls and PID-controllers. These functions are integrated into the ladder diagram and logic gate graphic programming languages in the form of function blocks which can be wired up on the screen. Such languages can be classified as high level, as the user does not require to have any hardware knowledge and can concentrate on the control task when programming. The main advantage of programmable controllers is the ease with which any function can be configured, altered or modified without rewiring; by simply entering a new program into the controller user memory. The "factory floor" programming language , which can be easily understood by non-specialized users, coupled with the automatic documenting of programs and comprehensive built-in fault diagnostics, has led to the impressive growth of programmable controllers.

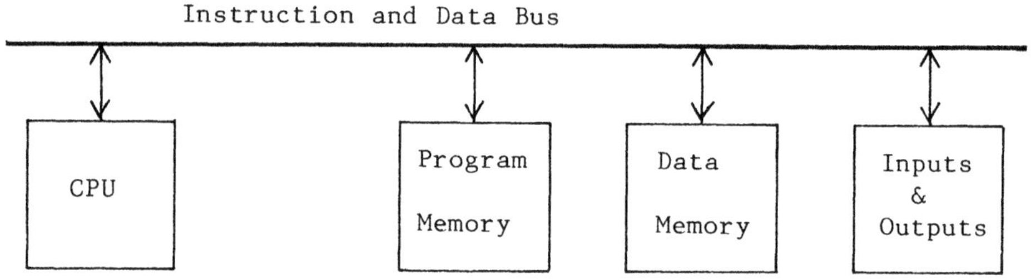

2a. Princeton (von Neumann) Architecture

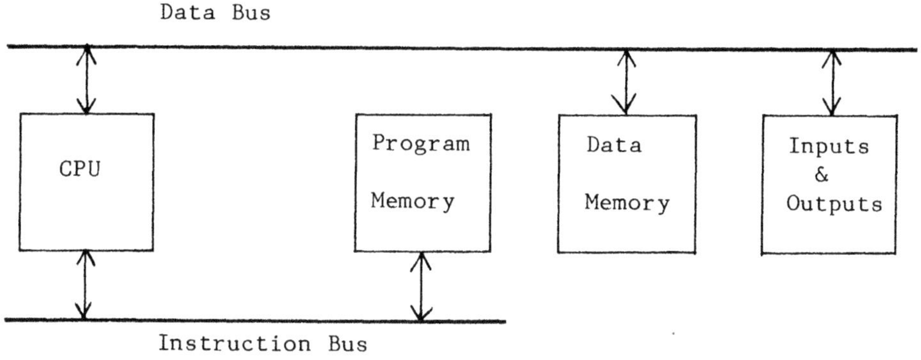

2b. Harvard Architecture

Fig.2 Princeton and Harvard Architectures

Microprocessor-based industrial controllers play an important role in the lower levels of hierarchical control systems. Powerful programmable controllers and microcomputers are increasingly replacing minicomputers at the real-time machine or process control levels. An important factor is the ruggedized construction of industrial microprocessors and in particular programmable controllers which do not require air-conditioned, dust-proof environments and can be mounted next to the machine or process subsection being controlled. Minicomputers are used in the intermediate levels e.g. for on-line production control and optimization. Other tasks include coordination of the machine-oriented controllers and data concentration and transmission to the higher levels. In many applications, a single minicomputer is able to handle a particular task at the machine-oriented control level. Other tasks can be solved using microprocessors in single or multiprocessor configurations. Such standard microprocessor systems are suitable for tasks involving data storage, arithmetic and slow logic or closed-loop control etc. Programmable controllers are better suited for real-time applications i.e. fast logic or slow closed-loop tasks with time constants down to 100 ms. On the other hand, fast closed-loop control applications such as the digital control of electrical drives require sampling times down to 1 ms. This is only possible with special controllers and architectures especially if high level languages are to be used. New types of "universal" industrial controllers with combined logic and fast closed-loop control functions are increasingly required in general industrial applications. Good examples of the applications of such controllers are process control and in robot control and numerical control systems for controlling the digital inputs and outputs and the drives simultaneously.

2. MICROPROCESSOR ARCHITECTURES

CPU architectures can be subdivided into three classes, i.e. Princeton, Harvard or special purpose. Fig.2a shows a typical Princeton machine architecture. This architecture is sometimes referred to as a von Neumann type of CPU and has the following characteristics:

1. Both programs and data share the available memory space and can be intermixed.

2. A single bus is used for instruction and data transfer to the CPU.

The second type of CPU is referred to as the Harvard machine. It is shown in Fig.2b. This type of machine architecture has two separate busses. The characteristics of the Harvard architecture are:

1. Separate memory modules for programs and data

2. Separate and independent busses for accessing instructions and data.

Standard 8 and 16 bit microprocessors such as the 8085, Z80, 8086, Z8000, 68000 are of the Princeton (von Neumann) type, memory and data can be intermixed on the single bus. One of the few standard 8 bit microprocessors with a Harvard architecture is the 8X300/305. Single chip microcomputers such as the TMS 1000, COP400, 8048, Z8 have hybrid architectures as they have only one bus for accessing programs and data. Data and programs can not however be intermixed in the same modules. Any desired type of architecture can be created with bit slice microprocessors. Bit slices are available in widths of 4 bits (Am2901), 8 bits (GP001) and 16 bits (Am29116). They can be connected in parallel to produce the desired processing bit width. New microprocessors developed for signal processing applications such as the TMS320 can be used in high performance industrial controllers.

The architectures of standard microprocessors, single chip microcomputers and bit-slice microprocessors are extensively covered in the available literature (see References 4 to 7). This chapter will concentrate on three different examples of architectures used in programmable controllers and on one innovative architecture designed for use in fast user-reconfigurable logic and digital closed-loop controllers for robots and numerical controls. The underlying architectural principle behind these high level machines is direct high level language execution.

3. PROGRAMMABLE CONTROLLERS

The main difference between programmable controllers and standard microcomputers is in software. Microcomputer systems are not restricted to one particular language and the user generally selects both the operating system and the language for each particular application. Programmable controllers on the other hand are packaged hardware plus software systems. The operating system and programming language are integral components of the programmable controller. Modern programmable controllers offer sophisticated programming techniques to the user in the form of ladder diagrams and logic gate plans (Fig.1). Once the user has entered his program graphically in the programming unit, its execution is supervised by the executive programs resident in the programmable controller. In contrast, when single board type microcomputer systems are used in real-time industrial applications, assembler level languages have to be used since general purpose languages are too slow or produce too much code. Detailed hardware knowledge is required by the user for assembler programming. This is not the case in programmable controllers, where the user does not require any specialized hardware knowledge for programming. The languages offered by programmable controllers have been developed for real-time control in industrial applications and hence are faster and more efficient for this purpose than the general purpose languages available for microcomputer systems. Conversely, the programmable controller languages are not optimal for general data processing applications as they lack functions for efficient filing, accessing and processing of large amounts of data.

Programmable controller languages are not standardized and differ from manufacturer to manufacturer. However the preliminary version of the DIN 19239 standard does specify a control language with three optional representations. The simplest from the point of view of hardware required for execution is the control list representation. Due to the low overhead required, the control list is found in the smallest programmable controllers which in some cases do not require separate programming units. More powerful programmable controllers offer the common ladder diagram representation. Not so common is the logic gate representation. Certain manufacturers offer programmable controller systems where the user can use any of the three representations.

The functions offered by programmable controller languages are continuously growing and top end programmable controllers approach the power of minicomputers and process computers in terms of execution speed, addressing range and number of functions. Fig.3 shows the functions available in programmable controllers. The hardware specification of a programmable controller determines its area of application. In general a figure of between 16 to 2048 each for inputs, outputs and flags (control coils) is not uncommon depending on the size of the programmable controller. Between 8 to 256 timers as well as counters are usually provided. The user program memory range varies between 2k and 48k instructions. Both RAM or EPROM user memory options are generally provided. Newer programmable controllers also have EAROMs (electrically alterable ROMs) for storing user programs. The cycle time required to process the user program is an important part of the specification as it determines the reaction time of the programmable controller. This can range between 1ms to 100ms for 1K logic statements or contacts/coils.

Software Functions:	Basic	Advanced	Process
	Boolean	Block transfer	Signalling
	Timers	Jump	Monitoring
	Counters	File	PID-control
	Data Move	Shift Registers	Communication
	Comparison	Sequencers	Logging
	Arithmetic	Floating point	Display
Hardware Functions:	**Small PCs**	**Large PCs**	
Inputs	16	4096	
Outputs	16	4096	
Timers	8	256	
Counters	8	256	
User Program	2K	48K	
Cycle time(per 1K)	100ms	1ms	

Fig. Functions of Programmable Controllers

Harvard, Princeton and special types of machine architectures are used in current programmable controllers. The older hardware state-machine based programmable controllers tended to have a Harvard machine architecture. Newer programmable controllers are increasingly based on standard or bit-slice microprocessors and tend to have von Neumann machine architectures. The use of standard microprocessors is increasing with the speed and range of functions available. At present standard microprocessors require additional hardware support to attain the executing speeds required in medium and high performance programmable controllers. Single chip microcomputers are in many cases adequate for small programmable controllers. Three different programmable controller architectures are described below. The first type is designed for execution of a graphic circuit-node type language. The second architecture is designed for execution of ladder diagrams and uses a hardware column solver to support the CPU. The third type is a direct execution microprogrammed architecture which uses a dual data bus for faster processing of list-based languages.

3.1 Circuit-Node Processor Architecture

The circuit-node programming language is based on the ladder diagram with the addition of of node numbers which uniquely define the interconnections between parallel and series contacts, coils or elements. Fig.4a shows a simple circuit-node ladder diagram and Fig.4b the corresponding instructions which are executed by the processor. The graphical circuit-node ladder diagram is converted into the appropriate sequence of instructions by the programming unit. Each instruction consists of 3 fields: the source node number, the contact or logic element designator and the destination node number. A special code (SC) is used in the destination node if the contact is any but the last contact of a parallel branch. A special code is also used for the source node when the instruction is used for setting an output. The algorithm for execution of this simplified version of a circuit-node language is shown in Fig.4c. Three steps are required for executing each instruction; one step for each instruction field. The architecture shown in Fig.5 is one example of a hardware solution of the algorithm.

The CPU in Fig.5 consists of components Y1 to Y5. The user instructions are addressed by counter Y1 and the corresponding instruction in the user memory Y9 is read into the instruction decoder Y2 via the instruction bus. Fields in the instruction are used for addressing the node number RAM Y3, the control coils (internal memories) in RAM Y6, the inputs (contacts) from Y7 and the outputs (coils) in Y8 as directed by Y2. Data is transferred between the CPU and the inputs and outputs via a single bit wide data input and output bus. The separate instruction and data busses indicate that this CPU is a Harvard machine. (Ref.18).

Referring to Fig.4c, step 1 is used for executing field 1 of the instruction. The node number RAM Y3 is addressed and the state of the source node, is read to the node processor input. Then in step 2, the

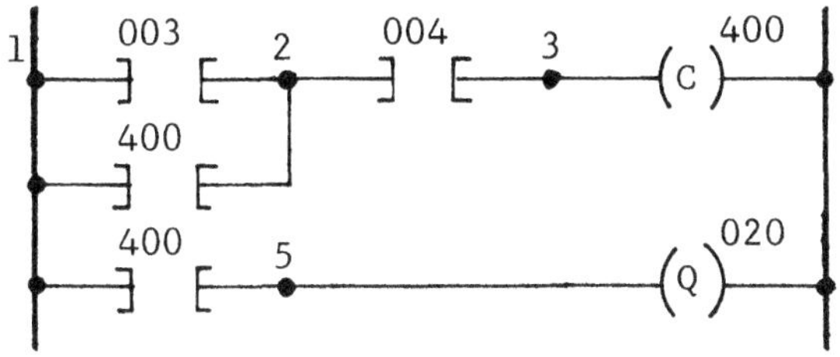

Fig. 4a Circuit-Node Diagram

Program Address	Field1 Source Node	Field2 Contact Number	Field3 Dest. Node
000	001	003	SC
001	001	400	002
002	002	004	003
003	003	SC	400
004	001	400	005
005	005	SC	020

Fig. 4b Circuit-Node Language

Step	Action of Circuit-Node Algorithm	
1	Read source node addressed by Field 1 to Node Processor	
2	Case: Special Code (SC) Select data bus output instead of node no. RAM for Step 3	Case: No Special Code Execute logic AND function with states of source node and contact addressed by field 2 as operands
3	Hold result of logic function in logic unit	Write result of logic function to node no. RAM or to data bus output depending on step 2

Fig. 4c Circuit-Node Algorithm

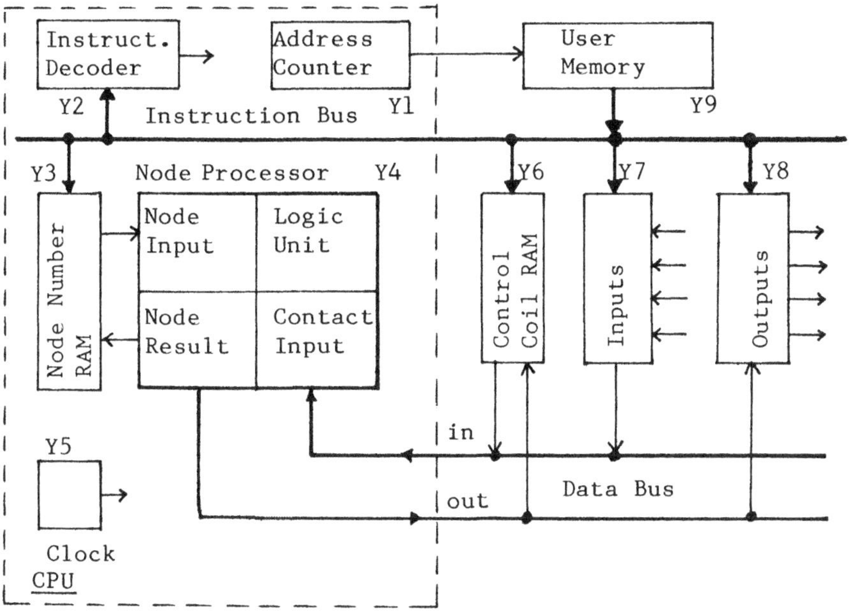

Fig. 5 Ciruit-Node Architecture

state of the contact or input (from Y6 or Y7) addressed in field 2 is AND-gated with the state of the source node from and held internally in the logic unit of the node processor. Finally in step 3, the state of F3 is stored in the addressed destination node in node number RAM Y3. This sequence represents the normal execution from node number to node number. Two special case of steps 2 and 3 are used for the execution of outputs and parallel branches. If a special code (SC) is programmed in field 2, then steps 2 and 3 are output instructions. In this case, the output is set directly to the result of the logic function. The decoder Y2 then selects the output data bus instead of the node number RAM for the subsequent step 3. This means that the result of the logic function is output via the data bus to the addressed output Y8 or coil RAM Y6, instead of writing it to the destination node in RAM Y3. If field 3 contains a special code, this indicates that an intermediate node of a parallel branch is being processed in step 3. The state of the preceding parallel contacts is held internally in the logic unit of the node processor and "carried through", until the final contact which does not have a SC field.

Although the operating principle and basic hardware for the execution of the circuit-node language is simple, this can be used as the basis for more sophisticated functions if more powerful hardware is used. The hardware CPU shown in Fig.6 can be incorporated totally into a program which is resident in a single chip microcomputer. In this way powerful programmable controllers with extensive functions in addition to boolean logic can be built. The single restriction is that only circuit-node type languages can be used with this type of architecture.

3.2 Column-Solver Processor Architecture

The column-solver architecture is designed for direct execution of ladder diagrams without circuit node numbers. A typical von Neumann architecture found in the majority of small and medium programmable controllers is shown in Fig.6. A 8X300 microprocessor Y2 is used as the major processing unit with microprogram memory Y1. The combined high level language instruction and data bus connects the processor to the column-solver Y3, programming unit interface Y4, scratch pad RAM Y5, status and control port Y6, coil RAM Y7, user instruction memory Y8 and the inputs Y9 and outputs Y10. (Refs. 19,20).

Each user instruction contains an opcode field which defines the function e.g. contact or coil and a reference field for the parameter. The ladder diagram displayed on the screen of the programming unit, is processed on a column by column basis. A "start of column" instruction at the beginning of each column contains information about the number of contacts in that column and the number of vertical interconnections as shown in Fig.7a. The processor scans the contacts in that column and combines the states of the contacts with the states of the inputs to those contacts to form an input power flow byte which is fed to the column solver input port (Fig. 7b). The vertical interconnections are fed to the "connect" port. A switch matrix consisting of logic gates is

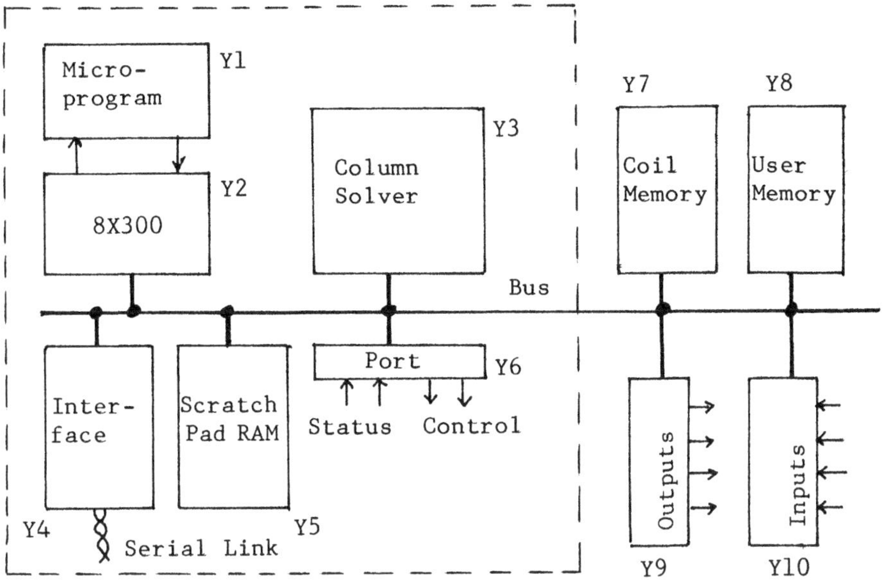

Fig. 6 Column-Solver CPU Architecture

Column 1		Column 2		Column 3	
Start CPC	ICC	Start CPC	ICC	Start CPC	ICC
Contact	003	Contact	004	Contact	400
Coil	400			Output	020
Coil	400				

(CPC = Contact position code ICC = Interconnection code)

Fig. 7a Ladder Diagram Language for Example in Fig. 4a

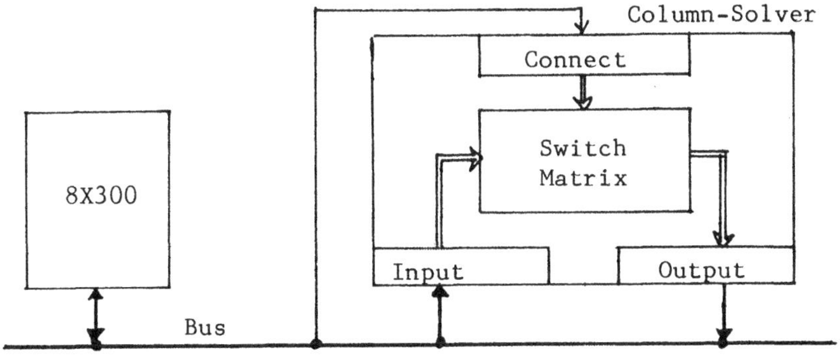

Fig. 7b Column-Solver Hardware (Y3)

used for switching the input power byte data to the output power byte as a function of the states of the contacts in that column. Power flow through vertical interconnections between contacts occurs via the switches controlled by the "connect" port. The output power byte defines the input conditions for the next column and can be read from the "out" port.

The column by column execution is very fast and can be extended to include more sophisticated functions such as timers, counters and arithmetic. This architecture is one of the most common in small to medium programmable controllers and can produce scan times down to 270ns per contact. It is only suitable for execution of ladder diagrams without node numbers.

3.3 Direct list-execution microprogrammed architecture.

The CPU interprets the user control list which is stored as code in the program memory. The main criterion for a programmable controller is the reaction time or cycle time (scan time) required for working through the user control list. Therefore the time required for the interpretation of each user language instruction in the control list must be kept as short as possible.

The cycle time definition includes not only the user program execution time but also the time required for transfer of the logic states of the input ports to the input image memory at the beginning of the cycle, for transfer of the output image memory states to the output ports at the end of the cycle and finally the time required for updating timer cells and for general program organization. The CPU uses the input and output images and not the actual ports, while executing the user program. This prevents errors due to repeated switching of the inputs and outputs within each cycle and allows the states of the outputs to be read by the CPU. (Refs. 21,22,23).

In a single data bus von Neumann architecture all important hardware functional blocks are connected to the single bus. These blocks consist of the input and output ports, the interfaces (e.g. to higher level computers, programming units or NC-systems), the memory for user program and operating system, the system data (e.g. input and output images) and user data RAM, the clock pulse generator for timer updating and finally the CPU itself. The microprogram memory shown in the dotted line in Fig.8 is not required if a general purpose microprocessor is used.

The single data bus architecture sets an upper limit to controller speed as all operations requiring data transfer have to be carried out using the slow data bus. The speed of the bus is limited by physical considerations (e.g. reflections, ringing etc.) as the bus is routed via printed circuit board backplanes within the enclosure and secondly because slow lower-cost memories are preferred due to cost considerations. The user program, operating system and the user and

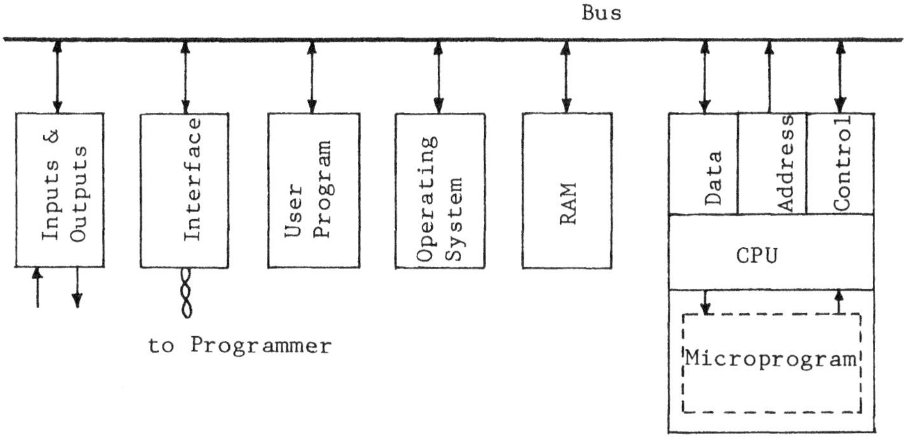

Fig.8 Single Data Bus Microprogrammed Architecture

system data memories typically have access times of 0.5 μs. This results in memory access operation time of about 1.5μs. At least 20μs are required for the execution of a binary logic instruction by a 16 bit MOS microprocessor. The interpretation of an AND INPUT instruction using an 8 bit microprogrammed microprocessor requires two memory accesses for fetching the 16 bit user instruction, one system data memory access for the operand from the process image and 10 interpretative microinstructions. This produces an execution time of 7 μs for a binary logic instruction if a microprogrammed microprocessor with a microinstruction cycle time of 250 ns is used.

The cycle time for single data bus designs is limited by the long access times which cannot be speeded up due to the bus. One alternative is to use a single card containing both the CPU and fast user program memory. If fast memories such as CMOS RAMs or bipolar PROMS are used for storing the user program, then the speed can be increased accordingly. For example a user memory with an access time of 100ns would result in an execution time of less than 3μs. The disadvantages of this approach is that it is unsuitable for EAROMs and EPROMS, which have significantly longer access times than 100ns. Secondly fast bipolar PROMS with 100 ns access times can only be used once and have to be thrown away if program modifications are necessary. Fast memories are expensive and increase system cost.

A different architecture offers faster cycle times with slow memories. Fig.9 shows a dual data bus architecture based on a microprogrammed microprocessor which allows binary instruction execution times of 3.5μs, even with slow memories by using a "bus coupling" processor for the user instruction fetch operation. Here the timer, system data RAM and CPU are connected to an internal fast data bus with an access time of approximately 50 nanoseconds. All other functional blocks are connected to the slow external data bus. The two buses are interconnected by the bus coupling unit, which essentially consists of an independent microprocessor. This architecture allows overlapping of the next user instruction fetch with the execution of the current user instruction.

The bus coupling unit (BCU) performs all user instruction fetches as well as read and write operations of data accessed via the slow data bus. The main processor provides the BCU with addresses via the internal data bus and commands via the microprogram BCU control bits. In linear user program execution new addresses are not required by the BCU which automatically increments its address counter after each access. A new address must only be provided by the main processor in case discontinuous program counter modifications (e.g. jumps or returns) are required by the user program. Since the input and output process images are stored internally in the fast system data RAM only two bytes for the user instruction have to be accessed externally per binary logic instructions. The execution time required for a binary logic instruction is 3.25μs. Eleven microinstruction are needed due to the additional process image byte access from the internal RAM. Hence the binary logic user instruction execution times can be reduced by a factor of two over a single data bus architecture.

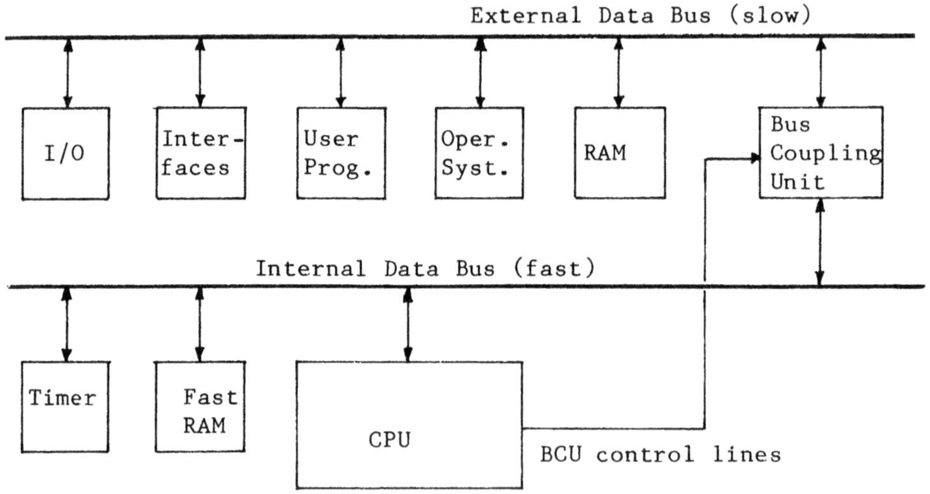

Fig.9 Dual Data Bus Microprogrammed Architecture

In summary the bus coupling unit halves the execution time by pre-fetching user instructions. It generates the necessary control signals for the slow external bus, thus relieving the fast main CPU. A simple bus coupling unit requires only a few components such as the 8X310 address director in addition to those required for the external interface of a bit-slice system. The increase in hardware cost is negligible. The direct list-execution microprogrammed architecture is the most versatile of the three architectures discussed, as all three language representations i.e. list, ladder and logic gate, can be used. The programming unit generates the intermediate list language for execution by the processor.

4. PROGRAMMABLE LOGIC AND CLOSED-LOOP CONTROLLERS

Many closed-loop control applications have slow plants such as in temperature control where sampling times of 100ms or more are satisfactory. In such cases standard microcomputers or programmable controllers can be used. Time constants as low as 5 ms are found in some key applications e.g. electrical drives, requiring sampling times down to 1ms. For such applications, special closed-loop controllers with greater high-speed arithmetic capability and additional hardware than found in standard microcomputers or programmable controllers are required. This hardware consists not only of arithmetic units but also special boards for A/D and D/A conversion as well as measuring systems and pulse-width modulated output power stages.

The driving forces behind the development of digital controllers are the advantages they offer over analog controllers. For instance no adjustments with screwdrivers and trimmers are necessary as the parameters can be accurately stored in tables in PROMs or entered via pocket terminals. Offset, drift and noise problems caused by operational amplifiers are absent. Digital controllers can be easily programmed to solve adaptive, non-linear and structure variable problems. Multi-loop and interlinked systems can also be easily realized. Built-in diagnostic programs can be used to reduce down time and to simplify trouble-shooting. Digital controllers are easy to put into operation and can be easily reconfigured on-site without a development system, if a suitable high level language is used. Such a language also makes structure and application modification easy without requiring any changes in the hardware.

Digital controllers also have some disadvantages over analog controllers. The measurement systems are more complex as the actual values from the real world are analog quantities and have to be converted to digital form. Programming bare hardware is difficult for the user without an embedded operating system and application-oriented high level language. This can increase the cost, which has to remain competitive with regard to an analog control system.

Before one can specify the requirements a digital controller has to fulfill in detail, one has to differentiate between dedicated single loop digital controllers and programmable closed-loop controllers. The dedicated single loop controller is designed to replace the conventional analog closed-loop controller found for instance in servo drives. In such controllers the program is fixed and can not be altered by the user. Only control parameters such as gain or integrating time constant can be set. The fact that a microprocessor executes the control algorithm instead of an analog amplifier is not directly apparent to the user except for easier adjustment of parameters using a pocket terminal instead of trimmers and the availability of digital displays. A serial or parallel digital interface for the input signals is also available. Dedicated units are usually designed using single chip microcomputers such as the 8051 and have to be very low cost as they have to compete with analog loop controllers. The trend in dedicated digital loop controllers for electrical drives is to integrate them into a package which includes the power stage of the chopper. In this way the microprocessor can be used for executing other functions in addition to the closed-loop control algorithms.

The programmable digital loop controller on the other hand, can be configured by the user as required, to fit different applications. For this purpose a high level language, which relieves the user from specialized hardware knowledge and machine level coding is necessary. Both simple and complex control configurations can be realized easily. In these controllers the user can alter both the program, i.e. the control structure, and the parameters to fit a wide variety of tasks. Since the hardware required for such systems is more complex than in the single loop dedicated controller, these units are designed to control more than one loop simultaneously. In addition they possess the capability of handling all logic control tasks associated with the loops. These tasks would otherwise have required a separate programmable controller in addition to the loop controller. By eliminating the separate programmable controller, the programmable digital loop controller becomes competitive in applications where multiple loops have to be controlled and where logic control is required.

A high level language digital loop controller of the programmable type is described below. It was designed for controlling drives and executing the logic tasks in numerical control and robot control systems. Therefore the requirements placed on the controller are primarily based on needs of these systems. The majority of numerical controls have 4 drives with some possessing 6 or more. Robots on the other hand generally have 6 drives for controlling the arm and manipulator. The fast servo drives found in such applications have time constants of 5 to 10 ms. Therefore sampling times of 1 ms are necessary in the closed-loop control algorithms. In order to control a drive directly without any analog controllers, 2 loops are required per drive; one current loop and one speed loop. Hence 12 loops have to be executed every 1 ms for 6 drives, allowing a maximum of 50μs for each PID-loop, i.e. a total of 600μs for the closed-loop control. The logic functions can be carried out in the remaining 400µs. Typical logic control found in numerical control and robot applications carries out tasks such as start-up, interlocking, reversing, safety monitoring etc. For this 128 digital inputs and 128 digital outputs are sufficient. Standard programmable controllers have execution times of 2µs per logic input. Taking this value for the digital loop controller, a maximum capacity of 2000 logic statements can be executed per 10 ms reaction time. This represents 500 3-input gate functions and was considered to be satisfactory. In addition hardware for actual value measurement and chopper control is required. The final and perhaps key requirement is an application-oriented high level language for configuring the closed-loop and logic control tasks of the digital loop controller.

A simple high level language was developed for the controller containing the key statements required for both closed-loop and logic control. The statements implemented can be subdivided into 3 groups, namely the logic control, closed-loop control and organizational groups. Only a few simple statements such as "PID controller" are required in addition to those found in programmable controllers for controlling electrical drives in numerical control and robot control systems.

4.1 High Level Language microprogrammed architecture

A high level computer language can be executed in various ways. Basically one differentiates between direct execution and indirect execution. In direct execution, the statements or instructions of the high level language are identical to the instruction set of the processor. Indirect execution on the other hand is based on the concept of an intermediate language, which itself can be the machine language of the processor. In this case the user program written in the high level language must first be converted to its equivalent in the machine language by a compiler before it can be executed by the CPU.

Since the instruction sets of standard microprocessors are not similar to the high level user language, each user statement must be converted to a sequence of assembler instructions before execution. If the conversion is carried out in the programming unit, one refers to it as compilation. It is also possible to load the user statements into the controller and jump to the appropriate subroutine written in the machine language of the CPU for each user statement. This is referred to as interpretation. Alternatively a microprocessor with a modifiable instruction set or a bit-slice processor can be used for direct interpretation of the user statements.

The choice - whether indirect or direct execution i.e. standard or custom microprocessor - depends on speed considerations. Generally speaking direct execution is faster than indirect execution. The final choice can only be made after benchmarks for typical statements have been programmed. Two benchmarks were programmed in machine code for the 8048, 8085, 8086 and 8X300 microprocessors. A hardware multiplier with a calculation time of 4μs was used in all systems. The first benchmark was the AND INPUT logic statement and the second benchmark was the PIC (16 bit proportional plus integral action controller) closed-loop statement. The results are shown in the table below:

Microprocessor	Execution times of statement: AND INPUT (logic)	PIC (closed-loop)
8048	92 μs	700 μs
8085	50 μs	299 μs
8086	20 μs	31 μs
8X300	1.5 μs	28 μs

The 8 bit microprocessors are obviously too slow for both the logic and closed-loop requirements. The 8086 was fast enough for the closed-loop benchmark but was too slow for the logic benchmark. Only the 8X300 was fast enough for both logic and closed-loop benchmarks giving estimated times of 28μs for a PI-controller algorithm and 1.5μs for the AND INPUT statement. Therefore a special direct execution architecture based on the 8X300 was chosen for the high level language digital loop controller. This microprocessor architecture was based on the principle of direct and optimal interpretation of the high level language. In contrast to standard microprogrammed systems, the machine architecture and resulting highly orthogonal microinstruction set were optimized for industrial control applications.

A block diagram of the CPU for the digital loop controller is shown in Fig.10. Although other bit-slice systems such as the 2901, 29116 etc. were considered, the 8X300 system offered the lowest cost and component count. Special features of this architecture are the fast internal bus connecting the master processor, slave processor and multiplier. All three units operate independently and parallel to each other, thus reducing execution time. For instance while the slave processor is fetching the next user high level language statement from the user program memory on the external bus, the master processor could be processing data or accessing the fast 1K x 8 RAM, and the multiplier could be calculating the product of two 16 bit numbers previously loaded by the master processor. The multiplier unit immediately starts calculating as soon as the operands have been loaded into its internal latches by the master processor. It flags end of calculation after 4μs to the master processor via an internal bus port. Although a more complex multiplier/divider unit or even a floating point unit could have been used, a simple 16 bit multiplier based on the Am25LS14 was selected due to its low cost. The slave processor is a logic engine which executes only four instructions i.e. memory read, memory write, user statement opcode fetch and no-operation. The address can be loaded into the slave processor by the 8X300 if required (e.g. for jumps) otherwise the address counter in the slave processor is automatically incremented or decremented under microprogram control. As soon as the external bus access operation is finished, the slave processor flags

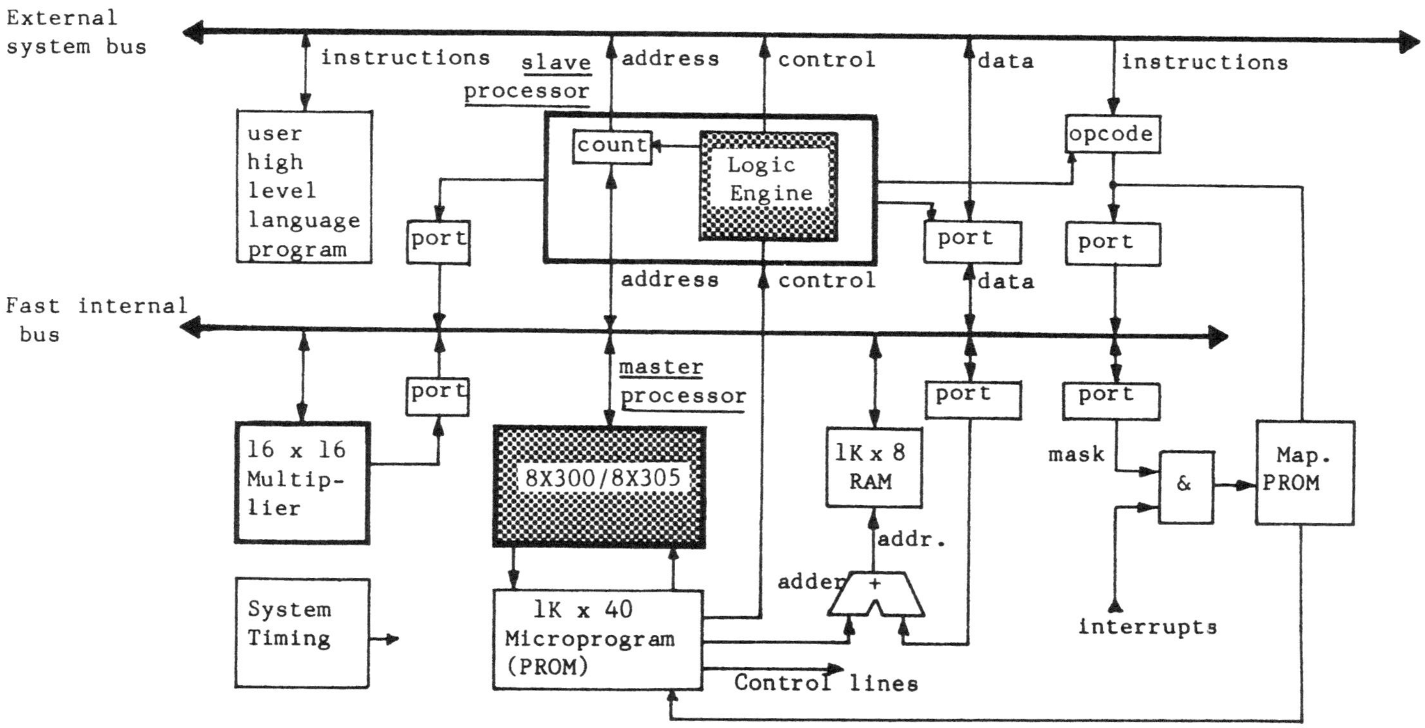

Fig.10 High Level Language CPU Architecture for Logic and Closed-Loop Controllers

the master processor via a port. In the case of a memory write operation, data must first be entered into the data port by the master processor. Conversely after a memory read operation data is available to the master processor in the data port. User statements are always stored in the opcode port. A bus from the opcode port is connected to a mapping PROM which generates the microprogram address for each high level user statement. In case of interrupts, the mapping jump is overlaid by a jump to the interrupt service routine in the microprogram. The interrupts can be masked by the master processor via a port on the internal bus.

One important feature of the architecture is the high speed 1K x 8 internal RAM. This RAM is addressed directly from an address field in each microinstruction. In this way the master processor gains 1024 extra high speed registers for calculations and storing bit fields or parameters. The address from the microinstruction field can be offset as required via an adder connected to a port, before entering the RAM. This significantly speeds up the CPU as vector addressing of the high level language can be simply decoded by writing the appropriate vector address corresponding to inputs, outputs and flags in the microinstruction address field and the actual number of the input, output or flag directly into the port. Note that the actual states of the inputs and outputs are not accessed directly. Instead a process image in the internal RAM is used, which is updated at the start of each user program cycle.

Another feature of the high level language loop controller is the real-time microprogrammed operating system, which controls all major functions of the HLL controller. It is responsible for start-up, supervision of user program execution and shut-down. In addition interrupts and faults are also handled by it. The interleaved execution of the logic and loop programs is an important feature of the HLL controller. The operating system was microcoded so that all CPU resources could be utilized and so that the implementation of extra statements exclusively for the operating system would be avoided.

The basic operation of the loop controller is supervised by the operating system. After power has been switched "on" or after a reset, an initialization procedure is executed in order to bring the hardware to a ready condition. The process image is then updated and cyclic execution of the user logic program is started. During the first cycle of the logic program after start-up, the closed-loop program is disabled as the logic conditions for all closed-loop controllers have to be brought to an initial state defined by the user in the user logic program. The closed-loop program is enabled for all subsequent cycles and is executed every 1ms. Fig.11 shows the multiplexed execution of the logic and closed-loop user programs. The logic program can be interrupted by system faults, by the 1ms clock for the closed-loop program and by a DMA request, in order of priority. The closed-loop program can only be interrupted by system faults. This ensures a constant sampling time. Whereas the complete closed-loop program is

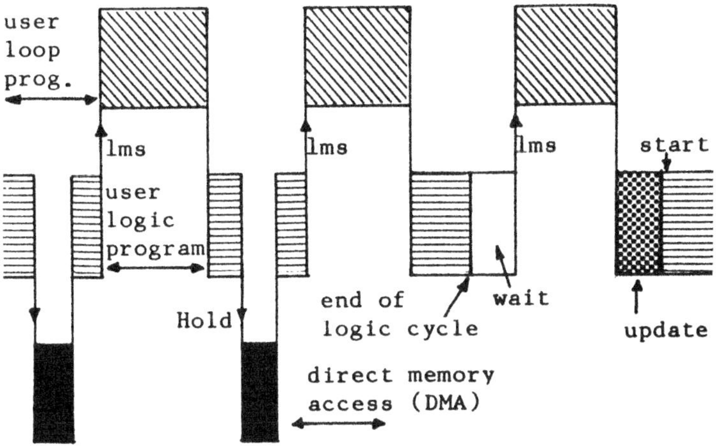

Fig.11 Multiplexed Execution of Logic and Closed-Loop Control Programs

executed at each 1ms interrupt, the logic program is asynchronous and its execution is continued after each 1ms interrupt, until the END PROGRAM statement is reached. Before updating the process image, a wait-loop is entered. Only after the next 1ms closed-loop interrupt has been processed, is updating carried out. This prevents the updating being interrupted by the closed-loop program. Transfer flags, which are updated with the process image, are used for synchronizing information flow between the logic control and closed-loop programs. A total microprogram of 1K words of 40 bits width was needed; this corresponds to 5K bytes when comparing with 8 bit processors. All logic statements together required 48 words. The PI-control microprogram required 268 words. The operating system took up 448 words.

5. ARCHITECTURAL CONVERGENCE

The evolution of programmable controllers from simple relay replacement units to general purpose industrial controllers and the levels of performance available in modern microprocessors has led to the convergence of programmable controller and standard microcomputer hardware. The original simple cyclic program execution without jumps found in the first programmable controllers, which was partly due to hardware restrictions, has today been replaced by modular structured procedure-type software organization in some programmable controllers.

Such software organization is equivalent to that found in structured general purpose programming languages for microcomputers, e.g. Pascal. Sophisticated operating systems oriented towards real time control are available in programmable controllers. These operating systems make it possible for the user to program and debug his application without having to have specialized computer training.

Standard microprocessors are themselves becoming more powerful. The increasing speeds of newer microprocessors allows their use in real time control as the instructions required for interpretation of a high level language can be executed at higher speeds. At present standard microprocessors are only suitable for small programmable controllers and are replacing special hardware circuits. In medium scale programmable controllers, some hardware support for the boolean logic instructions is necessary. Large scale programmable controllers generally use bit-slice or special microprocessors. Economic reasons dictate the use of standard microprocessors wherever possible at present. The developments in logic arrays and automated silicon chip design will make it also feasible to manufacture special purpose microprocessors for industrial controllers. Cyclic program execution is the major uniting feature of all programmable controllers. Each cycle represents one scan of the process or machine to be controlled. Although the cyclic program execution of programmable controllers can be interrupted for critical conditions, the underlying cyclic operation is maintained. Cyclic operation is just one of several possibilities for microcomputer systems. Other alternatives are interrupt driven, multiplexed or batch processing modes. The program execution mode is determined by the operating system and software and not the hardware.

6. CONCLUSION

The goals of increasing productivity and lowering energy consumption can only be realized by increasing the use of automation down to the lowest levels of industrial control. This can only be achieved by lowering the cost of industrial controllers as far as possible. The high functional capability of industrial controllers and the introduction of "universal" programmable controllers with sophisticated logic and closed-loop control already assures wide application. The availability of low cost and powerful microprocessors with suitable functions is one of the requirements. The other perhaps more important requirement is a user friendly programming language. This simplifies programming and debugging, thus minimizing the rising software costs, which already make up the major share of total costs for an installation. The evolution of programmable controller hardware from random logic to standard microprocessors points to a synthesis between programmable controllers and microcomputers. Programming units based on powerful standard microcomputers are extensively used for programming and documentation. This use of standard microcomputer systems for the programming units in programmable controllers is a first step in the synthesis. New multiprocessor structures and networks will have as much an impact on programmable controllers as they did on microcomputers. The combination of the simple man-machine user interface and real time speed of the programmable controller with the data processing capability and standard software available for microcomputers will determine the direction for new industrial control systems.

7. REFERENCES

1 Taylor,D.,Morgan,L.: "High-Level Languages for Microprocessor Projects", 1980, NCC Publications (UK)

2 Kompass,E.J.,Williams,T.J.: "Computer Software for Industrial Control", Annual Advanced Control Conference, Purdue University, 1981, Control Engineering

3 Skrokov,M.R.: "Mini- and Microcomputer Control in Industrial Processes", 1980, Van Nostrand Reinhold Co.

4 Osborne,A., Kane,G.: "16-Bit Microprocessor Handbook", 1981, OSBORNE/McGraw-Hill

5 Osborne,A., Kane,G.: "4 & 8-Bit Microprocessor Handbook", 1981, OSBORNE/McGraw-Hill

6 RCA Solid State Division, Data Sheet File no.1324 "EPIC GP001"

7 Advanced Micro Devices: "Build a Microcomputer", 1978, AMD Inc.

8 Andrews,M.: "Principles of Firmware Engineering in Microprogram Control" 1981, Pitman

9 Salisbury, A.: "Microprogrammable Computer Architectures", 1976, Elsevier

10 Giloi W.K.: "Rechnerarchitektur", 1981, Springer-Verlag

11 Bode,A.,Händler,W.: "Rechnerarchitektur", 1980, Springer-Verlag

12 Weitzmann,C.: "Distributed Micro/Minicomputer Systems", 1980, Prentice-Hall.

13 Schmid, H.: "Multi-Mikroprozessor systeme", 1982, Elektronik, Vol.2 pp.87-95, Vol.4 pp.55-62, Vol.7 pp.67-73.

14 Schmidt,G.,Swik,R.: "Softwarefragen beim Einsatz von Mikroprozessoren für MSR-Aufgaben", 1978, Regelungstechnische Praxis, Vol.5, pp.143-150.

15 Behrens,D.,Kotte,W.,Krimmling,H-J.,Rath,G.: "The Simatic S5

Stored Program Automation System", Siemens Power Engineering, Vol.1, No.5, May 1979, pp.145-149.

16 Anon.: "Survey of Programmable Controllers", Elektronik, 1980, Vol.25, pp.126-132.

17 Anon.: "Speicherprogrammierbare Steuerungsgeräte", VDI-Berichte, No.396, 1981, VDI-Verlag, Düsseldorf.

18 Kintner,P.M.: "Controller Programmable with Logic Instructions Representative of Circuit Nodes and Elements of a Ladder Diagram", Cutler-Hammer World Trade Inc., European Patent Specification EP 0 001 700 B1, 1982.

19 Hill,L.W.,Stoodley III,T.J.,Malcolm,R.: "Improved Programmable Controller", Gould Inc., European Patent Application EP 0 007 153 A2, 1980.

20 Lynch,F.,Hill,L.: "Low-power Schottky replaces MOS in PC", Control Engineering, Vol.26, 1979, No.6, pp.139-140.

21 Bischoff,H.,Müller,B.,Tendulkar,G.: "Automatisierungsgerät SIMATIC S5-130W", Siemens Energietechnik, Vol.2, No.9, Sept. 1980, pp.362-363.

22 Tendulkar,G.A.,Müller,B.: "Microprocessor Architecture for Industrial Controllers", pp.233-242, Microprocessor Systems: 6th Euromicro Symposium, 16-18 Sept.1980 London, North Holland Publishing Co.

23 Tendulkar,G.A.: "High Level Language Digital Drive Controller", IECON 82 Conf.Proc., Nov.15-19 1982, pp.196-201.

24 Flynn, M.J.: "Directions and Issues in Architecture and Language", Oct. 1980 Issue, Computer, pp.5-22, IEEE.

25 Maggioli,V.J.:"The Microcomputer/programmable Controller Syndrome", IEEE Trans.on Industry Applications, Vol.IA-18,No.3, May/June 1982,pp.300-305.

8. ACKNOWLEDGEMENTS

The author is presently with Siemens AG, Erlangen, West Germany in the Control and Automation Systems Department. Most of the research was done in Gerätewerk Erlangen. The contents of this chapter however do not necessarily have any correspondence with the products or views of Siemens AG.

CHAPTER 12

DIGITAL CONTROLLERS AND THEIR MICROPROCESSOR-BASED IMPLEMENTATION

Spyros G. Tzafestas
Electrical Engineering Department
University of Patras
Patras, Greece

1. INTRODUCTION

Process control computers are recognized to be the first big step in filling the gap between control theory and practice, whereas the introduction of microprocessors (μPs) is the second and bigger step [1-7]. Of course it must be noted that the transition to microprocessor-aided control is still slow due to several reasons such as the natural conservatism of industry, etc. Actually, the direct replacement of a controller by a microprocessor-based version does not always lead to considerably improved performance or very much smaller cost. The main advantage is that with a *microcontroller* (μP-based controller) one has increased flexibility in modifying its structure, and this is possible by the utilization of low-cost computer facility in small packages.

Microcontrollers (μControllers) are distinguished in *"general purpose μControllers"* which are designed for use in a large variety of process systems, and *"special purpose μControllers"*, each one of which is designed to work for a specific type of process. The greater part of activity (designs, reports, articles, seminars, workshops, etc) was oriented towards the special purpose application, covering the basic interfacing and programming knowledge needed for designing a special purpose μController from the beginning. A μController can be easily of the *distributed type* by employing appropriate data-highways for servicing the various devices at high data rates (larger than 1Mbit/sec) with negligible noise contamination (although the signals travel long distances in an intesively hostile environment). In future *fiber optics* will become the data high-way medium, provided that some current difficulties will be overcome and that prices will go very much down.

In distributed control the actual closed-loop system involves one μP for each loop or for each few loops. A central computer

S. G. Tzafestas (ed.), Microprocessors in Signal Processing, Measurement and Control, 231–279.

is used for supervisory control, information interchange, monitoring, and for communication with higher levels in the control hierarchy. Among the principal advantages of distributed control are: flexibility, increased reliability, centralized information storage, rich repertory of control algorithms, etc. Examples of such distributed process controllers are the ones produced by *Honeywell* (each remote station can control up to 128 loops) and by *Yokogawa-Japan* (each remote station can control up to 32 loops). In both cases coaxial cables are used for extended data bus. These companies also sell their products as small multi-loop *direct digital controllers* (DDCs). Some other typical small stand-alone controllers are the *Bristol μController* (8 DDC loops) and *Measurex* (4-6 DDC loops or 6-8 supervisory loops).

Stand alone single-loop μControllers [6] possessing the capability of a certain degree of adjustment or programmability are very useful both for their own or for implementing distributed control. Originally the computer was used for making adjustment to the set point of analog controllers, but gradually its use was turned to DDC, while retaining a sufficient amount of analog control set-up. Hence control programs were naturally developed imitating (simulating) classical analog control algorithms (mainly proportional+integral+derivative: PID algorithms). It is useful to remark that, due to that properly tuned PID controllers give generally satisfactory performance, there has not been much incentive for different control laws. That is why the development of more sophisticated controllers (state controllers, adaptive controllers, etc) is somehow lagging behind PID-type controllers.

Our main purpose in this chapter is to provide the basic equations for a variety of discrete-time (digital) controllers which are amenable to μP-aided implementation. These involve classical controllers, minimum variance/self-tuned controllers and some state feedback controllers. For completeness, a useful general discussion on microprocessors and microprocessor-based controllers is included, and some case studies are briefly outlined to illustrate the versatility and power of μP-based control.

2. MICROPROCESSORS AND MICROCONTROLLERS

In digital control the signals are both sampled and quantized (digitized). The main objective in designing a μController is to obtain a real-time controller. Due to the complexity and dimensionality of the systems to be controlled, this requirement needs special skills regarding the insight of the process, and knowledge of control methodology and microprocessor engineering. The control is accomplished by feeding back output or state signals suitably processed in real or non-real time as the case may be. Thus inherent in the design of a digital control system is the objective of improving the quality of signals (using filters), transforming signals (using coordinate transformations), estimating, discriminating, recognizing and transmitting signals. This constitutes the signal processing and communication part of any control system. Standard controllers (compensators and PID controllers) need relatively less signal processing operations, but newer controllers (state space controllers, self-tuned controllers, etc) need sophisticated signal processing operations (parameter estimators, Kalman filters, etc) requiring high computational capabilities [1-2].

2.1. General Description of Microcontrollers

A μController is an I/O machine which is easily programmed and includes analog-to-digital (A/D) and digital-to-analog (D/A) converters. Some of the channels are *digital* (for interrupts, displays, stepping devices or motors, etc) and other can be *analog*. The control algorithm is built by *hardware, software*, or both, inside the controller. The word length and the A/D, D/A accuracies are selected to meet the particular design requirements. The parameters of the control algorithm (law) can be changed by the designer, or by a central controller in an on-line fashion. Until a generally acceptable μController is commercially available, the control engineer is faced the problem to design his own μController by connecting the necessary components such as μPs, clocks, memories, A/Ds, D/As, and other LSI devices. The μController receives and processes analog and logic inputs and provides in real-time analog and logic commands. When very fast computation is needed, one can use special hardware multipliers or dividers. It is advisable that a μController be designed such as to be able to carry out (in addition to real-time control) off line information processing such as self-checking, reprogramming, etc.

The basic LSI building chips of μControllers are:

- The *clock* (synchronizes the whole system).
- The *μP* (executes programs and controls the activities of a bus organized system).
- *Memories* (such as ROM, RAM, PROM, EPROM for data and program storage).
- *Interfaces* (A/D, D/A converters, I/O ports, differential voltage buffers, etc).
- *Multipliers* (or dividers) for fast computation.
- *Buffers* (latches, tristates, MUXs, decoders, etc).

The actual assembly of μControllers may be found in several books and articles (see Chapter 12 of this volume). After the selection of the appropriate μP and the design of a μController, the next step is to program it for implementing the desired control algorithm. The final step is to verify and select (*"tune"*) the parameters of the controller program. This is usually done using an analog or digital simulator of the process under control. The programming is performed with the help of suitable I/O peripherals and extensive software equipment. Regarding the programming we just mention that *machine language* requires minimum peripherals but it is difficult to write, modify and debug the program, and the *assembly language* needs organization of the registers and memory by the programmer. The *high-level languages* (BASIC, FORTRAN, PLM, etc) do not require the tedious organization of the registers and memory, but suffer from smaller efficiency of the object code. They can be used for large programs or when the time for programming and debugging is limited.

The hardware part includes an I/O facility for communication, and enough memory for data and program storage. Also it includes some of the following: keyboard-display, keyboard-printer, keyboard-CRT, reader, magnetic tape, diskettes, etc. The software part includes: a monitor program for the supervision of the whole system (including I/O), the organization of the memory and the transfer of commands to the various programs. It also involves an *editor* (organizes any text of ASCII characters into lines and blocks), a *debugger* (analyzes the control algorithm), a *compiler*, a *cross-assembler* (translates programs into the machine language of another computer), and an *emulator* (i.e. a mix of hardware and software) to enable the user to debug in real time the software and the hardware (partially or totally).

2.2. Parallel-Organized Microcontrollers

The alternative to high-speed stand-alone μControllers (or processors) is to use multiple parallel processors operating simultaneously [7]. Two typical types of parallel processors exist:

(a) Single-Instruction Multiple Data Processor (SIMD, Fig. 1a).
(b) Multiple-Instruction Multiple Data Processor (MIMD, Fig. 1b).

In the *SIMD machine* there is one control unit (CU), many processing units (PU), and an interconnection network (I.N.). All pure control instructions are performed by the control processor. The arithmetic instructions are transmitted to the arithmetic processors where they are executed simultaneously by all PUs usually on different data. Arithmetic processors can be deactivated by the control processor and can exchange data via the interconnection network which can have various stru-

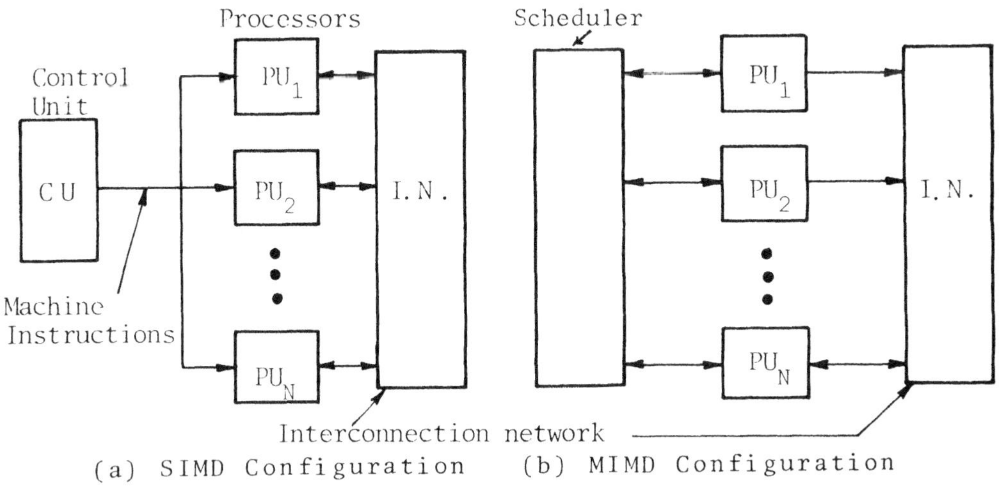

Fig. 1. The two types of parallel processing.

ctures . SIMD machines are appropriate (efficient) for matrix computation.

In a *MIMD machine* each processing unit fetches and executes its own instructions. Thus at any instant different processors may be executing completely different program segments. A MIMD processor uses special hardware and software signals to coordinate the activities of the individual processors. Parallel processors can be easily modified if a change in the control law exceeds the computational capabilities of the machine. In the case of a single processor one has to redesign a faster processor, i.e. usually a different system. Parallel processors can be expanded by simply adding more PUs to meet the added demands. Since modern control laws typically involve vector matrix computations, they are appropriate for parallel processors.

2.3. An Overview of Some Microprocessors

Here we shall briefly describe some recent trends in μP design and give some examples of recent μPs. Microprocessors and microcomputers based upon them are classified according to *word length* (4,8,12,16 bit machines), and to *technology* being used for their manufacturing (P-MOS:1971, N-MOS:1973, TTL:1974, I^2L:1975) [3-4].

Two popular 8-bit μPs are the *Zilog Z-80*, and the *Motorola 6809*. Z-80 works with binary 8080 programs and M6809 works with assembly language using a special assembler which takes 6800 programs and produces 6809 code. Motorola developed a group of programs written for 6800 and produced a powerful 8-bit processor which is suitable for implementing high level languages at some of the latest 16-bit machines. Another interesting machine is Intel 8088 implementing the same instruction set as the 16-bit 8086, but uses 8-bit external bus. The 16-bit data words are fetched in two bus cycles; 8088 is in the worst case half as fast as the 8086. However in typical controller implementation with a predominance of multiplications and division the speed difference is very small. The 8088 has the advantage of being capable to use any software (e.g. compiler or operating system) written for the powerful 8086 on a cheaper system with an 8-bit bus.

Many μP companies have now available complete microcomputers (a μP, a small amount of memory ROM and RAM, and some I/O lines) on a single chip [6]. The older chips (e.g. TMS-1000) were 4-bit machines. A typical new single-chip (5V supply) microcomputer is *Intel 8048* (8-bit CPU, 1kbytes ROM, 64 bytes RAM, 27 I/O lines, mask programmed at the factory). A pin-compatible EPROM version is also available for prototypes and reproduction systems. A similar single-chip machine is *MOSTEK 3870* which is based on Fairchild F8 (2-4 kbytes ROM, up to 128 bytes RAM, timer, 32 I/O lines). It can be mask programmed at the factory (when more than 250 are ordered) but an EPROM version is available bringing the ROM address and data bus to a 24-pin socket. A useful cut-down version of the 8048, the 8022 is also available which implements a subset of the 8048 instruction set and has onchip an A/D conventer.

The new microcomputer *Intel 2920* is suitable for signal processing and contains onchip a D/A converter which can be used in a successive-approximation way for A/D conversion under program control (9-bit precision). Also a PROM is involved onchip holding a 192-instruction program to allow regular sampling and processing of the analog signal (Maximum speed: 192 instructions X 400 nsec=76.8 μs, i.e. 13kHz sampling rate or 6.5kHz signal bandwidth). If less than 192 instructions are used faster rates can be obtained. For example a second-order filter uses 10 instructions an can be implemented at a rate of 250kHz (see Appendix of Chapter 5 of this volume).

Regarding the communication oriented μPs we mention *Mostek SCU-1* (serial control unit) which is designed to be used as a front-end communication controller operating under direction from a control processor and up to 255 of them can be accomodated in one system.

Presently, a number of *microcomputer development systems* (MDS) are available, which besides the basic microcomputer chips have a number of specialized support chips such as DMA (direct memory access) and ICE (In-circuit emulator) chips, floppy-disc controllers, etc. A DMA controller performs the protocol required to gain and release the bus for direct memory access transfer. As examples of MDSs we mention the *Intellec MDS* (Intel), the *DAI-MDS* (Data Applications International), and the *Motorola 6800 MDS*. All of them are self-contained modular microcomputer development systems. The standard Intellec MDS consists of four microcomputer modules, an interconnected printed circuit mother board, power supplies, a chassis and a front panel. The CPU module contains an Intel 8080 μP (2μs cycle, 78 instructions, unlimited subroutine nesting, vectored interrupt, DMA capabilities, a 16-line address bus which is associated with a bidirectional 8-line data bus). The monitor module contains the Intellec MDS monitor and all MDS peripheral interface hardware (clocks, control and data transfer circuitry to interface the peripherals). The memory is arranged into 65536 addressable 8-bit bytes. There are 16k of RAM modules. The software support involves the system monitor, the text editor and the macro-assembler.

Some new Intel MDS models are the *Model 120*, the Model 225, and the *Model 286* (16 bits). The model 120 uses for the CPU the MCS-80/85 μP families, 32k bytes RAM and 4k bytes ROM. Also it has self-test diagnostic capabilities, the ISIS-II disk operating system, and an Integral 250 k-byte floppy disk which is extendable to 2M bytes and software compatible with previous Intellec systems. The model 225 has an 8085A-2 CPU, 64k bytes RAM, 4k bytes ROM, supports PL/M, FORTRAN, BASIC, PASCAL and COBOL languages. The model 286 (Intellec Series III) is a high performance system solution designed specifically for iAPX 88/86 μP. It contains two host CPU's (e.g. the iAPX86 and 8085), 96k bytes of RAM, full range of iAPX 88/86-resident high-level languages, such as PL/M 88/86, PASCAL 88/86, and FORTRAN 88/86.

The *DAI* microcomputer development systems are called *digital control element (DCE) microcomputer systems*, and their modules have three features in common: single eurocard format, plug-to-plug compatibility, and software compatibility. The DCE-1 microcomputer system provides 4k EPROM spaces and 512 bytes of RAM, whereas DCE-2 has a RAM expandable to 2k. The DCE-X microcomputer system provides up to 64k bytes (8 bit) memory capacity. All these DCE's can be interchanged without any modification. Also a *Real-World* card (RWC) is available for the connection with the real world (analog signals, heavy currents, noisy signals, data communication, contact closures, etc.). The designer can simply connect these RWC's in any combination, up to 15, along with a DCE card and a power supply module into the parallel wired eurocard rack. Some of the available real-time cards are: high speed analog data acquisition (RWC-V8/16), analog voltage output (RWC-V08), serial communication interface (RWC-CCE), serial time monitor (RWC-SLD), standard current driver (RWC-SCD), position and temperature measurement (RWC-PTM), and scientific-mathematics module (RWC-MATH). Other RWC cards are currently under development. Regarding the *Motorola MDSs* we mention the M6800 EXORciser, the EXORciser II MDS (also based on M6800), the EXORterm MDS (based on M6800 and M6809), and the EXORmacs MDS based on M68000. The last one is a state-of-the art MDS for designing and developing advanced 16-bit μP-based systems using Motorola's μP families, microcomputers and peripheral parts. The internal card set accomodates 15 cards, five of which are included in the basic system (MPU/MMU, DEbug, FDC and 256k dynamic RAM). The basic system may involve either a 32 MD hard disc or a 1MB floppy disk storage. The software includes a multi-task operating system, M6800 resident PASCAL compiler and structured MACRO Assembler, a CRT editor, a firmware-resident debugger, and self-test diagnostics. Also for all Motorola MDSs there are available the following software tools: M68000/6800/6809 resident FORTRAN, M68000/6809 resident MPL, M6800 resident basic interpreter, M6800 resident COBOL, M68000 Cross Macro Assemblers, M68000 Cross PASCAL Compiler, M6800/6801 Cross Macro Assembler, and M6800/6809 MDOS operating system/CRT editor/Macro assembler.

We close this section with a few comments on 16-bit microprocessors. The first 16-bit μP was *Texas' TMS9900* which was followed by *Intel 8086*, *Zilog Z-8000*, and *Motorola M68000*. These μPs are presently very strong competitors. The 64k bytes address space of 8-bit μPs is somehow restrictive, expecially for large high-level language and multiprogramming applications. However allowing an address space larger than 64k bytes implies that addresses can, in general, occupy more than two bytes (16 bits). For this reason the problem has been overcome in various ways. Only the M68000 is fully and permanently working with a large address space. The Intel 8086 allows the programmer to work with a 64k byte "window" in a much larger address space-up to 1 M byte. The Z-8000 is provided in two versions; the *"unsegmented"* version (with 64k byte address space) and the *"segmented"* version which can work in cooperation with another memory management chip allowing address spaces of up to 8 M bytes. The M68000 has a maximum address space of 16 M bytes. Protected areas of store can be set-up in both the Z-8000 and M68000 with the aid of separate memory management chips. Regarding Z-8000 an alternative memory management board was developed by *Central Data Corporation* which is actually more powerful than Zilog's own product [5]. Both Z-8000 and M68000 have dual operation modes, i.e. *system mode* (supervisor mode) and *normal mode* (user mode). These are used when a supervisory (system) program wants to exert protection and control over user (normal) programs. The Z-8000 has sixteen 16-bit registers and one more register for use on occurence of an interrupt. The M68000 has sixteen 32-bit registers which from a logical point of view work in 32-bit architecture. The addresses occupy 24 bits, but 16-bit addresses can be specified whenever desired, so that small systems to not pay high penalty by having to handle 24-bit addresses.

2.4. Comparison of Computational Characteristics of Some Microcomputers

As it will be clear from the analysis to be presented in the next sections, the implementation of discrete-time controllers needs a few basic mathematical operations, namely *addition, multiplication by scalar, multiplication by a variable, division, delay, matrix* operations, *conditional branch* operations, and some *program control* operations. The characteristics of two 8-bit and four 16-bit representative microcomputers are summarized in Tables 1 and 2 [2]. Addition and subtraction are performed by a single instruction on the basic data word, and unit delay is implemented by moving data out of (delayed value) and into (current value) of CPU register.

TABLE 1

Execution Times of Arithmetic Operations

ARITHMETIC OPERATOR*	EXECUTION TIMES (μsec)*					
	8 BIT NMOS		16 BIT NMOS		16 BIT CMOS/SOS	16 BITBIPOLAR
	8048	Z-80	9900	LSI-11	ATMAC	ATAC
ADDITION	2.5	1.0	4.66	3.5	.28	0.25
MULTIPLICATION	190.$^{+}$	91.0^{+}	17.3	44.0	11.5^{+}	5.5
DIVISION	360.0^{+}	113.0^{+}	36.0	78.0	18.6^{+}	11.2
DELAY	5.0	2.0	9.33	7.0	.70	.5

(+) Average subroutine Time, (*) Fixed-point register operations in μP basic data word length at maximum CPU clock speed.

The 16-bit μPs have special instructions for multipication and division, but in 8-bit μPs one must use binary routines. For high speed operation *special LSI hardware function units* (SHFU) have to be used (for multiplication/division, exponentials, trigonometric functions, square root, logarithms, correlation, floating point arithmetic, convolution, FFT etc). These special function chips are interfaced to a microcomputer as peripherals on the I/O bus, but in some cases they are interfaced directly to a dedicated CPU port.

Some examples of improving speed using SHFUs are the following.

TABLE 2

Microcomputer Throughput for Controller Applications (Fixed-point Data Operations, Average of P addressing Modes).

INSTRUCTION (OPERATION) AND EXECUTION %	EXECUTION TIMES (μsec)					
	8 BIT NMOS		16 BIT NMOS		16 BIT CMOS/SOS	16 BIT BIPOLAR
	8048	Z-80	9900	LSI-11	ATMAC	ATAC
AND	3.33	2.8	4.7	17.43	.28	.625
OR	3.33	2.8	4.7	10.18	.28	.625
XOR 40%	3.33	2.8	4.7	10.18	.28	.625
TEST BIT 40%	5.0	2.8	4.0	9.65	.28	1.0
BIT CLEAR/SET	6.25	3.93	4.0	10.18	.28	.625
COMPLEMENT	2.5	1.0	4.0	7.25	.28	1.27
LOAD 30%	4.06	3.28	4.7	9.3	.35	.604
STORE 30%	3.92	3.39	4.7	9.3	.35	1.42
THROUGHPUT IN KOPS (KILO-OPERATIONS PER SECOND	251	325	220	101	3100	1081

(Execution times computed on the basis of maximum clock specification).

I. Z-80 (8 bits): Using an AMD 9511 APU (working at 4MHz) a 16-bit fixed-point multiplication needs 42 μsec compared with 92 μsec achieved by software subroutine.
II. ATMAC (16 bits): Using a TRW MPY-16AJ (16x16 bits) multiplier, a 16-bit multiplication needs 1 μsec compared with 11,5 μsec obtainedusing the machine's own instructions.

Program control operations (logical, test bit, I/O, load/store memory, branch operations, etc) are needed, which can be executed using different addressing modes (direct, indirect, relative and immediate). Table 2 provides their execution times, along with the operation time of a representative mixture of them needed in a typical controller. In this special instruction mix 40% is spent for the logical operations (AND, OR, Test bit) and the remaining 60% for the load and store (I/O). The result is expressed in "*throughput*" (operations per second) which is a convenient figure of merit.

More complex functions can be generated by combining basic arithmetic and program control operations, using either software subroutines or hardware or both. As an indication we mention that the execution times for fixed point sum of products for fractional data are 107 μsec for Z-80, 13 μsec for ATMAC and 9 μsec for ATAC. Using an 8-bit CMOS/SOS accumulator/multiplier in ATMAC μComputer the operation time reduces to 2.7 μsec/step. Using a look-up table (with equally spaced independent variables) a 16-bit fixed-point fraction on a Z-80 needs 194 μsec/step.

3. CLASSICAL DIGITAL CONTROLLERS

3.1. System Discretization

Digital systems are described by discrete (pulse, sampled-data) transfer functions which depend on the complex variable $z=\exp(sT)$ where $s=a+j\omega$. Given the transfer function $G(s)$ of a continuous-time system its conversion to digital form $G_A(z)$ can be made by a variety of methods [8-9]. Some of them, which are most frequently used are the following.

Pure Z-transform Discretization: Given

$$G(s)=\sum_{i=1}^{n} g_i/(s+\pi_i) \tag{1a}$$

then

$$G_A(z)=\mathcal{Z}G(s)=\sum_{i=1}^{n} g_i/(1-z^{-1}e^{-\pi_i T}) \tag{1b}$$

where T is the sampling period. This method preserves the impulse response and stability, but not the frequency response of G(s). Obviously partial fraction of G(s) is needed.

Z-transform with Zero Hold: Here

$$G_A(z)=\mathcal{Z}\left[\frac{1-e^{-sT}}{s}G(s)\right]=(1-z^{-1})\,\mathcal{Z}G(s)/s \tag{2a}$$

for example, if $G(s)=\pi/(s+\pi)$, then

$$G_A(z)=y(z)/u(z)=z^{-1}(1-e^{-\pi T})/(1-z^{-1}e^{-\pi T}), \tag{2b}$$

which in difference-equation form is written as

$$y_k=e^{-\pi T}y_{k-1}+(1-e^{-\pi T})u_{k-1},\quad y_k=y(kT),\quad u_k=u(kT) \tag{2c}$$

This method preserves the stability, but not the impulse and frequency responses of G(s). Partial fraction of G(s) is also needed.

Backward Difference Method: We replace s is G(s) by D=d/dt and use the backward difference approximation $D=(1-z^{-1})/T$ where z^{-1} is to be considered as the unit delay operator i.e. $z^{-1}y_k=y_{k-1}$. Then

$$G_A(z)=G(s)\Big]_{s=(1-z^{-1})/T} \qquad (3)$$

For example a 1st-order compensator (lead or lag) $G_c(s)=A(s+\mu)/(s+\pi)$ gives $G_{cA}(z)=A_1(z+\mu_1)/(z+\pi_1)$, $A_1=A(1+\mu T)/(1+\pi T)$ with $\mu_1=-1/(1+\mu T)$ and $\pi_1=-1/(1+\pi T)$. This is equivalent to the difference equation

$$y_k=-\pi_1 y_{k-1}+A_1(u_k+\mu_1 u_{k-1}) \qquad (4)$$

This method needs very small T for acceptable accuracy. It preserves the stability but not the impulse and frequency responses. No partial fraction expansion of G(s) is needed.

Tustin Method: We use the transformation $s=\frac{2}{T}(1-z^{-1})/(1+z^{-1})$ and obtain

$$G_A(z)=G(s)\Big]_{s=\frac{2}{T}(1-z^{-1})/(1+z^{-1})} \qquad (5)$$

Stability is preserved, but not the impulse and frequency responses. No partial expansion is needed. This method is equivalent to trapezoidal integration. Small sampling periods T lead to high accuracy. Applying this method to the above 1st-order compensator leads again to the difference model (4) with $A_1=A(2+\mu T)/(2+\pi T)$, $\mu_1=(\mu T-2)/(\mu T+2)$ and $\pi_1=(\pi T-2)/(\pi T+2)$.

3.2. Classical Design of Digital Compensators

Three methods exist, namely: transfer function manipulation, discrete root-locus, and design in the Tustin plane.

Transfer Function Manipulation: The typical configuration of a digital closed-loop system controlled by the digital compensator $G_{cA}(z)$ is as shown in Fig. 2 where F(z) is the digital transfer function of the continuous process under control.

The closed-loop digital transfer function is

$$T(s)=y(z)/r(z)=G_{cA}(z)F(z)/\{(1+G_{cA}(z)F(z)\} \qquad (6)$$

Given the desired, physically realizable, closed-loop transfer function T(z), one can determine $G_{cA}(z)$ by solving (6), i.e.

$$G_{cA}(z) = T(z)/F(z)(1-T(z)) = \left(\sum_{i=0}^{p} a_i z^{-i}\right) / \left(\sum_{i=0}^{q} b_i z^{-i}\right) \qquad (7)$$

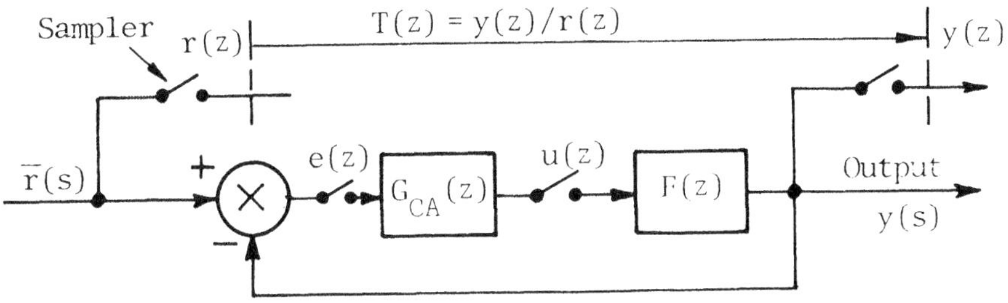

Fig. 2. Typical closed-loop system with digital compensator.

Clearly, $G_{cA}(z)$ in (7) is physically realizable if $b_o \neq 0$. Also poles and zeros of F(z) outside the unit circle cannot be cancelled by zeros and poles of $G_{cA}(z)$. These two facts restrict considerably the closed-loop transfer functions which can be achieved. Common closed-loop requirements are:i) zero steady-state error (step, ramp, parabolic, etc), dead beat (finite-settling time) response, and ripple-free response.

Discrete Root Locus Method: Suppose that in Fig.2, $G_{cA}(z)=KD(z)$ where K is a variable gain. For K>0 the "root locus" is characterized by the following two conditions:

$$|D(z)F(z)| = 1/K \text{ and } \angle D(z)F(z) = 180^o \qquad (8)$$

The first is the amplitude condition and the second the phase condition.

As an example consider a system with forward path $G(z)=K(z-z_1)/(z-1)(z-z_2)$. The root locus is determined by $1+KG(z)=0$ and is shown in Fig.3. Drawing the damping line $\alpha/\omega=0.4$ one can find the value of K for which the closed-loop poles lie on this line. Note that G(z) is the transfer function of the system $\bar{g}(s)=A/s\tau_1(1+s\tau_2)$ determined by the $\mathcal{Z}$-transform/zero-order hold method.

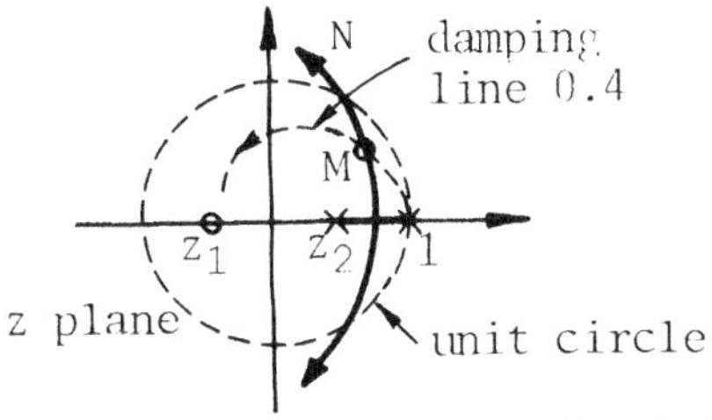

Fig.3.Root locus of G(z)=D(z)F(z)

Design in the Tustin Plane: The method of Bode diagrams cannot be applied in the z-domain since the relation $z=\exp(sT)$ is transcedental. For this reason we use the bilinear mapping

$$w=\frac{z-1}{z+1}=\frac{1-z^{-1}}{1+z^{-1}} \quad \text{or} \quad w=\frac{1-e^{-j\omega T}}{1+e^{-j\omega T}}=j\,tg\frac{\omega T}{2} \tag{9}$$

This is Tustin's mapping except for the factor 2/T. Setting $w=\upsilon+jv$ we see that $v=tg(\omega T/2)$ or $\omega=(2/T)tg^{-1}v$. To determine $G_{cA}(z)$ in Fig. 2 by the Bode method we work as follows.

i) Replace z in F(z) by the inverse of (9) namely $z=(1+w)/(1-w)$ and find $F^*(w)$.
ii) Draw the Bode diagrams of $F^*(w)$ in the usual way.
iii) Choose a $G^*_{cA}(w)$ as in the s-plane.
iv) Replace w in $G^*_{cA}(w)$ by (9), i.e. $w=(z-1)/(z+1)$ to find the required digital compensator $G_{cA}(z)$.

Similar steps are followed when using the Tusting plane $w^*=(2/T)w$, which has the advantage that gives similar quantities as the s-plane.

Application examples of these methods can be found in standard textbooks on "Discrete-Time Control".

3.3. Digital PID Controllers

Proportional plus Integral plus Derivative (PID) controllers are being used in process industry for more than four decades started with the pneumatic ones. The PID control algorithm has the general form (see Fig. 2):

$$u(t)=K_p e(t)+K_i\int_0^t \varepsilon(\tau)d\tau+K_d d\varepsilon(t)/dt \tag{10a}$$

where u(t) is the output of the controller, $\varepsilon(t)=r(t)-y(t)$ the error signal, and K_p, K_i, K_d the proportional, integral, and derivative gains, respectively. The transfer function of the controller is

$$G_c(s)=\bar{u}(s)/\bar{e}(s)=K_p+K_i/s+sK_d \tag{10b}$$

which has the digital equivalent (obtained by the backward difference method):

$$G_{cA}(z)=U(z)/E(z)=K_p+TK_i/(1-z^{-1})+(K_d/T)(1-z^{-1}) \tag{10c}$$

The time-domain representation of (10c) is

$$u_k = K_p e_k + TK_i (1+z^{-1}+\ldots) e_k + (K_d/T)(1-z^{-1}) e_k$$

$$= K_p e_k + TK_i \sum_{j=0}^{k} e_j + (K_d/T)(e_k - e_{k-1}) \qquad (11a)$$

If the integral in (10a) is approximated by a triangular-hold, and the derivative by a two-period average, the PID equation (11a) gives

$$u_k = K_p e_k + \frac{TK_i}{2} \sum_{j=1}^{k} (e_j + e_{j-1}) + \frac{K_d}{2T}(e_k - e_{k-2}) \qquad (11b)$$

When using a digital PID controller {viz.(11a) or (11b)} the problem is to *"tune"* (select) the gains K_p, K_i and K_d, and the sampling period T. Usually, the sampling rate can be adjusted by factors of 2 with the aid of a clock.

In incremental (velocity) form the PID equation (11a) takes the form:

$$\Delta u_k = K_p(e_k - e_{k-1}) + TK_i e_k + \frac{K_d}{T}(e_k - 2e_{k-1} + e_{k-2}) \qquad (11c)$$

This form, where $\Delta u_k = u_k - u_{k-1}$, is appropriate when the actuators following the controller have integrator like characteristics such as stepping motors, etc. To reduce large changes Δu_k after fast changes in the set-point input r_k, r_k is not included in the derivative term, i.e. we use the form $(K_d/T)(2y_{k-1} - y_k - y_{k-2})$. Further reduction of the controller output changes can be obtained, if the set point variable is only present in the integral term, in which case we have the equation

$$\Delta u_k = K_p(y_{k-1} - y_k) + TK_i e_k + \frac{K_d}{T}(2y_{k-1} - y_k - y_{k-2}) \qquad (11d)$$

Tuning procedures for analog PID controllers can be found in many papers (e.g. [10]), and for digital controllers in [11]. Some improvements to the numerical integration schemes for overcoming various practical difficulties were given in [12].

3.4. Microprocessor-Aided Implementation

Digital compensators and PID controllers have been implemented either by hardware [13-14] or by software [15], the latter offering more generality and flexibility. In the market there are available several variations of PID controllers designed using various techniques such as minimum squared error, square-root design, etc. The updating of the sampling rate according to process deviation is also possible [16].

Classical Compensators: A μP-based system (2k store, 16 bits) for the implementation of digital compensators for a-c electromechanical (synchro) systems was designed and constructed at the Polytechnic of Central London [17]. Similar set-ups were studied at Patras University in final year projects. Since the μP operates on digital signals synchro-to-digital (S/D), D/S, A/D and D/A converters were used. Available devices have up to 14 bits and conversion time about 100 μsec. Although the converter may provide 14 bits (say) to the processor, an 8-bit processor can be used by reading the 14-bit data in two operations and working in double precision to maintain the required accuracy. Of course a 16-bit μP can perform this task with fewer operations. The μP must be selected so as to meet the speed of response and the accuracy required. Denoting the input position by X, the output to the motor by E, and the measured output position by Y a typical sequence of operations is: (1) Select converter 1, (2) Input X, (3) Scale X, (4) In range? If not go to (1), (5) Calculate X1 (transformed value of X), (6) Select converter 2, (7) Input Y, (8) Scale Y, (9) Compute E=X1-Y, (10) Select converter 3, (11) Output E, and (12) Go to step (1). Most of these operations need one instruction, but the calcualation of X1 (step 5) may be lengthy.

μP-Implementation of PID Controllers: From among the various commercially available μP-based industrial controllers we mention here the *MICON* (Microcomputer Controller) of Process Systems Inc. which can be used in conjunction with a *DigiCon* (Direct Digital Control) valve [18]. Using a DigiCon valve one can easily control currently available analog final control elements via D/A converters and E/I amplifiers as required. The MICON is based on the Intel 4004 μP and has the general organization of Fig. 4. Besides the factory programmed ROM, a RAM, a field PROM, and an I/O shift register is provided. The basic instruction execution needs 8 or 16 periods of a 667 kHz clock. Up to 63 controllers can be connected to the I/O bus of a supervisory or optimizing minicomputer via an interface control and parallel addressing. All internal operations of MICON are digital with the exception of the software-hardware mix that performs the A/D conversion in the feedback path. The control action is computed using eqn. (11a), with u_k representing the valve opening ($u_o=0$). The integral term I_k is computed using the recursive equation $I_k=I_{k-1}+TK_ie_k$, $I_o=0$. As an option the predictive algorithm $u_{k+1}=u_k(r)/y_k$ is included, where u_k is the present valve opening, r is the set-point (reference value) and y_k is the present value of the output variable. This option is used whenever rapid stable reaction is important, usually when the error exceeds 2%.

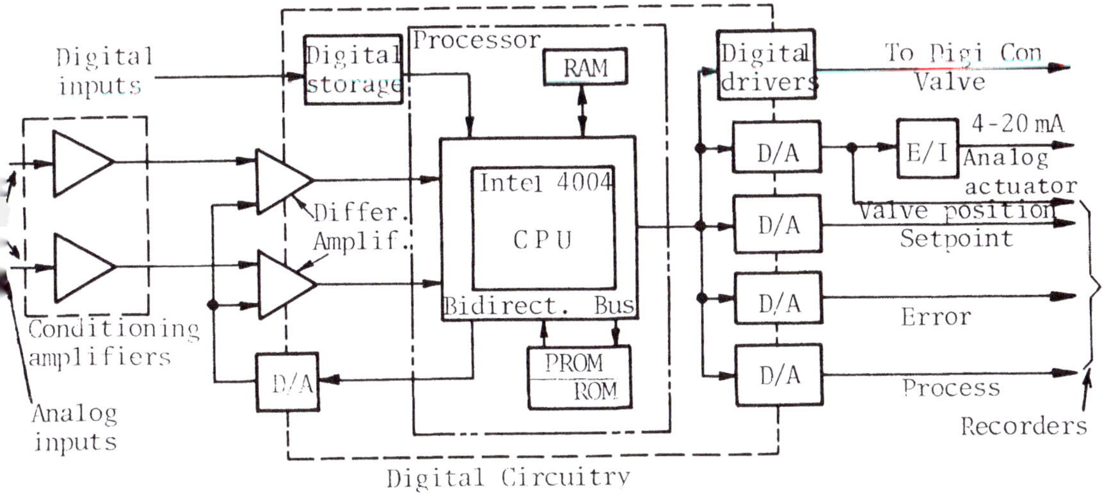

Fig. 4. The MICON:PID and Predictive algorithms are involved in PROM.

Two laboratory μP-based PID controllers have been designed by Reed and Mergler [19] and Aylor, Ramey and Cook [20].

The first is based on the positional PID algorithm (11a) with $K_i=K_p/\tau_i$ and $K_d=K_p\tau_d$, where τ_i is the integral (reset) time and τ_d the derivative time of the equivalent analog controller. The implementation was made by using the M6800 μP, three PIA's (peripheral interface adaptors), 4096 8-bit words of RAM, a 1024-word ROM containing a system development monitor ROM, and 2048 words of EPROM. The I/O operations are considerably simplified by the PIA chip which is appropriate to interface directly the CPU. A data direction register and a control register determine the behavior of the I/O lines and the action of the control lines associated to two bidirectional bit-programmable ports. Peripherals include a teletype, A/D and D/A converters, thumbweel switches, and a time base oscillator which sends a synchronization signal every 10 msec. The general hardware set-up is shown in Fig. 5. The ROM contains a monitor program (for data entry and programming via the teletype). The EPROM contains the control algorithms as well as I/O routines, evaluation routines and interrupt handling routines. All data and intermediate results of the control algorithm are placed in the RAM. The data which exist on the thumbweel switches passes directly onto the data bus when the processor sends the appropriate address on the address bus. The controller gains are entered through the switches in decimal form (XX.XX) as $K_p, K_i'=TK_i=TK_p/\tau_i$ and $K_d'=K_d/T =K_p\tau_d/T$, i.e. precalculation of K_i' and K_d' is needed. Setpoint is also entered in decimal form.

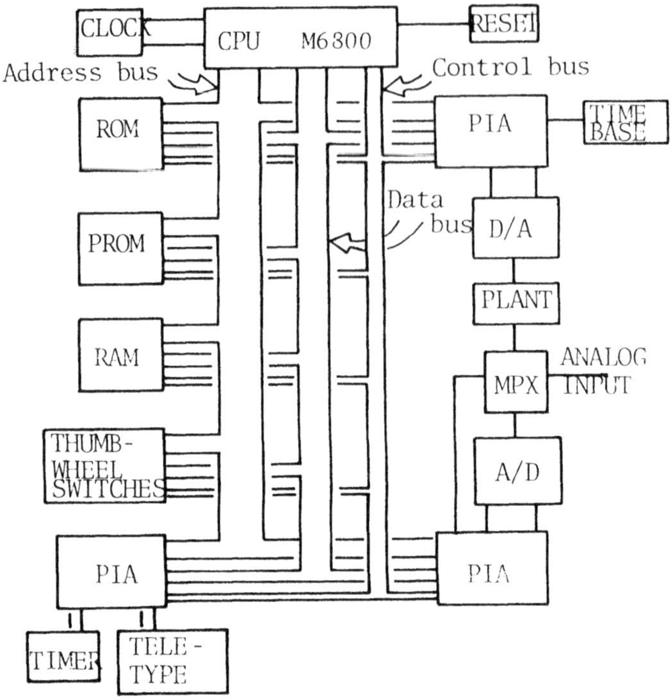

Fig. 5. Hardware set-up of the μP-based Controller.

The system can also evaluate the step response by computing every 1 msec the three criteria $IAE=\Sigma|\varepsilon_k|$, $ISE=\Sigma\varepsilon_k^2$ and $ITAE=\Sigma k|\varepsilon_k|$. The trial and error procedure of the controller is:

(i) Output the accumulated (IAE, ISE, ITAE) values to the teletype.

(ii) Obtain new gains and set point from the thumbweel switches and compute again the criteria.

(iii) Compare the results for a series of equal step inputs and adjust the gains K_p, K_i' and K_d' so as to minimize the accumulated errors.

The controller was tested using a two-time-constant analog circuit (τ_1=0.4 sec, τ_2=0.6 sec) and needed an average of 7 msec to perform the PID calculations. The computation of the above three criteria need two additional multiplications i.e. 4 msec more. Various tuning techniques were tried (Ziegler-Nichols, Kalman, Dahlin) with performance ranged from very good to unstable. Since the controller computation time is less than 10 msec and many chemical processes have time constants of the order of seconds or ten seconds, the controller can be used in a multi-loop control system. Of course by incorporating a hardware multiplier the computational time of the controller can be reduced by a factor of 5 to 10.

Now we turn our discussion to the PID predictor of [20] which is an extension of the preceding controller. The starting point is the controller equation (11a). The main problem is that this algorithm can be implemented if the time required to perform the computations and output of the result is small compared to the sampling time. If this is not so, one must predict the value of e_k at time k-1 (using past values e_{k-1}, $e_{k-2}, \ldots$) and carry out the computations for e_k during the time interval $[(k-1)T, kT]$. The simplest way for doing this prediction is to use the straight-line assumption $(1-z^{-1})e_k/T = (1-z^{-1})e_{k-1}/T$ which leads to the equation

$$e_k = 2e_{k-1} - e_{k-2} \tag{12}$$

Introducing (12) into (11a) yields, after some manipulation

$$u_k = K_1 e_{k-1} + K_2 e_{k-2} + K_3 e^*_{k-1} \tag{13a}$$

where e^*_{k-1} can be generated by the recursive equation

$$e^*_{k-1} = e^*_{k-2} + e_{k-1}, \quad e^*_{k-1} = \sum_{j=0}^{k-1} e_j \tag{13b}$$

and

$$K_1 = 2K_p + 2K_i T + K_d/T$$
$$K_2 = -(K_p + K_i T + K_d/T), \quad K_3 = K_i T \tag{13c}$$

For systems that have by themselves integrating characteristics, e.g. a motor $\Theta_m(s)/\upsilon(s) = K_m/s(1+s\tau_m)$, one can omit the integral (summation) term of the controller. In practice one must be very careful when omitting the integral term since the motor may have nonideal characteristics (such as stall current, etc) and thus a large steady-state error may occur. In case where the integral term is omitted we obtain the PD equation

$$u_k = K_1 e_{k-1} + K_2 e_{k-2} \tag{14a}$$

$$K_1 = 2K_p + K_d/T \quad \text{and} \quad K_2 = -(K_p + K_d/T) \tag{14b}$$

Clearly, omitting the integral term improves the computer throughput.

Suppose that the digital PD controller (14a,b) controls a

motor $G_m(s)=K_m/s(1+s\tau_m)$ via a D/A (zero-order hold) device. The transfer function of the elements of the system are:

Controller

$$G_{cA}(z)=(K_1z+K_2)/z^2$$

Zero-Order Hold and Motor

$$G_{mA}(z)=\frac{K(z-1)}{z}\left[\frac{Tz}{\alpha(z-1)^2}-\frac{z}{\alpha^2(z-1)}+\frac{z}{\alpha^2(z-e^{-\alpha T})}\right]$$

Therefore the closed-loop transfer function (assuming unity feedback) is

$$T(z)=G_{cA}(z)G_{mA}(z)/[1+G_{cA}(z)G_{mA}(z)]$$

For stability all roots of the characteristic equation

$$C(z)=1+G_{cA}(z)G_{mA}(z)=0$$

must lie inside the unit z-circle.

In practice the gain values must be chosen, for a given value of the ratio $\beta=T/\tau_m$, such that to obtain a stable closed-loop system. Such a design example is included in [20]. The motor parameters are $\alpha=1/\tau_m=13.48$ sec^{-1} and $K_m=0.818$ rad/volt/sec^{-2}. It is assumed that the necessary software has a throughput rate of 400 Hz, and that it is desired to control by a single controller three such motors. Thus by appropriate multiplexing of the software algorithm, one must obtain an effective throughput rate of (400/3)Hz=133 Hz or T=7.5 msec. Here $\beta=\alpha T=0.101$ and for stable operation the selection technique of [20] leads to $K_1=14218$, $K_2=11552$ (a possible stable set).

If e_{k-1} and e_{k-2} come from the output of an 8-bit A/D converter and K_1, K_2 are represented by 16-bit words, then the multiplications and additions indicated in (14a) would require 24-bit words. Of course the resulting value u_k can be truncated again to 8 bits, the D/A converter size. Clearly, the system throughput can be maximized if all I/O and controller computations are kept in binary instead of floating- point arithmetic.

3.5. Further Design Aspects of Digital Controllers

PID Controllers: For small sampling periods the parameters

K_p, K_i and K_d of (11a,b) can be found using the corresponding parameters K_p, τ_i and τ_d of the analog controller determined by standard analog tuning procedures, namely $K_p = K_p$, $K_i = K_p/\tau_i$ and $K_d = K_p\tau_d$. However for longer sampling times the parameters have to be selected directly in discrete form. The transfer function of a general discrete controller is

$$G_{cA}(z) = \frac{u(z)}{e(z)} = \frac{\beta_o + \beta_1 z^{-1} + \dots + \beta_p z^{-p}}{\alpha_o + \alpha_1 z^{-1} + \dots + \alpha_q z^{-q}} \triangleq \frac{\underline{\beta}(z^{-1})}{\underline{\alpha}(z^{-1})} \tag{15}$$

which is realizable only when $\alpha_o \neq 0$ (say $\alpha_o = 1$). Note that usually $\beta_o \neq 0$. Similarly the digital transfer function of a linear process is

$$F(z) = \frac{y(z)}{u(z)} = \left[\frac{b_o + b_1 z^{-1} + \dots + b_n z^{-n}}{1 + a_1 z^{-1} + \dots + a_n z^{-n}}\right] z^{-\mu} \triangleq \frac{B(z^{-1})}{A(z^{-1})} z^{-\mu} \tag{16}$$

where $z^{-\mu}$ is the pure time delay involved in the process. When there is no delay we have $\mu = 0$.

In order for the steady-state error (offset) to step disturbances to be zero the controller must have a pole at $z=1$ (final value theorem of $\mathcal{Z}$ transform). Hence G_{cA} must have the form

$$G_{cA}(z) = (\beta_o + \beta_1 z^{-1} + \dots + \beta_p z^{-p})/(1 - z^{-1}) \tag{17}$$

which leads to the time domain algorithm

$$u_k = u_{k-1} + \beta_o e_k + \beta_1 e_{k-1} + \dots + \beta_p e_{k-p} \tag{18}$$

For $p=1$ we obtain a PI controller, and for $p=2$ a PID one. As we have seen (viz (12)) to make faster the controller one can use the linear prediction equation $e_k = 2e_{k-1} - e_{k-2}$, in which case (18) becomes

$$u_k = u_{k-1} + (2\beta_o + \beta_1)e_{k-1} + (\beta_2 - 1)e_{k-2} + \beta_3 e_{k-3} + \dots + \beta_p e_{k-p} \tag{19}$$

The parameters $\beta_o, \beta_1, \dots, \beta_p$ must be selected so as to obtain good performance. To this end three general methods exist:

i) *Tuning rules* based on several criteria.
ii) *Trial and error* starting with a low gain, increasing it

until loop damping is small, and then decreasing it according to some heuristic rule.

iii) *Parameter optimization* which in general needs numerical techniques. A standard criterion is $J=\Sigma_{k=0}^{N}\{e_k^2+\Theta(u_k-\bar{u}_k)^2\}$ where $e_k=r_k-y_k$ and $\bar{u}$ is the steady state of u for step changes. The parameters are determined from the conditions $\partial J/\partial\beta_i=0$, $i=0, 1,\ldots,\beta_p$.

Deadbeat Controllers: A deadbeat controller is designed such that after a step change in the set-point the new steady state is achieved in a finite-settling time. Suppose that $\mu=0$ in (16), and consider a step change $r_k=1$ $(k=0,1,2,..)$ in the set-point (reference) input. The requirement for minimum settling time implies that $y_k=r_k=1$ and $u_k=u_n$ for $k\geqslant n$ where n is the order of the process under control. Then for $b_o=0$ in (16) one obtains

$$r(z)=1/(1-z^{-1})$$

$$y(z)=y_1z^{-1}+\ldots+y_{n-1}z^{-(n-1)}+1\{z^{-n}+z^{-(n+1)}+\ldots\}$$

$$u(z)=u_o+u_1z^{-1}+\ldots+u_n\{z^{-n}+z^{-(n+1}+\ldots\}$$

Thus

$$\frac{y(z)}{r(z)}=\alpha_1z^{-1}+\ldots+\alpha_nz^{-n}=\underline{\alpha}(z^{-1});\frac{u(z)}{r(z)}=\beta_o+\beta_1z^{-1}+\ldots+\beta_nz^{-n}=$$
$$=\underline{\beta}(z^{-1}) \quad (20)$$

$$\alpha_1=y_1,\ \alpha_2=y_2-y_1,\ldots,\alpha_n=1-y_{n-1};\ \sum_{i=1}^{n}\alpha_i=1 \quad (21)$$

$$\beta_o=u_o,\beta_1=u_1-u_o,\ldots,\beta_n=u_n-u_{n-1};\ \sum_{i=0}^{n}\beta_i=u_n=\frac{1}{F(1)}=\frac{1}{K} \quad (22)$$

From (6) (viz sec. 3.2) and the first of (20) it follows that

$$T(z)=\underline{\alpha}(z^{-1})=(\alpha_1z^{n-1}+\ldots+\alpha_{n-1}z^{-1}+\alpha_n)/z^n \quad (23)$$

Also

$$F(z)=\frac{y(z)}{u(z)}=\frac{y(z)/r(z)}{u(z)/r(z)}=\frac{\underline{\alpha}(z^{-1})}{\underline{\beta}(z^{-1})} \quad (24)$$

Hence {viz (7)}:

$$G_{cA}(z)=\underline{\beta}(z^{-1})/\{1-\underline{\alpha}(z^{-1})\}$$
$$=(\beta_o+\beta_1z^{-1}+\ldots+\beta_nz^{-n})/(1-\alpha_1z^{-1}-\ldots-\alpha_nz^{-n}) \quad (25)$$

The parameters are found from

$$F(1)=\underline{\alpha}(1)/\underline{\beta}(1)=(\alpha_1+...+\alpha_n)/(\beta_o+...+\beta_n)=K=1/u_n$$
(see (24))

$$F(1)=B(1)/A(1)=(b_o+...+b_n)/(1+a_1+...+a_n) \quad \text{(see (16))}$$

They are:

$$\beta_o=1/(b_1+...+b_n)=u_o \quad \text{(recall that } b_o=0)$$

$$\alpha_1=b_o\beta_o,\ \alpha_2=b_1\beta_o,...,\alpha_n=b_n\beta_o$$
$$\beta_1=a_1\beta_o,\ \beta_2=a_2\beta_o,...,\beta_n=a_n\beta_o \tag{26}$$

Note from (23) that the closed-loop transfer function has an nth pole at the origin $(z=0)$. If the process under control contains μ units of pure delay, the above procedure can readily be applied by writing $F(z)$ as

$$F(z)=(b_1z^{-(1+\mu)}+...+b_nz^{-(n+\mu)}/(1+a_1z^{-1}+...+a_nz^{-n})$$
$$=\frac{\hat{b}_1z^{-1}+...+\hat{b}_\mu z^{-\mu}+\hat{b}_{\mu+1}z^{-(\mu+1)}+...+\hat{b}_{\hat{n}}z^{-\hat{n}}}{1+\hat{a}_1z^{-1}+...+\hat{a}_nz^{-n}+\hat{a}_{n+1}z^{-(n+1)}+...+\hat{a}_{\hat{n}}z^{-\hat{n}}}$$

with $\hat{b}_1=\hat{b}_2=...=\hat{b}_\mu=0,\hat{b}_{\mu+i}=b_i\ (i=1,2,..,n)$, $\hat{a}_i=a_i\ (i=1,2,...,n)$, $\hat{a}_{n+i}=0\ (i=1,2,...,\mu)$. Here the minimum settling time requirement implies that

$$y_k=r_k=1 \text{ for } k\geqslant\hat{n} \quad \text{and } u_k=u_n \text{ for } k\geqslant n$$

The resulting controller has the form

$$G_{cA}(z)=(\beta_o+\beta_1z^{-1}+...+\beta_nz^{-n})/(1-\alpha_{\mu+1}z^{-(\mu+1)}-...$$
$$..-\alpha_{\mu+n}z^{-(\mu+n)})$$

4. MINIMUM-VARIANCE AND SELF-TUNING CONTROLLERS

4.1. Single-Input Single-Output Controllers

Minimum variance controllers regulate the system so as the output variables of the system under control are kept as close as possible to their reference values and the corresponding process products have desired specifications. For linear systems one can apply linear stochastic theory [21-24].

Here we shall present the *generalized minimum variance con-*

troller [24] which is designed such that the variance of a linear combination of the output and control variables, namely

$$J=E\{y_{k+\mu+1}^{2}+\gamma u_{k}^{2}\} \tag{27}$$

is minimized, where $E\{.\}$ is the stochastic expectation operator, μ is the process pure delay time, and γ a weighting factor. The term γu_k^2 is put in order to limit excessive input changes.

The SISO process under control has the form shown in Fig.6.

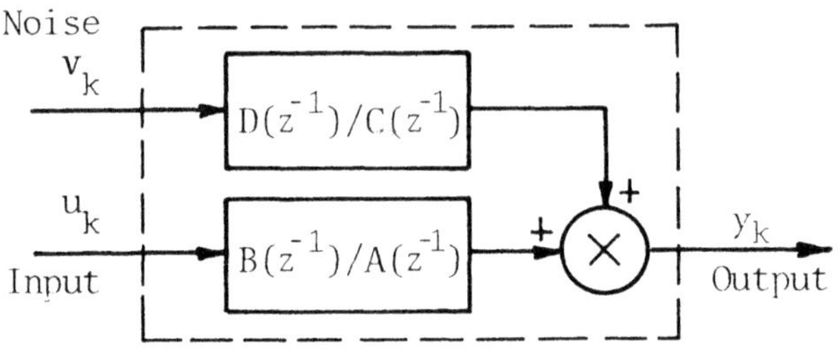

Fig. 6. Stochastic system regulated by a minimum variance controller.

The actual output y_k of the process is

$$y_k=F(z)u_k+F_w(z)v_k \tag{28}$$

where v_k is a white noise sequence with $E\{v_k\}=0$, $E\{v_k v_k\}=1$ and $E\{v_k v_{k+j}\}=0$ for $j\neq 0$, $F(z)$ is the transfer function of the process under control given by (16) with $b_o=0$, and $F_w(z)$ is a linear digital filter of the form

$$F_w(z)=\frac{w(z)}{v(z)}=\lambda\frac{D(z^{-1})}{C(z^{-1})}=\lambda\frac{1+d_1 z^{-1}+\ldots+d_n z^{-n}}{1+c_1 z^{-1}+\ldots+c_n z^{-n}} \tag{29}$$

The problem is to *select the controller $G_{cA}(z)$ such that J in (27) is minimized.*

The first term of (27) is equal to

$$y_{k+\mu+1}=\frac{B(z^{-1})}{A(z^{-1})}zu_k+\lambda\frac{D(z^{-1})}{C(z^{-1})}z^{\mu+1}v_k \tag{30a}$$

where the noise term can be written as

$$\lambda\frac{D(z^{-1})}{C(z^{-1})}z^{\mu+1}v_k=\lambda\{U(z^{-1})z^{\mu+1}+\frac{H(z^{-1})}{C(z^{-1})}\}v_k \tag{30b}$$

$$U(z^{-1})=1+\upsilon_1 z^{-1}+\ldots+\upsilon_\mu z^{-\mu};H(z^{-1})=h_o+h_1 z^{-1}+\ldots\ldots+h_{n-1}z^{-(n-1)} \tag{30c}$$

Here $U(z^{-1})z^{\mu+1}$ represents the part of the noise w_k that cannot be controlled by u_k, and $H(z^{-1})/C(z^{-1})$ is the part that can be controlled. The coefficients υ_i and h_i can be found by solving (30b), i.e. $D(z^{-1})=C(z^{-1})U(z^{-1})+H(z^{-1})z^{-(\mu+1)}$. Introducing (30a) into (27) and setting $\partial J/\partial u_k=0$ gives

$$\{z^{\mu+1}y_k-\lambda U(z^{-1})z^{\mu+1}v_k\}b_1+\gamma u_k=0 \tag{31}$$

Now from (28) one finds that

$$\lambda v_k=\frac{C(z^{-1})}{D(z^{-1})}y_k-\frac{C(z^{-1})A(z^{-1})}{D(z^{-1})B(z^{-1})}z^{-\mu}u_k \tag{32}$$

Thus the desired minimum variance controller $G_{cA}(z)=u(z)/y(z)$ is found to be

$$G_{cA}(z)=-\frac{A(z^{-1})H(z^{-1})}{zB(z^{-1})C(z^{-1})U(z^{-1})+(\gamma/b_1)A(z^{-1})D(z^{-1})} \tag{33}$$

where $H(z^{-1})=D(z^{-1})-C(z^{-1})U(z^{-1})$.

Setting $\gamma=0$ we obtain the *pure minimum output variance* controller. In many cases the process and noise characteristic polynomials are the same i.e. $C(z^{-1})=A(z^{-1})$. If we have zero pure delay ($\mu=0$), then $U(z^{-1})=1$ and $H(z^{-1})=C(z^{-1})-D(z^{-1})$. In this case the $\gamma=0$ controller is

$$G_{cA}(z)=-\frac{A(z^{-1})\{D(z^{-1})-C(z^{-1})\}}{zB(z^{-1})C(z^{-1})} \tag{34}$$

This is precisely the controller implemented in [25].

In [26] the following model

$$\begin{aligned}&y_{k+\mu+1}+a_1^*y_k+\ldots+a_n^*y_{k-n+1}\\&=b_o^*(u_k+b_1^*u_{k-1}+\ldots+b_{\mu+n-1}^*u_{k-\mu+1-n})+w_{k+\mu+1}^*\end{aligned} \tag{35}$$

is employed which is appropriately derived from (28). Obviously (35) is again of the form (28) with

$$A(z^{-1})=1-z^{-(\mu+1)}A^*(z^{-1});A^*(z^{-1})=a_1^*+\ldots+a_n^*z^{-(n-1)}$$

$$B(z^{-1})=z^{-1}B^*(z^{-1});B^*(z^{-1})=b_o^*\{1+b_1^*z^{-1}+\ldots+b_{\mu+n-1}^*z^{-(\mu+n-1)}\}$$

$$C(z^{-1})=A(z^{-1}),\ D(z^{-1})=1$$

Therefore the controller (34) is directly applicable with $A(z^{-1})=C(z^{-1})$ and gives

$$G_{cA}(z)=-\frac{D(z^{-1})-A(z^{-1})}{zB(z^{-1})}z^{\mu+1}=-\frac{A^*(z^{-1})}{B^*(z^{-1})} \tag{36}$$

This is exactly the controller implemented in [26].

All the above controllers do not satisfy the zero-offset condition $\lim_{z\to 1} G_{cA}(z)=\infty$ in the case of a proportional acting process. To make these controllers zero-offset ones one must add a pole at z=1 (i.e. an integral term). More freedom results if the controller transfer function is multiplied by the PI term $G_{PI}(z)=1+\lambda/(z-1)$. Clearly for $\lambda\neq 0$ we have

$$\lim_{z\to 1} G_{c,total}(z)=\lim_{z\to 1} G_{PI}(z)G_{cA}(z)=\infty$$

If in spite of the addition of the z=1 pole, this condition is not satisfied, we add more poles at z=1.

Alternatively, one can start directly to design a zero-offset controller by minimizing the criterion

$$J'=E\{(y_{k+\mu+1}-r_k)^2+\gamma(u_k-\bar{u}_k)^2\}$$

where $\bar{u}_k=\{A(1)/B(1)\}r_k=r_k/K$ is the value of u_k for $y_k=r_k$ (the zero offset case). Working as above we find the zero-offset controller

$$u_k=G_{c1}(z)y_k+G_{c2}(z)r_k$$

where

$$G_{c1}(z)=zH(z^{-1})\{D(z^{-1})-C(z^{-1})\}/M(z^{-1})$$

$$G_{c2}(z)=A(z^{-1})D(z^{-1})(1+\gamma/b_1K)/M(z^{-1})$$

$$M(z^{-1})=zB(z^{-1})C(z^{-1})U(z^{-1})+(\gamma/b_1)A(z^{-1})D(z^{-1})$$

4.2. Multi-Input Multi-Output Controllers

When the process under control has multiple inputs and multiple outputs we use the matrix model [27,28]

$$\underline{A}(z^{-1})\underline{y}_k=\underline{B}(z^{-1})z^{-\mu}\underline{u}_k+\underline{D}(z^{-1})\underline{v}_k \tag{37}$$

where $\underline{y}_k$, $\underline{u}_k$, $\underline{v}_k$ are vectors, and $\underline{A}(z^{-1})$, $\underline{B}(z^{-1})$, $\underline{D}(z^{-1})$ matrix-valued polynomials of appropriate sizes:

$$\underline{A}(z^{-1})=\sum_{i=0}^{n}\underline{A}_i z^{-i},\quad \underline{B}(z^{-1})=\sum_{i=0}^{n}\underline{B}_i z^{-i},\quad \underline{C}(z^{-1})=\sum_{i=0}^{n}\underline{C}_i z^{-i}$$

with $\underline{A}_o=I$ (unit matrix). Here, the generalized variance criterion to be minimized takes the form

$$J=\underline{E}\{\tilde{\underline{y}}^T_{k+\mu+1}\tilde{\underline{y}}_{k+\mu+1}+\tilde{\underline{u}}^T_k\underline{\Gamma}\tilde{\underline{u}}_k\} \tag{38}$$

where $\tilde{\underline{y}}_{k+\mu+1}=\underline{y}_{k+\mu+1}-\underline{r}_k$, $\tilde{\underline{u}}_k=\underline{u}_k-\bar{\underline{u}}$, and $\bar{\underline{u}}_k=B^{-1}(1)A(1)\underline{r}_k$ is the offset steady state value of $\underline{u}_k$.

To determine the controller we work as in the single-input single-output case [23,28], i.e. we write

$$\underline{y}_{k+\mu+1}=\underline{A}^{-1}(z^{-1})\{\underline{B}(z^{-1})z\underline{u}_k+\underline{H}(z^{-1})\underline{v}_k\}+\underline{U}(z^{-1})z^{\mu+1}\underline{v}_k$$

with

$$\underline{U}(z^{-1})=I+\underline{U}_1z^{-1}+\ldots+\underline{U}_\mu z^{-\mu};\ \underline{H}(z^{-1})=\underline{H}_o+\underline{H}_1z^{-1}+\ldots+\underline{H}_{n-1}z^{-(n+1)}$$

$$\underline{D}(z^{-1})=\underline{A}(z^{-1})\underline{U}(z^{-1})+z^{-(\mu+1)}\underline{H}(z^{-1})$$

Solving $\partial J/\partial\underline{u}_k=0$ gives

$$\begin{aligned}\underline{u}_k=&\{\underline{U}(z^{-1})\underline{D}^{-1}(z^{-1})\underline{B}(z^{-1})z+(\underline{B}_1^T)^{-1}\underline{\Gamma}\}^{-1}.\\ &.\{I+(\underline{B}_1^T)^{-1}\underline{\Gamma}\ \underline{B}^{-1}(1)\underline{A}(1)\underline{r}_k\\ &-\underline{A}^{-1}(z^{-1})\underline{H}(z^{-1})\underline{D}^{-1}(z^{-1})\underline{A}(z^{-1})\underline{y}_k\}\end{aligned} \tag{39}$$

For $\underline{\Gamma}=0$, we obtain the pure minimum output variance controller

$$\underline{u}_k = \frac{B^{-1}(z^{-1})z^{-1}}{1-z^{-(\mu+1)}}\{A(z^{-1})(\underline{r}_k-\underline{y}_k)+|\underline{D}(z^{-1})-\underline{H}(z^{-1})|\hat{v}_k\} \quad (40a)$$

$$\hat{v}_k = \underline{D}^{-1}(z^{-1})\{\underline{A}(z^{-1})\underline{y}_k-\underline{B}(z^{-1})z^{-\mu}\underline{u}_k\} \quad (40b)$$

4.3. Self-Tuning Controller

Self-tuning (or *self-optimizing*) controllers constitute a subclass of adaptive controllers and involve three stages, namely: (i) *parameter identification*, (ii) *controller computation*, and (iii) *adjustment of the controller*. These controllers are able to adapt their parameters, when the process characteristics vary, in an automatic way by means of an estimator [22-28].

A survey of parameter estimation (identification) techniques, appropriate for the model (28) are presented in [29]. These techniques are:

- Conventional least squares
- Generalized least squares
- Modified least squares
- Maximum likelihood
- Bayesian procedure
- Simplified estimators

Just for illustration we present here the conventional least squares estimator for the model (35), which is rewritten in the form

$$\underline{\tilde{y}}_k = M_k\Theta + w_k^*, \quad \tilde{y}_k = y_k - b_o^* u_{k-\mu-1} \quad (41)$$

where

$$M_k = [-y_{k-\mu-1} - \dots - y_{k-\mu-n}; b_o^* u_{k-\mu-2}, \dots, b_o u_{k-2\mu-n}] \quad (42a)$$

$$\Theta^T = [a_1^*, a_2^*, \dots, a_n^*; b_1^*, b_2^*, \dots, b_{\mu+n-1}^*] \quad (42b)$$

Applying the sequential (on-line) least-squares estimator {eqn. (3) of [29] with $P_k^{-1} = \Sigma_k$} one obtains

$$\hat{\Theta}_{k+1} = \hat{\Theta}_k + \frac{\Sigma_k M_{k+1}^T(\tilde{y}_{k+1} - M_{k+1}\hat{\Theta}_k)}{1 + M_{k+1}\Sigma_k M_{k+1}^T}, \quad \hat{\Theta}_o = 0 \quad (43a)$$

$$\Sigma_{k+1} = \Sigma_k - \Sigma_k M_{k+1}^T M_{k+1}\Sigma_k / (1 + M_{k+1}\Sigma_k M_{k+1}^T), \quad \Sigma_o = \sigma_o I \quad (43b)$$

where σ_o is a large number.

Since w_k^* is a sequence uncorrelated with M_k, this estimator is unbiased. If the coefficient b_o^* is also unkown we carry out the multiplications in (35) and estimate the parameters b_o^*, $b_o^*b_1^*,\ldots,b_o^*b_{\mu+n-1}^*$. If the process M_k is stationary, then we can use the simplified estimator {eqn (51) of [29]}:

$$\hat{\theta}_{k+1}=\hat{\theta}_k+\frac{1}{k+1}SM_{k+1}^T(\tilde{y}_{k+1}-M_{k+1}\hat{\theta}_k)$$

where $S=j_o\Sigma_k$ for some $j_o<k$. Clearly, this estimator does not need the recursive computation of the matrix gain Σ_k for any k.

Having determined by some estimator the estimates $\hat{\theta}_k$ of the process parameters, one can implement the minimum variance controller. For example, the controller (36) becomes

$$\hat{G}_{cA}(z)=-\hat{A}_k^*(z^{-1})/\hat{B}_k^*(z^{-1}) \tag{44}$$

Similar self-tuned controllers can be deviced for multi-input multi-output systems as well. That the self-tuning controller converges to the truly optimal one, follows from the efficiency and convergence of the estimator to the true parameter values.

A useful and fast suboptimal controller for single-input multi-output systems is implemented in [26]. In this case if $y_k^1, y_k^2,\ldots,y_k^p$ are the p-outputs, the feedback is closed using a weighted sum of the form:

$$y_k=\sum_{j=1}^{p}\rho_j y_k^j \tag{45}$$

Since the outputs are interdependent, minimizing the variance of y_k does not in general give the truly minimum variance of each y_k^j. In fact, choosing constant weights ρ_j may lead some outputs far from the desired set-points. The situation is improved by assigning dynamic weights of the type

$$\rho_j=c_jV_k^j/V_k,\quad V_k=\sum_{j=1}^{p}c_jV_k^j \tag{46a}$$

where c_j are preselected constants and V_k^j is the variance of y_k^j determined recursively by

$$V^j_{k+1}=V^j_k+\{(y^j_k)^2-(y^j_{k-N-1})^2\}/(N+1) \tag{46b}$$

Choosing the ρ_j's in this way ensures to a certain extend that a dynamical search about the minimum of the variance of each individual output is made. This technique can easily be extended to multi-input and multi-output systems [26].

5. DIGITAL STATE FEEDBACK CONTROLLERS

This class of controllers is based on the state-space representation of the system, namely

$$\underline{x}_{k+1}=\underline{A}\ \underline{x}_k+\underline{B}\ \underline{u}_k,\quad \underline{y}_k=\underline{C}\ \underline{x}_k+\underline{D}\ \underline{u}_k \qquad (x_o \text{ given}) \tag{47}$$

where $\underline{u}_k$ is the p-dimensional control vector, $\underline{x}_k$ is the n-dimensional state vector, $\underline{y}_k$ is the m-dimensional output vector, and $\underline{A},\underline{B},\underline{C},\underline{D}$ are matrices of appropriate dimensions. The block diagram representation of this model is shown in Fig. 7.

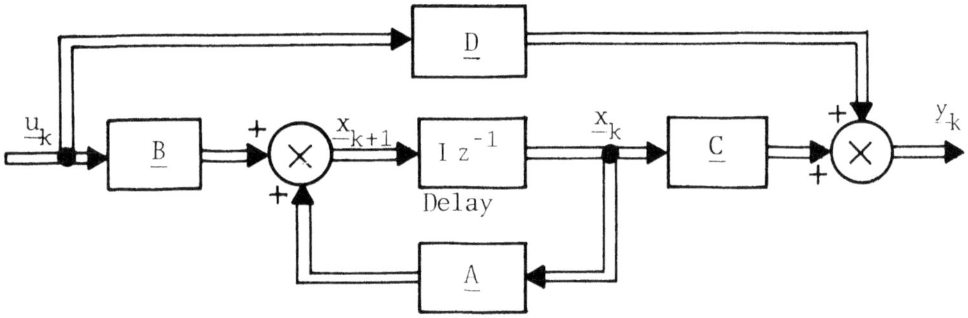

Fig. 7. Block diagram of state-space model.

The discretization of a continuous-time state-space model can be done using the methods described in Sec. 3.1.
The solution of (47) is

$$\underline{x}_k=\underline{A}^k\underline{x}_o+\sum_{i=0}^{k-1}\underline{A}^{k-i-1}\underline{B}\ \underline{u}_i,\quad \underline{y}_k=\underline{C}\ \underline{x}_k+\underline{D}\ \underline{u}_k \tag{48}$$

and the corresponding transfer function is

$$\underline{F}(z)=\underline{C}(z\underline{I}-\underline{A})^{-1}\underline{B}+\underline{D} \tag{49}$$

where $(z\underline{I}-\underline{A})^{-1}=\mathrm{adj}(z\underline{I}-\underline{A})/|z\underline{I}-\underline{A}|$ and $|z\underline{I}-\underline{A}|=\det(z\underline{I}-\underline{A})$.

Integral control can be inserted by augmenting the state vector with an additional state x_{n+1}. In the continuous-time case this is given by $x_{n+1}=\int x_1 dt$ or $dx_{n+1}/dt=x_1$, and so in the discrete-time case we have $x_{n+1,k+1}=x_{n+1,k}+Tx_{1,k}$.

Here, the following state feedback controllers will be briefly described:

(i) Linear quadratic controller
(ii) Eigenvalue controller
(iii) Dead beat controller
(iv) Model Matching Controller
(v) Controller with Observer
(vi) Minimum variance controller
(vii) Model-following controller

5.1. Optimal Linear Quadratic Controller

This is a proportional-type negative-state-feedback time-varying controller and has the form [34].

$$\underline{u}_{N-k}=\underline{G}_{N-k}\ \underline{x}_{N-k},\quad k=1,2,\ldots,N \tag{50a}$$

where

$$\underline{G}_{N-k}=-\{\underline{R}+\underline{B}^T\underline{P}_{k-1}\underline{B}\}^{-1}\underline{B}^T\underline{P}_{k-1}\underline{A} \tag{50b}$$

$$\underline{P}_k=(\underline{A}+\underline{B}\ \underline{G}_{N-k})^T\underline{P}_{k-1}(\underline{A}+\underline{B}\ \underline{G}_{N-k})+\underline{G}_{N-k}^T\underline{R}\ \underline{G}_{N-k}+\underline{Q} \tag{50c}$$

with $\underline{P}_o=\underline{Q}_o$. This controller minimizes the quadratic cost function

$$J=\frac{1}{2}\underline{x}_N^T\underline{Q}_o\underline{x}_N+\frac{1}{2}\sum_{k=0}^{N-1}(\underline{x}_k^T\underline{Q}\ \underline{x}_k+\underline{u}_k^T\underline{R}\ \underline{u}_k) \tag{51a}$$

where $\underline{Q}\geqslant 0$, $\underline{Q}_o\geqslant 0$, $\underline{R}>0$ are weighting matrices. The resulting optimal cost is

$$J_{N-k}^o(x_{N-k})=\frac{1}{2}\underline{x}_{N-k}^T\underline{P}_k\underline{x}_{N-k} \tag{51b}$$

To accomodate in the model the constant set-point (reference) variable $\underline{r}_k$ and a disturbance $\underline{w}_k$ we use a state space generating model for $\underline{r}_k$ and $\underline{w}_k$, and utilize as total input the error $\underline{e}_k=\underline{r}_k-\underline{y}_k-\underline{w}_k$ and as state vector the state vectors of the original and two subsidiary models [30]. An alternative way is to introduce integral output control i.e. $x_{n,k+1} = x_{n,k} + \underline{E}\,\underline{y}_k$ where $\underline{E}$ is a diagonal matrix.

5.2. Eigenvalue Controller

Here we shall describe an indirect technique in which the system is assumed in its Input-Luenberger canonical form [30], namely

$$\underline{x}_{k+1} = \underline{A}\,\underline{x}_k + \underline{B}\,u_k,\quad \underline{A} = [\underline{A}_{ij}],\ i,j = 1,2,\ldots,p \tag{52a}$$

where $\underline{x}_k$ is n-vector, $\underline{u}_k$ is p-vector, and

$$\underline{A}_{ii} = \begin{bmatrix} 0 & 1 & \cdots & 0 \\ 0 & 0 & 1 \cdots & 0 \\ \hline -(a_{ii})_o & \cdots & & -(a_{ii})_{\sigma_i - 1} \end{bmatrix} = \begin{bmatrix} \bar{A}_{ii} \\ \hline \tilde{A}_{ii} \end{bmatrix}, \tag{52a}$$

$$\underline{A}_{ij} = \begin{bmatrix} 0 \\ \hline -(a_{ij})_o, \cdots -(a_{ij})_{\sigma_{j-1}} \end{bmatrix} = \begin{bmatrix} \bar{A}_{ij} \\ \hline \tilde{A}_{ij} \end{bmatrix} \tag{52b}$$

$$\underline{B}_i = \begin{bmatrix} 0 \ldots 0 \ldots 0 \\ \cdots\cdots\cdots \\ 0 \ldots 0 \ldots 0 \\ \hline 0 \ldots 1 \ldots 0 \end{bmatrix} = \begin{bmatrix} \bar{B}_i \\ \hline \tilde{B}_i \end{bmatrix} \tag{52c}$$

where σ_j is defined in Luenberger [31].

The problem is to select the gain matrix of the linear static state feedback controller $\underline{u}_k = \underline{F}\,\underline{x}_k$ so as to drive the eigenvalues of the system, i.e. the solution of $|z\underline{I} - \underline{A}_c| = 0$, where $\underline{A}_c = \underline{A} + \underline{B}\,\underline{F}$ is the closed-loop matrix, to desired positions $\lambda_1^d, \lambda_2^d, \ldots, \lambda_n^d$.

Let $\underline{\Lambda}$ be any matrix that has the desired eigenvalues. Then clearly $\underline{A} + \underline{B}\,\underline{F}$ has the same eigenvalues $\lambda_1^d, \ldots, \lambda_n^d$ as Λ, if there exists a nonsingular linear transformation $\underline{T}$ such that $\underline{A} + \underline{B}\,\underline{F} = \underline{T}\,\underline{\Lambda}\,\underline{T}^{-1}$. To solve this equation for $\underline{F}$ and $\underline{T}$ we write it in the decomposed form

$$\begin{bmatrix} \bar{A} \\ \hline \tilde{A} \end{bmatrix} + \begin{bmatrix} \bar{B} \\ \hline \tilde{B} \end{bmatrix} \underline{F} = \begin{bmatrix} \bar{T} \\ \hline \tilde{T} \end{bmatrix} \underline{\Lambda}\,\underline{T}^{-1},\quad \underline{\Lambda} = \begin{bmatrix} \Lambda_1 & & 0 \\ & \ddots & \\ 0 & & \underline{\Lambda}_p \end{bmatrix} \tag{53}$$

where $\bar{A}=[\bar{A}_{ij}]$, $\tilde{A}=[\tilde{A}_{ij}]$, $T=[T_{ij}]$, dim Λ_i=dimA_{ii}, dimT_{ij}= dimA_{ij}, and

$$T_{ij}=\begin{bmatrix}\bar{T}_{ij}\\ \hdashline \tilde{T}_{ij}\end{bmatrix},\quad \bar{T}_{ij}=\begin{bmatrix}(T_{ij})_1\\ \vdots\\ (T_{ij})_{\sigma_j-1}\end{bmatrix},\quad \tilde{T}_{ij}=(T_{ij})_{\sigma_i}$$

Then (53) gives

$$F=-\tilde{A}+\tilde{T}\,\Lambda\,T^{-1} \tag{54a}$$

with the condition $\bar{A}\,T=\bar{T}\,\Lambda$ or $\bar{A}_{ii}T_{ij}=\bar{T}_{ij}\Lambda_j$ $(j=1,2,\ldots,p)$ which has the solution

$$T_{ij}=\text{col.}\left[\pi_{ij},\pi_{ij}\Lambda_j,\ldots,\pi_{ij}\Lambda_j^{\sigma_j-1}\right] \tag{54b}$$

where $\pi_{ij}=[\pi_{i1},\pi_{i2},\ldots,\pi_{ip}]$ is an arbitrary σ_j-row vector. Using (54b) in (54a) we finally obtain

$$F=-\tilde{A}+\pi\,\Lambda^*T^{-1},\quad \pi=\begin{bmatrix}\pi_1 & & 0\\ & \ddots & \\ 0 & & \pi_p\end{bmatrix},\quad \Lambda^*=\begin{bmatrix}\Lambda^{\sigma_1}\\ \vdots\\ \Lambda^{\sigma_p}\end{bmatrix} \tag{55}$$

For examples see [30,32].

5.3. Dead-Beat Controller

For dead-beat control F in $u_k=Fx_k$ must be chosen such that $A_c=(A+B\,F)^n=0$ where $n=\dim A_c$.

Since here

$$|zI-A_c|=\alpha_n+\alpha_{n-1}z+\ldots+\alpha_1 z^{n-1}+\alpha_o z^n=0$$

the Cauley-Hamilton theorem gives

$$\alpha_n I+\alpha_{n-1}A_c+\ldots+\alpha_1 A_c^{n-1}+\alpha_o A_c^n=0$$

From the deadbeat property $A_c^n=0$, it follows that $\alpha_1=\alpha_2=\ldots=\alpha_n=0$, and the characteristic equation becomes

$$|zI-A_c|=\alpha_o z^n=0$$

This means that a control loop with dead beat behavior is characterized by a multiple-pole at $z=0$ of order n.

For the calculation of modal (eigenvalue) dead-beat controllers the reader is referred to [30,p.51]. The deadbeat controller drives the system from any initial state $x_o \neq 0$ to the zero final state $x_n=0$ (state deadbeat controller). Note that the deadbeat controller of Section 3.5. leads to the same closed-loop characteristic equation $z^n=0$, and drives the system to the zero state after n steps, too.

5.4. Model Matching Controller

The model matching controller $\underline{u}_k=\underline{F}\ \underline{x}_k+\underline{G}\ \underline{\upsilon}_k$ is such that the closed-loop system

$$\underline{x}_{k+1}=(\underline{A}+\underline{B}\ \underline{F})\underline{x}_k+\underline{B}\ \underline{G}\ \underline{\upsilon}_k\ ,\ \underline{y}_k=(\underline{C}+\underline{D}\ \underline{F})\underline{x}_k+\underline{D}\ \underline{G}\ \underline{\upsilon}_k \qquad (56)$$

has the same input-output transfer function as that of a desired model (prototype):

$$\underline{\hat{x}}_{k+1}=\underline{\hat{A}}\ \underline{\hat{x}}_k+\underline{\hat{B}}\ \underline{\upsilon}_k,\ \underline{y}_k=\underline{\hat{C}}\ \underline{\hat{x}}_k+\underline{\hat{D}}\ \underline{\upsilon}_k \qquad (57)$$

Let $\underline{T}$ be a nonsingular similarity state transformation matrix. Then using the identity

$$\{z\underline{I}-(\underline{A}+\underline{B}\ \underline{F})\}^{-1}=\underline{T}\{zI-T^{-1}(A+BF)T\}^{-1}\ T^{-1}$$

we see that the transfer function of (56), namely

$$\underline{H}_c(z)=(\underline{C}+\underline{D}\ \underline{F})\{z\underline{I}-(\underline{A}+\underline{B}\ \underline{F})\}^{-1}\underline{B}\ \underline{G}+\underline{D}\ \underline{G},$$

is also the transfer function of the system

$$\begin{aligned} x'_{k+1}&=\underline{T}^{-1}(\underline{A}+\underline{B}\ \underline{F})\underline{T}x'_k+\underline{T}^{-1}\underline{B}\ \underline{G}\ \underline{\upsilon}_k \\ \underline{y}_k&=(\underline{C}+\underline{D}\ \underline{F})\underline{T}\ \underline{x}'_k+\underline{D}\ \underline{G}\ \underline{\upsilon}_k \end{aligned} \qquad (58)$$

Thus the model matching controller design problem reduces to that of choosing $\underline{F}$, $\underline{G}$ and $\underline{T}$ such that [33]:

$$\underline{T}^{-1}(\underline{A}+\underline{B}\ \underline{F})\underline{T}=\underline{\hat{A}}\ ,\ \underline{T}^{-1}\underline{B}\ \underline{G}=\underline{\hat{B}} \qquad (59a)$$

$$(\underline{C}+\underline{D}\ \underline{F})\underline{T}=\underline{\hat{C}}\ ,\ \underline{D}\ \underline{G}=\underline{\hat{D}} \qquad (59b)$$

Let us assume that the system under control is given in the

input Luenberger form, and that (for simplicity) the prototype has the same dimensionality as the controlled system. Then decomposing the relation (59a) as {see definitions (52a-c), (53)}:

$$\bar{\underline{A}}+\bar{\underline{B}}\ \underline{F}=\bar{\underline{T}}\ \hat{\underline{A}}\ \underline{T}^{-1},\quad \bar{\underline{B}}\ \underline{G}=\bar{\underline{T}}\ \hat{\underline{B}} \tag{60a}$$

$$\tilde{\underline{A}}+\tilde{\underline{B}}\ \underline{F}=\tilde{\underline{T}}\ \hat{\underline{A}}\ \underline{T}^{-1},\quad \tilde{\underline{B}}\ \underline{G}=\tilde{\underline{T}}\ \hat{\underline{B}} \tag{60b}$$

Clearly $\tilde{B}=I$, and so (60b) gives

$$\underline{F}=-\tilde{\underline{A}}+\tilde{\underline{T}}\ \hat{\underline{A}}\ \underline{T}^{-1},\quad \underline{G}=\tilde{\underline{T}}\ \hat{\underline{B}} \tag{60c}$$

Introducing (60c) in (60a) and (59b), noting that $\bar{B}=0$, gives

$$\begin{aligned} &\bar{\underline{A}}\ \underline{T}=\bar{\underline{T}}\ \hat{\underline{A}},\quad \bar{\underline{T}}\ \hat{\underline{B}}=0 \\ &(\underline{C}-\underline{D}\ \tilde{\underline{A}})\underline{T}+\underline{D}\ \tilde{\underline{T}}\ \hat{\underline{A}}=\hat{\underline{C}},\quad \underline{D}\ \tilde{\underline{T}}\ \hat{\underline{B}}=\hat{\underline{D}} \end{aligned} \tag{61}$$

If the linear algebraic system (61) has a solution $\underline{T}$ with $|\underline{T}|\neq 0$, then $\underline{F}$ and $\underline{G}$ in (60c) are the desired model matching controller matrices $\underline{F}$ and $\underline{G}$.

The extension to the case where the dimensionalities of the prototype and the controlled system are different is straightforward [33].

5.5. Controller With Observer

When the state of the system is not available, for closing the feedback control $\underline{u}_k=-\underline{F}\ \hat{\underline{x}}_k$ one can use an approximate (observed, reconstructed) state $\hat{x}_k$ generated by an observer of the type [35]:

$$\hat{\underline{x}}_{k+1}=\underline{A}\ \hat{\underline{x}}_k+\underline{B}\ \underline{u}_k+\underline{H}(\underline{y}_k-\hat{\underline{y}}_k),\quad \hat{\underline{y}}_k=\underline{C}\ \hat{\underline{x}}_k$$

which results in the error equation

$$\tilde{\underline{x}}_{k+1}=(\underline{A}-\underline{H}\ \underline{C})\tilde{\underline{x}}_k$$

The observer gain $\underline{H}$ must be chosen such that the matrix $(\underline{A}-\underline{H}\ \underline{C})$ has desired eigenvalues inside the unit circle, or $(\underline{A}-\underline{H}\ \underline{C})^n=0$ i.e. deadbeat, or such that to minimize a quadratic criterion. By comparing the characteristic equation of the closed-loop state feedback controller and the observer one can directly see the duality of the respective gains $\underline{F}$ and $\underline{H}$. Using this duality we conclude that to design the observer gain matrix $\underline{H}$ for the system $(\underline{A},\underline{C})$ one has to design the controller gain $-\underline{F}$ in $\underline{u}_k=-\underline{F}\ \underline{x}_k$ for (A,B) i.e.

Controller: $A, B, -F$
Observer : A^T, B^T, H^T

To reduce the order of the observer one can only reconstruct the nonmeasured states. More on observers and their μP-based Implementations can be found in the chapters of *Sinha*, and *Holmberg* and *Orava* of this book.

5.6. Minimum Variance State Controllers

Consider the stochastic state-space model

$$\underline{x}_{k+1} = \underline{A}\ \underline{x}_k + \underline{B}\ \underline{u}_k + \underline{B}'\underline{v}_k, \quad \underline{y}_k = \underline{C}\ \underline{x}_k + \underline{v}_k$$

where $\underline{v}_k$ is a zero-mean Gaussian and white noise sequence, and the generalized variance criterion

$$J_{k+1} = E\{\underline{x}_{k+1}^T \underline{Q}\ \underline{x}_{k+1} + \underline{u}_k^T \underline{R}\ \underline{u}_k\}, \quad \underline{Q} \geqslant 0, \quad \underline{R} > 0$$

which, if minimized, provides the corresponding state feed-back minimum variance controller [23].

In fact, setting $\partial J_{k+1}/\partial \underline{u}_k = 2\underline{B}^T\underline{Q}(Ax_k + Bu_k + \underline{B}'\underline{v}_k) + 2\underline{R}\ \underline{u}_k = 0$, we find

$$\underline{u}_k = -(\underline{B}^T\underline{Q}\ \underline{B}+\underline{R})^{-1}\underline{B}^T\underline{Q}(\underline{A}\ \underline{x}_k + \underline{B}'\hat{\underline{v}}_k) \tag{62}$$

where the reconstructed noise $\hat{v}_k$ is $\hat{v}_k = y_k - Cx_k$.

If the dead time is not included in the system matrix $\underline{A}$, the state equation is $\underline{x}_{k+\mu} = \underline{A}\ \underline{x}_k + \underline{B}u_k + \underline{B}'\underline{v}_k$, and the controller (62) simply takes the form

$$\underline{u}_k = -(B^T\underline{Q}\ \underline{B}+\underline{R})^{-1}\underline{B}^T\underline{Q}(\underline{A}\ \hat{\underline{x}}_{k+\mu} + \underline{A}^{\mu}\underline{B}'\hat{\underline{v}}_k) \tag{63a}$$

with the predicted state

$$\hat{\underline{x}}_{k+\mu} = E\{\underline{x}_{k+\mu} \mid k-1\} = \underline{A}^{\mu}\underline{x}_k + \sum_{i=0}^{\mu-1} \underline{A}^{\mu-i-1}\underline{B}\ \underline{u}_{k-\mu-i} \tag{63b}$$

Another version is obtained by minimizing the output-vector variance criterion

$$I_{k+\mu+1} = E\{\underline{y}_{k+\mu+1}^T \underline{S}\ \underline{y}_{k+\mu+1} + \underline{u}_k^T\ \underline{R}\ \underline{u}_k\}$$

Clearly, due to the relation $\underline{y}_k = \underline{C}\ \underline{x}_k + v_k$, the resulting con-controller has exactly the form (63a) with $\underline{Q} = \underline{C}^T\underline{S}\ \underline{C}$.

5.7. Implicit Model-Following Controller

Continuous-time model-following controllers are fully studied in Landau's book [36]. Here we will be concerned with a discrete-time implicit model following controller [37].

The system under control is

$$\underline{\xi}_{k+1}=\underline{A}_p\,\underline{\xi}_k+\underline{B}_p\underline{u}_k,\quad \underline{y}_k=\underline{C}\,\underline{\xi}_k \tag{64}$$

and an implicit prototype (model) is

$$\hat{\underline{\xi}}_{k+1}=\hat{\underline{A}}\,\underline{\xi}_k+\hat{\underline{B}}\,\hat{\underline{u}}_k,\quad \hat{\underline{y}}_k=\underline{C}\,\hat{\underline{\xi}}_k,\quad \hat{\underline{u}}_{k+1}=\underline{\Gamma}\,\hat{\underline{u}}_k \tag{65}$$

where $\hat{\underline{u}}_k$ is an uncontrolled command signal.

It is desired to make the system output follow the prototype output, and hence a performance index, involving the output error $\tilde{\underline{y}}_k=\underline{y}_k-\hat{\underline{y}}_k$, of the type

$$J_N=\frac{1}{2}\sum_{k=0}^{N}L_k$$

$$L_k=\tilde{\underline{y}}_k^T\underline{Q}_o\tilde{\underline{y}}_k+\underline{u}_k^T\underline{R}\,\underline{u}_k=\tilde{\underline{\xi}}_k^T\underline{Q}_1\tilde{\underline{\xi}}_k+\underline{u}_k^T\underline{R}\,\underline{u}_k \tag{66}$$

where $\tilde{\underline{\xi}}_k=\underline{\xi}_k-\hat{\underline{\xi}}_k$ and $\underline{Q}_1=\underline{C}^T\underline{Q}_o\underline{C}$, should be minimized.

The problem determined by (64)-(66), can be reformulated in the compact form

$$\underline{x}_{k+1}=\underline{A}\underline{x}_k+\underline{B}\,\underline{u}_k,\quad J_N=\frac{1}{2}\sum_{k=0}^{N}L_k \qquad \underline{x}_k=\begin{bmatrix}\underline{\xi}_k\\ \hat{\underline{\xi}}_k\\ \underline{u}_k\end{bmatrix}$$
$$L_k=\underline{x}_k^T\,\underline{Q}\,\underline{x}_k+\underline{u}_k^T\underline{R}\,\underline{u}_k \tag{67}$$

$$\underline{A}=\begin{bmatrix}\underline{A}_p & 0 & 0\\ \hat{\underline{A}} & 0 & \hat{\underline{B}}\\ 0 & 0 & \underline{\Gamma}\end{bmatrix},\ \underline{B}=\begin{bmatrix}\underline{B}_p\\ 0\\ 0\end{bmatrix},\ \underline{Q}=\begin{bmatrix}\underline{Q}_1 & -\underline{Q}_1 & 0\\ -\underline{Q}_1 & \underline{Q}_o & 0\\ 0 & 0 & 0\end{bmatrix}$$

This is precisely a standard optimal-quadratic control problem {see Section 5.1.}, and the resulting controller has the form (50a), namely

$$\underline{u}_{N-k}=\underline{G}_{N-k}\underline{x}_{N-k}=\left[\underline{G}^1_{N-k}\,\vdots\,\underline{G}^2_{N-k}\,\vdots\,\underline{G}^3_{N-k}\right]\underline{x}_k$$
$$=\underline{G}^1_{N-k}\underline{\xi}_{N-k}+\underline{G}^2_{N-k}\hat{\underline{\xi}}_{N-k}+\underline{G}^3_{N-k}\hat{\underline{u}}_{N-k} \tag{68}$$

where the particular gain matrices $\underline{G}^{i}_{N-k}$ (i=1,2,3) are determined from (50b), i.e.

$$\underline{G}^{1}_{N-k}=-(\underline{B}^{T}_{p}\underline{P}_{11}\underline{B}_{p}+\underline{R})^{-1}(\underline{B}^{T}_{p}\underline{P}_{11,k-1}\underline{A}_{p}+\underline{B}^{T}_{p}\underline{P}_{12,k-1}\hat{\underline{A}})$$

$$\underline{G}^{2}_{N-k}=0 \tag{69}$$

$$\underline{G}^{3}_{N-k}=-(\underline{B}^{T}_{p}\underline{P}_{11}\underline{B}_{p}+\underline{R})^{-1}(\underline{B}^{T}_{p}\underline{P}_{12,k-1}\hat{\underline{B}}+\underline{B}^{T}_{p}\underline{P}_{13,k-1}\underline{\Gamma})$$

For an example regarding the application of the controller (68)-(69) to the lateral axis stability augmentation problem see [37].

6. EXAMPLES OF MICROPROCESSOR-BASED CONTROLLERS

In this Section we shall present a number of additional microprocessor-based (implemented) controllers which fall in the categories described in Sections 3 to 5.

6.1. DC Servo μController

A useful μP-based digital controller for DC servo systems was designed and built around the 12-bit single-chip IM6100 microprocessor which recognizes the instruction set of the *Digital Equipment Corporation* (DEC) PDP-8 minicomputer [38]. This system called *UNICO System IV CNC* (computer numeric control) offers the advantages of a programmable controller with the power and flexibility of a general purpose minicomputer. The hardware and software of the system are modularly designed to allow their configuration modification in accordance to particular problem requirements. The PDP-8 instruction set is simple but powerful and has extensive software, and flexible I/O instructions. The 12-bit μ-processor used has some advantages over the 8-bit and 16-bit processors.

A typical System IV CNC involves a CPU, an EOC (extended option control), a RAM, a PDI (parallel data interface), a DIK (data input keyboard), a DOD (data output display), a PTR (paper tape reader), a (CEM) cable extender module, and the power supply. Up to twenty one additional I/O modules can also be plugged into the parallel bus in reserved areas, namely the DIS (data input station) and DOS (data output station). The EDB (extended data bus) is used for remote operation of parallel data devices and for expanding the parallel data I/O capacity. The software involves the RTS-II (real time operating system) the AP (application program) and the

DP (diagnostic programs).

Each DC servo drive consists of a DC motor used to drive the load in several control applications, an SCR (silicon controlled rectifier) amplifier used for power amplification, a transformer used to isolate the drive from the line voltage, and a drive interface module used to control the amplifier. An extended description of the above units can be found in [38]. This programmable μController can be used to impement a variety of desired servo control algorithms.

6.2. Self-Tuning and Self-Selecting Controllers

A universal self-tuning μController based on the control law (38), the sequential estimator (43a,b), and the weighted output-sum feedback determinated by (45) and (46a,b), was provided in [26]. Two different implementations were made; one based on the M6800 8-bit μP and another based on the Intel 8085 8-bit μComputer

In the first (8-bit computation), all variables x were assumed to be in the interval $-1.0 \leqslant x \leqslant +1.0$ (i.e. FF to 7F hexadecimal). Since this gives rise to overflow when computing the inverse of a number such as $K=1/(1+M_{k+1}\Sigma_k M_{k+1}^T)$ {see (43a)}, the following approximation is used

$$K=1.0-M_{k+1}\Sigma_k M_{k+1}^T$$

where 1.0 equals to 7F hexadecimal. In this way the overflow difficulty is overcome, since there is no overflow in multiplication.

The flowchart of the self-tuning controller is shown in Fig. 8. The initialization routine clears and initializes the RAM, introduces the desired weighting feedback factors ρ_j {viz. (45) and (46a,b)}, and initiates the data aquisition system. In order for the estimator (43a,b) to have enough data to provide accurate values of the system parameters $\underline{\theta}$ {viz (43a,b)}, the identification routine begins a number of sampling periods after the start-up of the regulator algorithm (routine) from rest. During the time between the initial start-up and the control signal provided by the self-tuning algorithm (typically 10 sampling intervals), a fixed preselected control is fed by the controller to the process under control. The overall program needs about 0.5 k bytes of ROM, and 80 bytes of RAM for storing intermediate values, and is completed in less than 25 msec, which is sufficient for many processes (except the very fast ones).

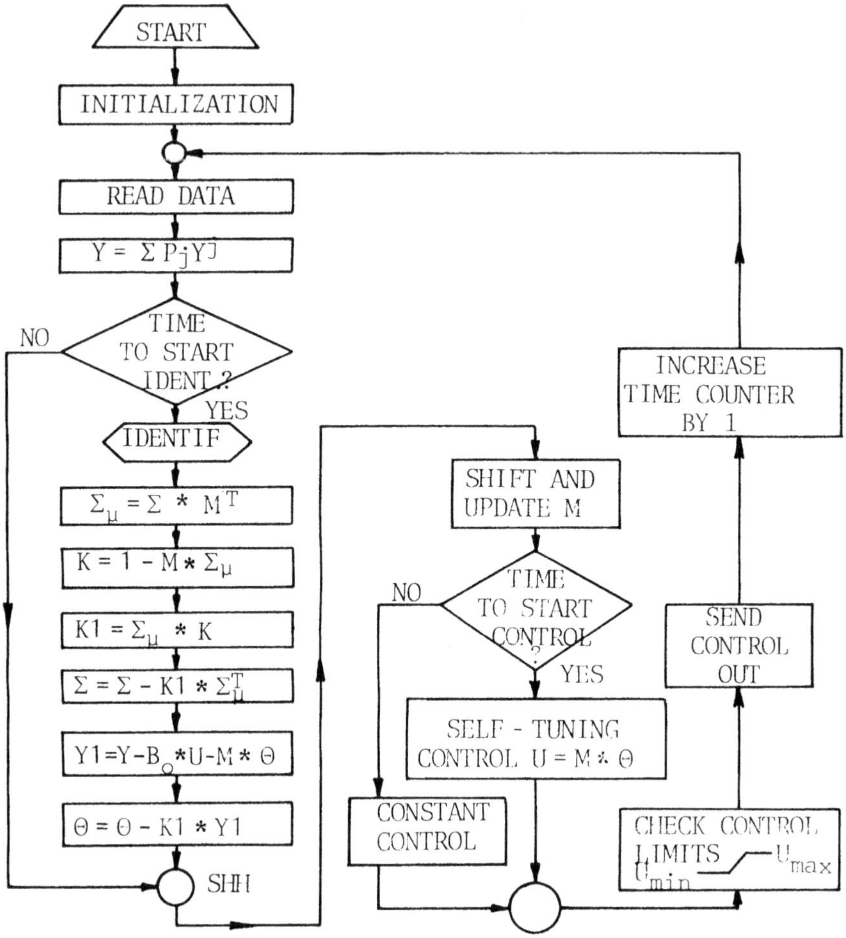

Fig. 8. Flowchart of a self-tuning digital controller.

Since the 8-bit computation restricts the achievable accuracy of the results, the controller program was rewritten using 32-bit floating-point arithmetic around the Intel 8085 μP. In this case the floating-point numbers lie in the range $\pm 3.4 \times 10^{38}$ and the integers in the range $\pm 215 \times 10^{7}$. The input data enters in 16-bit form and the output in an 8-bit word. The details of this implementation can be found in [26]. Less than 4kbytes of ROM and 0.5 kbytes of RAM are needed. The overall computation time is about 230 msec, which is too long for many practical cases (electric generators, radar systems, etc). To overcome this difficulty a modification of the controller algorithm was implemented. In this version the control is computed and output for every sampling cycle, but the identification routine is performed every 15 samples.

Clearly, the identification of the system parameters at a slower rate than the control has the additional advantage of not taking into account temporal disturbances, and thus convergence to the true values without oscillations is achieved. This improves considerably the performance of the overall controller. The computation time of this version is less than 35 msec.

A second case study regarding the μP-based implementation of self-tuning and self-selecting controllers is described in [39]. Firstly, a useful stability property (in the Lyapunov sense) of linear self-tuning regulators is established, which justifies to a certain extend the separation of the adaptive control problem into an identification and a controller design problem.

Secondly, the gain adaptation problem through identification is treated, and thirdly the gain adaptation is achieved by adjusting the model (prototype) parameters using the controller. The process is then adjusted to the model. This is actually a model reference technique. The cases of a sinusoidal perturbation system, and of the multiparameter adjustment problem are also discussed. Finally a self-selecting PI controller is studied and designed.

The implementation of the above controllers was made using a small hybrid computer system involving an *EAI Mini Ac* analog computer, an extension containing several adaptation (multiplier-integrator-multiplier) units, and an *Intellec 80* microcomputer development system (with A/D-D/A converters, interrupt unit, and clock). About 30 standard *macros* have been developed, which help to code the adaptive controllers right from the block diagrams. These Macros are expanded by Intel's Macro Assembler. When short sampling intervals are used, interrupt driven programs must be employed. In [39] a set of examples (model-reference control of a generator, state variable filter identification, self-selecting control, self-tuning controllers, etc) are included.

6.3. Optimal and Multilevel Controllers

We start by describing the results of [40] regarding the μP-based optimal control of an induction motor-speed system. As it is shown, the three basic issues which must be taken into account in the system design are: the input dead time due to signal processing in the μP, the input variable constraint due to avoiding commutation failure of the inverter, and the state variable constraint due to limited capacities of thyristors or diodes.

The static Scherbius induction motor system modelled and employed in the optimal μP-based control study is shown in Fig.9.

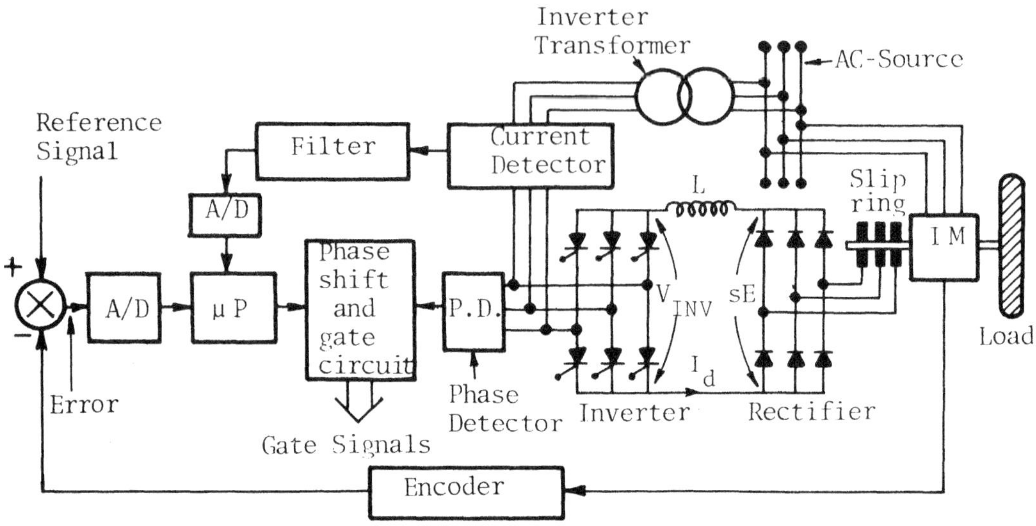

Fig. 9. Structure of the speed control system.

It consists of a wound rotor induction motor, a static rectifier, a 3-phase separately excited inverter and an inverter transformer. The control variable of the induction motor is the phase angle γ of the separately excited inverter. For high system reliability a cycloconverter can be used in place of the rectifier and inverter set.

The continuous-time system state-space model is $\dot{\underline{Y}}(t)=\underline{A}'\underline{Y}(t)+\underline{B}_1'\upsilon(t-T_A)+\underline{B}_2'\underline{d}(t)$, where $\underline{Y}=[y_1,y_2]^T$, $\underline{d}=[d_1,d_2]^T$ (the system disturbance),

$$\underline{A}'=\begin{bmatrix} 0 & 4q/\tau_m \\ -1/\tau_e & -(1+q)/\tau_e \end{bmatrix}, \underline{B}_1'=\begin{bmatrix} 0 \\ a/\tau_e \end{bmatrix}, \underline{B}_2'=\begin{bmatrix} -1/\tau_m & 0 \\ 0 & 1/\tau_e \end{bmatrix}$$

Here τ_e is the electrical time constant, q is the circuit constant, τ_m is the mechanical time constant, $y_1=\omega/\omega_s$ (ω,ω_s are angular velocity and synchronous angular velocity of the motor respectively), y_2=i=direct choke current/maximum choke current, $d_1=\tau_L$=load torque, $d_2=1.0$, and $a\upsilon=s_o(\beta V_{INV})$, s_o being the no load slip $s=(\omega_s-\omega)/\omega_s$. The dead time T_A, which is the processing time of the controller to generate the control input, is assumed small compared to the sampling period.

The input and state constraints are assumed of the form

$$V_{min} \leq \upsilon \leq V_{max}, \quad Y_{2min} \leq y_2 \leq Y_{2max}$$

Discretizing the system model (with sampling period $T \geq T_A$) yields

$$\underline{Y}_{k+1} = \underline{A}\underline{Y}_k + \underline{B}_1 \upsilon_{k-1} + \underline{B}_2 d_k, \quad \underline{W}_k = \underline{C}\,\underline{Y}_k, \quad \underline{C} = I$$

The reference (desired) signal is $\underline{R}_k = [R_{1,k}, 0]^T$ and the error signal is $\underline{e}_k = \underline{R}_k - \underline{W}_k$. Introducing new state variables as

$$\underline{X}_{1,k} = \underline{e}_{k-2}, \underline{X}_{2,k} = \underline{e}_{k-1} - \underline{e}_{k-2}, \quad \underline{X}_{3,k} = (\underline{e}_k - \underline{e}_{k-1}) - (\underline{e}_{k-1} - \underline{e}_{k-2})$$

$$\underline{Z}_{1,k} = (\underline{R}_k - \underline{R}_{k-1}) - (\underline{R}_{k-1} - \underline{R}_{k-2}), \underline{Z}_{2,k} = (\underline{R}_{k+1} - \underline{R}_k) - (\underline{R}_k - \underline{R}_{k-1})$$

$$\underline{D}_k = (\underline{d}_k - \underline{d}_{k-1}) - (\underline{d}_{k-1} - \underline{d}_{k-2}), \, u_k = (\upsilon_k - \upsilon_{k-1}) - (\upsilon_{k-1} - \upsilon_{k-2}),$$

and observing that $\underline{e}_k = \underline{X}_{1,k} + 2\underline{X}_{2,k} + \underline{X}_{3,k}$, one easily arrives at the model

$$\underline{X}^o_{k+1} = \underline{A}_o \underline{X}^o_k + \underline{B}_{o1} u_{k-1} + \underline{B}_{11} \underline{Z}_k + \underline{B}_{o2} \underline{D}_k$$

where $\underline{Z}_k = [\underline{Z}^T_{1,k}, \underline{Z}^T_{2,k}]^T, \quad \underline{X}^o_k = [X_{1,k}, X_{2,k}, X_{3,k}]^T,$

$$\underline{A}_o = \begin{bmatrix} I & I & 0 \\ 0 & I & I \\ 0 & 0 & \underline{A} \end{bmatrix}, \underline{B}_{o1} = \begin{bmatrix} 0 \\ 0 \\ \underline{B}_1 \end{bmatrix}, \underline{B}_{11} = \begin{bmatrix} 0 & 0 \\ 0 & 0 \\ -A & I \end{bmatrix}, \underline{B}_{o2} = \begin{bmatrix} 0 \\ 0 \\ -\underline{B}_2 \end{bmatrix}$$

Assuming that desired signals and sisturbances are step and/ or ramp signals yields $\underline{Z}_{1,k} = 0$, $\underline{Z}_{2,k} = 0$ and $\underline{D}_k = 0$.

In order to deal with the state constraints two new state variables $x_{p10,k}$ and $x_{p11,k}$ are introduced as

$$x_{p10,k+1} = x_{p10,k} + (-y_2 + Y_{2max})^2 H(h_1), x_{p10,0} = 0$$

$$x_{p11,k+1} = x_{p11,k} + (y_{2,k} - Y_{2min})^2 H(h_2), x_{p11,0} = 0$$

$$h_1 = -y_2 + Y_{2max}, h_2 = y_2 - Y_{2min}$$

$$H(h_1) = \begin{cases} 0 \text{ if } h_1 \geq 0 \\ 1 \text{ if } h_1 < 0 \end{cases}, H(h_2) = \begin{cases} 0 \text{ if } h_2 \geq 0 \\ 1 \text{ if } h_2 < 0 \end{cases}$$

Finally, in order to take into account the input dead time, the system state equations are rearranged by considering three more state variables $x_{7,k}=\upsilon_{k-3}$, $x_{8,k}=\upsilon_{k-2}$, and $x_{9,k}=\upsilon_{k-1}$.
The final state space model, so obtained, is

$$\underline{X}_{k+1}=\underline{A}\ \underline{X}_k+\underline{B}\upsilon_k$$

where $\underline{X}_k=\left[\underline{X}_1^T,\underline{X}_2^T,\ \underline{X}_3^T,\ x_7,x_8,x_9,x_{p10},x_{p11}\right]_k^T$, $\underline{B}^T=\left[00000000100\right]^T$, and $\underline{A}$ has obvious definition. The cost function involves two terms. The first is $\underline{X}_k^T\underline{Q}\ \underline{X}_k=e_{1,k}^2$ and the second depends on the conditions for h_1,h_2, and $h_{1\upsilon}=-\upsilon+V_{max}$, $h_{2\upsilon}=\upsilon-V_{min}$. Actually, nine realizable cases were considered, and the corresponding linear quadratic state controllers {viz.(50a,b)} were derived [40].

Simulation studies were pursued for a system with the following parameters, $\tau_m=0.28$ sec, $\tau_e=0.101$ sec, $q=1.27$, $T_A=10$ msec, $T=10$ msec, $a=2.41$ and $-0.5\leq\upsilon\leq 0$. The conclusions drawn from these studies can be summarized as follows.

(i) The response of the induction motor will be greatly affected if the input deadtime, corresponding to the signal processing time in the µP, is neglected (even if this time is as small as 10 msec).
Although this controller processing time can be neglected in process control (chemical, thermal systems, etc), in power electronics applications, having time constants about ten times the sampling period, special attention should be given to it.

(ii) The PID controller processing time in these experiments was about 5.8 msec in the Intel 8080A µP, and 4.2 msec in the Intel 8085 µP.

(iii) The constraints on the input variable value (controlling phase angle of the inverter or the cycloconverter) and on the state variables have to be taken into account for successful control of power electronic systems and electrical machines.

We close our discussion on µP-based optimal control by mentioning the important implementation results of [7] and [41].

The first work [7] describes the implementation of a 2-level multi-µP operated controller, based on the *SIMD* and *MIMD* configurations studied in Section 2.2. The controller implemented has the general form $\underline{u}_k=\underline{G}_k\underline{x}_k+\underline{N}(\underline{x}_k,k)$, where $\underline{G}_k$ in the time varying gain matrix and $\underline{N}(\underline{x}_k,k)$ is an additive nonlinear feedback vector. Several applications of this controller were studied. One of them is the control of the *MIT-Scheinman arm* in the Advanced Automation Research Laboratory of the Purdue University. The elements of $\underline{G}_k$ and $\underline{N}(\underline{x}_k,t)$ are time varying and in particular those of $\underline{N}(\underline{x}_k,t)$ require special function generators for, say, powers, sines, exponentials, etc. These functions could be computed in parallel by a group of processors connected in the MIMD configuration. The design of the controller is based on a hierarchical division of the computational work into elementary functions and control law computation. One group of processors compute the basic functions (powers, exponentials, calculus operations, special functions) and the other does the actual computation of the controller outputs on the basis of the control law equations. This second group uses the elementary functions from the processing section. In complex cases and very short sampling periods a SIMD μP system could be used for fast matrix multiplication. To obtain faster, multiplication/division a special hardware fast multiplier could be used (e.g. a TRW MPY-16AJ multiplier) A reconfigurable design which allows modification of the structure of the whole controller system, under program control from MIMD or SIMD and vice versa can be employed to further increase flexibility and processor utilization..

The second work [41], presents a 4-level hierarchical method for the joint problem of system identification and optimization [42], which improves convergence and divides the computational effort among the higher levels. The application of the method to a *penicilin fermementation process* was made using a microprocessor implementation.

7. CONCLUDING REMARKS

In this chapter we discussed several aspects of microprocessor-based controllers, and presented the basic controller algorithms including the classical compensators and PID controllers, minimum-variance and self-tuning controllers, and state-feedback controllers. From among the large variety of existing results regarding the μP-based implementation of these controllers we picked-up and outlined a few which they are representative, but of course non-exhaustive.

Three particular problems which have to be resolved when implementing μP-based controllers are:

(i) Selection of the appropriate microprocessor.
(ii) Selection of the sampling period.
(iii) Selection of the appropriate interface.

A systematic technique for selecting the best microprocessor that satisfies desired design criteria is described in |43|. This technique is based on the construction and optimization, with respect to the μP type, of an appropriate system index expressing in a unified mathematical way the design criteria imposed.

Regarding the selection of sampling period we refer the reader to [44] where the problem is investigated in the framework of μP-based motor control systems, but can be appropriately used in other systems as well. Finally, regarding the interface selection and design the reader is referred to [45] where a comprehensive study and presentation of current μP interfacing techniques is provided.

Some further general comments and conclusions about μP-aided industrial control are given in [46]. A methodology for estimating the execution-time bound and memory requirements of control algorithms coded in Pascal and implemented with a specified μP is developed in [47]. The method is based on some modifications of the Pascal P-compiler, which provides the P-code and, together with it, a subsidiary sequential file of records to help timing and storage estimation.

Work is in progress by the writer and his co-workers at Patras University and N.R.C. Demokritos towards the end of implementing state-space multivariable feedback controllers of the PID, eigenvalue assignment/model matching, self-tuning, and model-reference types using the MDSs available at these institutions (Intellec MDC, DAI-MDC and M6800 MDC).

REFERENCES

1. Tzafestas, S.G. and Alexandridis, N.A.:1978, "Microprocessor-Based Approach to Control Engineering Problems", Proc. 1978 SCSC Conf., Newport, Calif., July 24-26, pp.488-498.
2. Tao, T.F.;Bar Yehoshua, D. and Martinez, R.:1977, "Applications of Microprocessors in Control Problems", Proc. 1977 JACC, WA21:pp.8-16.
3. Witten, I.H. (1981) "The New Microprocessors", Proc.IEE, Vol.128, pt.E, No.5, pp. 197-204.
4. Jones, M.R. and Pooch, U.W.:1976, "Evolution and Design of Microcomputers", Proc. Symp. on Trends and Appls 76: Micro and Mini Systems,Gaithersburg, Maryland.
5. Roloff, J.J.:1980, "Managing Memory to Unloose the Full Power of Microprocessors", Electronics, Vol.53, pp.130-134.
6. Patterson, D.A. and Sequin, C.H.:1980, "Design Considerations for Single-Chip Computers of the Future", IEEE Trans., Vol.C-29, pp.108-116.
7. Graham, J. and Saridis, G.N.:1980, "Two-level Multi-Micro-

processor Digital Controller", Automatic Control Theory and Applications, Vol. 8, No.1, pp.30-36.
8. Tzafestas, S.G.:1971, "Advanced Techniques of Digital-Model Design and Real-Time Simulation for Direct Digital Control", Proc. Intern. AICA Symp. on Simul. of Complex Syst., Tokyo, Aug. (IMACS Publ.).
9. Tzafestas, S.G.:1980, "Digital Signal Processing Techniques" Lecture Notes, Control Syst. Lab., Patras Univ.
10. Tzafestas, S.G.:1977, "Tuning Two-and Three-Term Industrial Controllers", Tech. Paper, Control Syst. Lab., Patras Univ.
11. Dahlin, E.B.:1968, "Designing and Tuning Digital Controllers (I and II)", Instrum. Technol., Vol. 41, pp. 77-83 and 87-91.
12. Bristol, E.H.:1977, "Designing and Programming Control Algorithms for DDC Systems", Control Engng, Vol. 24, pp. 24-26.
13. Tzafestas, S.G. and Cosmatos, P.:1973, "Logical Design of a Digital PID Controller", Tech. Demo, N.R.C. Demokritos, Athens.
14. Tzafestas, S.G. and Frangakis, G.:1971, "Variable Digital Design for P+I Controllers", Control and Instrumentation, pp. 45-47.
15. Tzafestas, S.G.:1973, "Computer-Aided Control of Chemical Systems", Technika Chronika, Tech. Chamber of Greece, Athens, pp. 707-721.
16. Khandheria, J. and Shanta, J.P.:1977, "Adaptive Sampling Increases Sampling Rate as Process Deviations Increase", Control Engng, pp. 33-35.
17. Burke, G.R.: 1976, "Microprocessors in Control Systems", Microprocessors, Vol.1, pp. 38-40.
18. Huff, E.C. and McArthur, K.G.:1975, "A Microprocessor-Based Digital Process Controller", IEEE Trans. on IECI, Vol. IECI-22, pp. 345-354.
19. Reed, M. and Mergler, H.W.:1977, "A Microprocessor-Based Control System", IEEE Trans. on IECI, Vol. IECI-24, pp. 253-257.
20. Aylor, J.H., Ramey, R.L. and Cook, G.:1980, "Design and Application of a Microprocessor PID Predictor Controller", IEEE Trans. on IECI, Vol. IECI-27, pp. 133-137.
21. Astrom, K.J.:1970, "Introduction to Stochastic Control Theory", Academic Press, N.Y.
22. Astrom, K.J. et.al.:1977, "Theory and Applications of Self-Tuning Regulators", Automatica, Vol.13, pp. 457-476.
23. Iserman,:1980, "Digital Control Systems", Springer.
24. Clarke, D.W. and Gawthorp, P.J.:1975", Self-Tuning Controller", Proc. IEE, Vol. 122, pp. 929-934.
25. Williams, D.C. and Delaney, D.G.:1972, "On-line Design and Implementation of Digital Compensators", I.Chem. E. Symp. Series No.35, pp. 6:14-6:19.
26. Sheirah, M., Malik, O. and Hope, G.:1982, Self-Tuning

Microprocessor Universal Controller", IEEE Trans. Ind. Electronics, Vol. IE-29, pp. 31-37.

27. Tzafestas, S.G.:1966, "Estimation of Parameters of Multi-Input and Multi-Output Linear Dynamic Systems", M.Sc.Thesis, Imperial College, London.
28. Keviczky, L.and Hetthessy, J.:1972, Self-Tuning Minimum Variance Control of MIMO Discrete Systems,Auto.Contr.Theory Appl.Vol.5
29. Tzafestas, S. G.:1970, "Some Computer-Aided Estimators in Stochastic Control Systems Identification, Int.J.Control, Vol.12, pp.385-399.
30. Tzafestas, S.G.:1982, "Optimization and Control of Dynamic Operational Research Models" (Ch.1), North-Holland, Amsterdam.
31. Luenberger, D.G.:1967, "Canonical Forms for Linear Multivariable Systems", IEEE Trans., Vol.AC-12, pp.290-293.
32. Paraskevopoulos, P.N. and Tzafestas, S.G.:1975, "New Results in Feedback Modal Controller Design", Int.J.Control, Vol. 21, pp. 911-928.
33. Tzafestas, S.G.:1975, "Feedback Transformation of Linear Sequential Machines", Proc. IEE, Vol.122, pp. 857-859.
34. Tzafestas, S.G.:1976, "Automatic Control Systems", Vol.3, Ch.3 (Optimal and Eigenvalue Control), Vagionakis Booksellers, Athens.
35. Tzafestas, S.G.:1972, "State Observer Design for Linear Sequential Machines", Int.J. Systems Sci., Vol.4, pp.13-25.
36. Landau, Y.D.:1979, "Adaptive Control:The Model Reference Approach", Marcel Dekker, N.Y.
37. Al-Bahrani, H. and Healey, M.:1972, "A Discrete Implicit Model-Following Control Law", Int.J. Control, Vol.16, pp. 1051-1056.
38. Perdikaris, G.A., Beck, T.L. and Loendorf, W.R.:1979, "A Microprocessor-based System for DC Servo Control", Tech. Report, Eng. Sci. Div., Wisconsin Univ.
39. Glattfelder, A.H., Huguenin, F. and Schaufelberger, W.: 1980, "Microcomputer-Based Self-Tuning and Self-Selecting Controllers", Automatica, Vol.16, pp.1-8.
40. Tsuchiya, T.:1980, "Basic Considerations on Applications of Optimal Control and the Microprocessor to the Induction Motor Speed Control System, "Int. J. Control, Vol.31, pp. 285-302.
41. Fawzy, A.S., Hinton, O.R. and Hanna, F.K.:1981, "A Four-Level Hierarchical Approach to the Joint Problem of System Identification and Optimization and Its Application to a Batch Fermenter Using Microprocessors", IEEE Trans. Syst. Man Cybern, Vol. SMC-11, pp. 227-231.
42. Mahmoud, M., Hassan, M. and Singh, M.:1980, "A New Hierarchical Approach to the Joint Problem of Systems Identification and Optimization", Large Scale Systems:Theory and Appl., Vol. 1, pp. 159-166.
43. Maymon, E. and Tabak, D.:1978, "Nonlinear Aspects and Se-

lection of Microprocessor Controllers", Int.J. Systems Sci., Vol.9, pp.827-835.

44. Wang, H.C.:1981, "Sampling Period and Stability Analysis for the Microcomputer-Based Motor Control Systems", IEEE Trans. IECI. Vol. IECI-28, pp. 98-102.
45. Zaks, R. and Lesea, A.:1979, "Microprocessor Interfacing Techniques", Sybex Inc.
46. Fagenbaum, J.:1982, "Industrial Controls", IEEE Spectrum, pp. 61-64.
47. Rajulu, R.G. and Rajaraman, V.:1982, "Execution-Time Analysis of Process Control Algorithms on Microprocessors", IEEE Trans. Industr. Electronics, Vol. IE-29, pp. 312-319.

Part V

APPLICATION TO CONTROL AND AUTOMATION

CHAPTER 13

SPEED CONTROL OF MOTOR DRIVES VIA MICROPROCESSORS

Fumio Harashima
Institute of Industrial Science
University of Tokyo
Tokyo, Japan

The advance in LSI technology has made it possible to utilize microprocessors economically and efficiently in motor drive control systems. This chapter describes the state of the art on speed control systems of motor drives via microprocessors. The recent progress of microprocessor-based motor control systems is, first, briefly reviewed, and the basic concept of the systems is presented. Then, the applications of modern control theory are discussed aiming speed accuracy, time-optimal dynamic response and robustness. Finally three practical examples of microprocessor-based motor speed control systems are shown.

1. INTRODUCTION [1],[2]

Recent advance in LSI technology has awakened a new interest in digital control of motor drive systems. The application of microprocessors enlarges the selection range of control algorithm and increases the performance of motors drive systems considerably. The addition of computational capacity enables the application of sophisticated control algorithms to the motor control systems. Figure 1 shows the basic functional diagram of a motor speed control system. In this Figure, the power supply feeds controlled electrical power to the motor through the

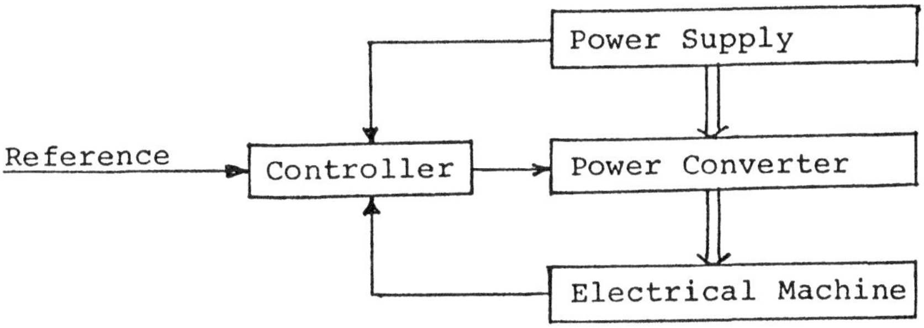

Fig. 1 Basic functional diagram of motor speed control system

S. G. Tzafestas (ed.), Microprocessors in Signal Processing, Measurement and Control, 283–298.

power converter, and the motor operates in a desired manner. The supply voltage, motor speed, motor current, ect. are detected by proper sensors, and the output signals from sensors are sent to the controller as feed back signals. The controller generates the control signal of power converter based on the reference signal and feed back signals. Conventionally, analog controllers based on operational amplifiers have been used as the controllers. The control algorithm used in analog controllers is, in most cases, limited to so-called PI control. A microprocessor-based digital controller, however, makes it possible to apply more sophisticated control algorithms such as modern control theory, and gives high control performance to the system.

Section 2 of this chapter presents the basic concept of microprocessor-based motor speed control system. The advantages of microprocessors applied in motor speed control systems are discussed and then feasible control algorithms are classified into six categories. They are digital PI control, digital PLL control, finite settling time control, minimum settling time control, model reference control, and adaptive control. These control algorithms are compared with each other in terms of control performance such as speed accuracy, dynamic response, robustness and so on. It is pointed out that the addition of computational capability by microprocessors enables the applications of modern control theory to motor speed control systems. The realization of the sophisticated control algorithms by microprocessors is also discussed.

Section 3 shows the detailed description on the applications of modern control theory to motor speed control. Three examples are presented aiming speed accuracy, time-optimal dynamic response and robustness, respectively. Each control system configuration, hardware implementation and microcomputer software are shown. The basic control performance of each system is described together with the experimental data.

Section 4 - 6 are devoted to practical applications of microprocessor-based motor speed control systems. Section 4 describes an example of 16-bit microprocessor-based dc motor control systems which have already been widely used in industry. The detailed description on this system is given, and the advantages of microprocessor-based systems over conventional analog systems are demonstrated.

In Section 5, an induction motor speed control system by microprocessor is shown. Microprocessor-based field oriented control of an induction motor has given rise to a very high performance of speed control.

Section 6 demonstrate a microprocessor-based synchronous motor speed control system. A converter-fed synchronous motor which employ margin angle control strategy has been realized by microprocessor, and shows a very high performance as a variable speed drive.

Section 7 is a conclusive remark. The present state of the art on microprocessor-based motor speed control system is summarized, and the future trend is discussed.

2. BASIC CONCEPT OF MICROPROCESSOR-BASED SPEED CONTROL SYSTEM [2],[3]

Figure 2 shows a typical example of microprocessor-based motor speed control system. This system consists of a rotor fed by a power supply through a power converter, sensors such as speed sensor (rotary encoder) and current sensor, and microcomputer as the controller. A control system using microprocessor presents following features:

(1) gives flexibility in control algorithm,
(2) has sampled-data operation,
(3) minimizes hardware restriction, and
(4) permits small size configuration of low cost.

(1) Flexibility of control algorithm

The complicated mode change of the control system such as start/stop and constant running can be realized by the software of the microcomputer. In addition, the control sequence, control parameters and even control algorithms can be also changed by the change of software without any change of hardware configuration.

(2) Sampled-data operation

A microprocessor-based control system is basically discrete-time and digital system. For this reason it is important to estimate the influence of sampling interval and round-off error on the control performance. The direct replacement of analog control algorithm by digital controller

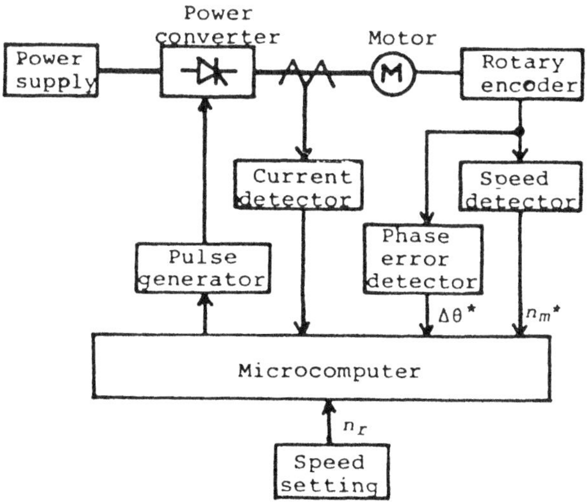

Fig. 2 Microcomputer-based motor control system

may deteriorate the control performance. The bit number and the sampling interval should be carefully determined in designing a microprocessor-based control system. On the other hand, a finite settling time control can be easily realized in a discrete-time system.

(3) Minimum hardware restriction

Sophisticated control algorithms can be implemented by the use of microprocessors. For example, modern control theory has begun to be applied in motor drive control system in order to optimize the control performance.

(4) Small size configuration of low cost

Small size and minimum parts number of a controller increase the reliability of a control system. The advance of LSI manufacturing process has made the cost of microprocessor-based controller cheaper than the cost of conventional analog controller.

Feasible control algorithms in motor drive system using microprocessor are listed in Table 1. For the evaluation of the performance of a motor control system, the accuracy, the speed of response and the sensitivity of the system can be considered as primary indices. In Fig. 3 several control algorithms of motor drives are classified according to accuracy, response and robustness. The PI control is conventional method and widely used in industrial applications. The Phase-Locked Loop control system gives more precise speed accuracy than that of the PI control system. With respect to dynamic response, however, the speed settling time of a PLL controlled motor is long since it is based on integral control. The minimum-time settling control achieves very fast response when the system configuration is optimally designed to achieve the deadbeat response. The minimum-time settling control system is a

Table 1 Control algorism of motor drives

No.	Control algorithm	Main object of control	Required operation mode*	Examples
1	Digital PID	good dynamic performance	N	many
2	Digital PLL	high accuracy	N,L	motor speed control
3	Finite time settling ctl.	short settling time	N,M,L	motor speed control
4	Minimum time settling ctl.	minimum settling time	N,M	motor speed control
5	Model reference control	low sensitivity to parameter	N,M,L	sliding mode control of motor
6	Adaptive control	adaptivity to param. fuluctuation	N,M,L	problem to be considered

* N=numerical operation, M=memory operation, L=logical operation

high gain feedback control system, so the performance of this system is sensitive to the deviation of motor parameters. The sliding mode control is introduced to obtain a robust control performance to the parameter deviation and/or the load torque disturbances.
Adaptive control is the final goal of motor drive control system. The flexibility of control algorithms in microprocessor-based controller makes it possible to realize high accuracy, fast dynamic response and robustness of the control system. For this purpose, high performance microprocessors are required.

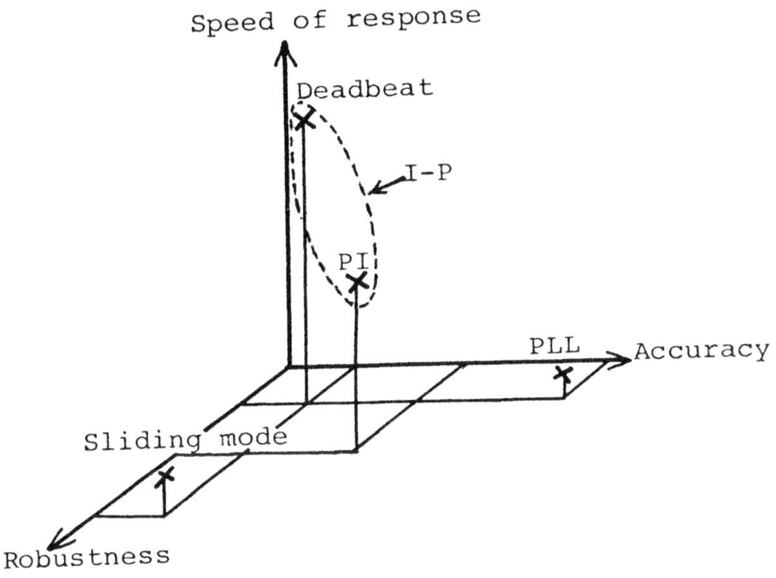

Fig. 3 Performance Evaluation of Control Algorithm of Motor Drive

3. APPLICATION OF MODERN CONTROL THEORY

3.1. Accurate Speed Control by PLL [3],[4],[5]

Figure 4 shows the block diagram of the digital PLL speed control system. One of the restriction to get high speed accuracy of the system in due to the quantization which causes discrepancy between detected phase error and actual phase error. The quantization error leads to a small oscillation of motor speed and phase error around their final values, as illustrated in Fig. 5. In order to analyse the small oscillation caused by the quantization, the following items are assumed;
(1) The time delay (τ) is equal to the sampling period (T_S).
(2) The waveforms of the oscillation are symmetrical. The period of oscillation is equal to $2KT_S$. In this case, the motor speed fluctuation is larger than that in non-symmetrical oscillation.

The following maximum values of motor speed fluctuation $2\Delta n_o$ (peak-to-peak) is obtained as the result of analysis using above assumptions, Fig. 4 and Fig. 5:

$$2\Delta n_o = K_m K_p \Delta\Theta_q \tanh\left(\frac{KTs}{2T_m}\right) \qquad (1)$$

where K denotes the maximum integer number which satisfied the following inequality:

$$(k-4)\frac{Ts}{Tm} - 2\left[1 - \frac{2\exp\{(k-2)Ts/Tm\}}{1+\exp(KT_s/T_m)}\right] < 0 \qquad (2)$$

Figures 6 shows experimental and simulation results of the oscillations mentioned above. By eq.(1) and (2), as are example, the following values are calculated:

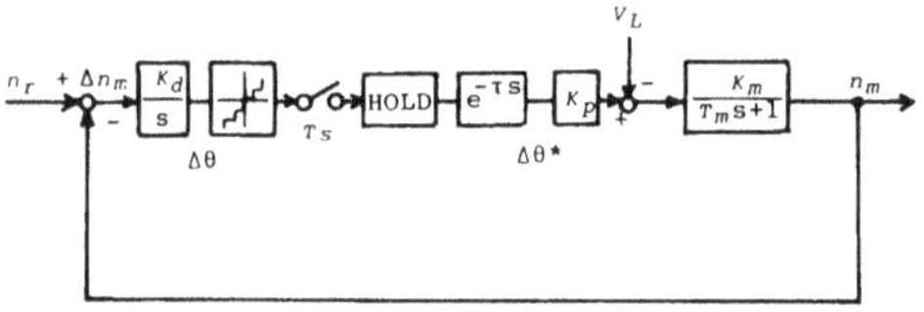

Fig. 4 Block diagram of digital PLL speed control system

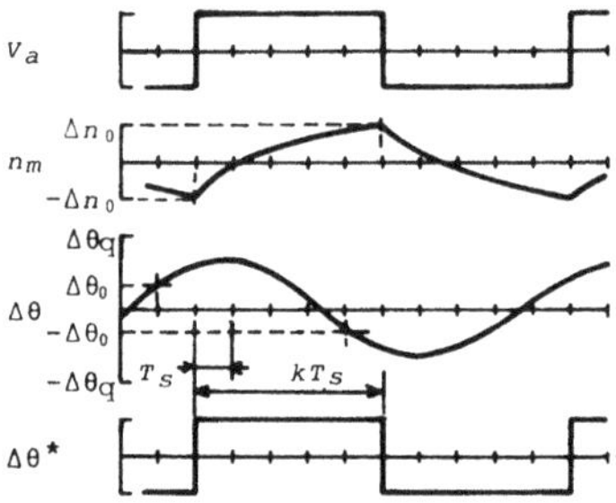

Fig. 5 Waveforms in limit-cylce due to quantization

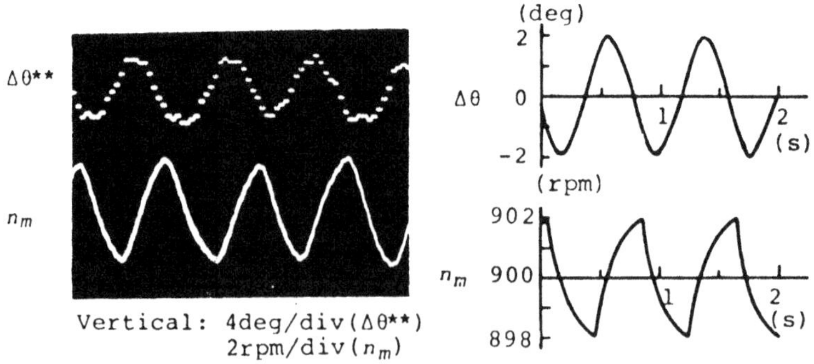

Vertical: 4deg/div($\Delta\theta^{**}$)
2rpm/div(n_m)
Hor.: 0.5 s/div
$\Delta\theta^{**}$: by precise phase error detector

(a) Experimental results (b) Simulation results

Fig. 6 Waveforms of motor speed and phase error in limit-cycle due to quantization

The period of the oscillation; $2KT_s = 0.9$ (s)
The fluctuation of the motor speed; $2\Delta n_o = 4.1$ (rpm)

Both, the period and the fluctuation, are in good agreement with the experimental results of Fig. 6.

The speed accuracy of 0.028% can be realized by using a 8-bit microprocessor.

3.2. Dead Beat Control of Motor Speed [6],[7]

A deadbeat performance means that a system responds to a stepwise input change in the quickest manner without overshoot.
It is extremely difficult to design an analog control system with a deadbeat response. On the other hand, such a performance can easily be realized by a digital control system. There are two methods to achieve the deadbeat response with a digital feedback controller. One is the introduction of a compensator in the forward control loop. The application of a PI controller and a digital filter as a forward loop compensator are contained in this type. Another type is a feedback compensation and the I-P controller is classified in this type. The feedback compensation is superior to the forward loop compensation in the controllability of the system.

The design of deadbeat control system using the I-P control algorithm is now going to be shown. The rotary encoder is employed as the position and speed sensor. Fig. 7 shows the hardware configuration of a microprocessor-based motor speed control system. In order to realize a good dynamic response of the system, the following consideration must be made. The detected average speed by counting up the number of encoder output pulses for a sampling period is not identical to the instantaneous speed and it contains the detection lag time at the sampling instance. This lag time is about half a sampling period.

A microprocessor-based digital control system inherently contains a control lag for processing the control signal. The execution time for processing the control program of this system is not so short as compared with a sampling period. Therefore, this control lag also affects the response and the stability of the system.

A predictive state observer is introduced to solve the two above mentioned problems. The block diagram of the system using predictive state observer is presented in Fig. 8. The predictive state observer gives the estimated valued of the instantaneous speed and position on the next sampling instance before the actual time is reached. This operation of the predictive state observer compensates both the detection lag and the control lag. Fig. 9 shows the experimental results. The expected deadbeat response is practically obtained by the P-I control system with the predictive state observer.

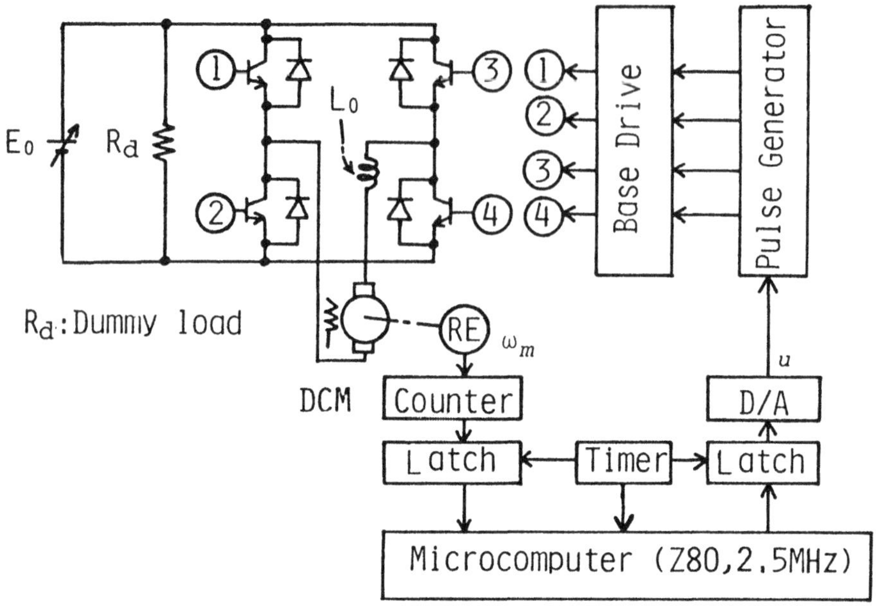

Fig. 7 Configuration of motor speed control system using microprocessor

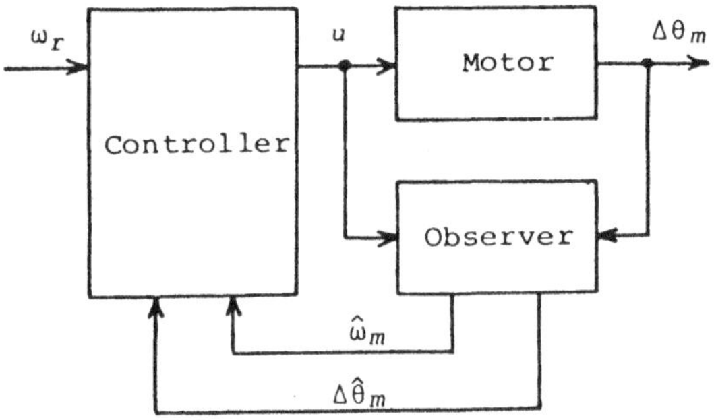

Fig. 8 Simplified Block Diagram of Speed Control System with Observer

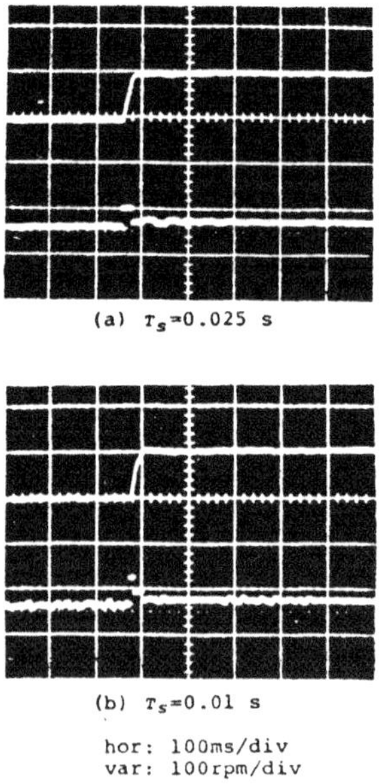

Fig. 9 Step Response of I-P Control System with Predictive State Observer

3.3. Robust Control by Sliding Mode [8],[9]

In many control applications, the control engineer is required to design a system that will be insensitive to parameter variations and disturbances. This can be a difficult task, especially in cases where the system parameters change over a wide range. A method of control called sliding mode control is suggested to overcome these difficulties, in which the representative point of the system is constrained to move (or slide) along a predetermined hyperplane or hyperplanes in state space. In order to achieve such a sliding regime, the control law is required to have a discontinuous nature, i.e. the system structure needs to be changed in time, resulting in a variable structure system (VSS). In this way, insensitivity to parameter variations and disturbances can be ensured. Such a control sheme is robust and simple to be designed since it does not require an exact modelling of the system. Only the bounds of the system parameters need to be known.

An application of sliding mode control to a position servo system is shown below. The hardware details of the system used for experimental investigations are shown in Fig. 10. A 24 V, 50W d.c. servo motor is driven by a PWM power MOSFET chopper amplifier operating at 10 kHz. A 10 bit digital shaft encoder is used to sense the output position while a d.c. tachogenerator coupled directly to the servo motor provides an analog signal for the output speed. A 10 bit tracking type A/D converter provides the digital signal required. Another 10 bit A/D converter is used to detect the current through an isolation amplifier. The microcomputer used is a Z-80 based system with a 2.5 MHz clock. A gear train with a gear ratio of 1/450 is inserted between the servo motor and the output shaft encoder.

Fig. 11 shows the equivalent block diagram of the system. The servo motor is modelled as a first order system neglecting the electrical time constant. Fig. 12 shows a similation results of phase plane trojectories. These similation as well as the experiment demonstrates that the establishment of a sliding mode lowers the sensitivity of a system with respect to parameter variations and disturbances.

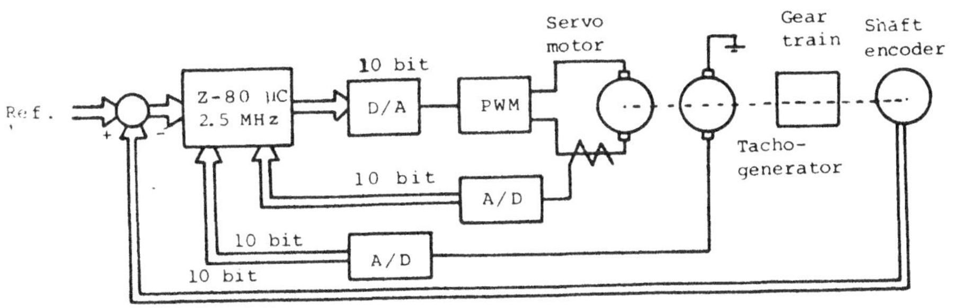

Fig. 10 The hardware details of the system

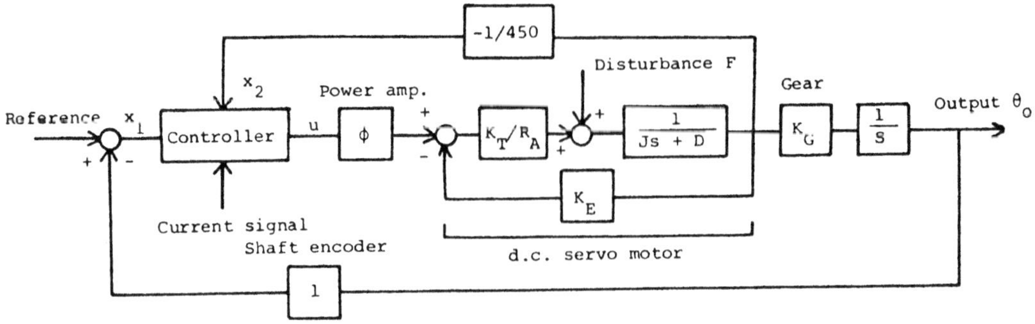

Fig. 11 Block diagram of the system

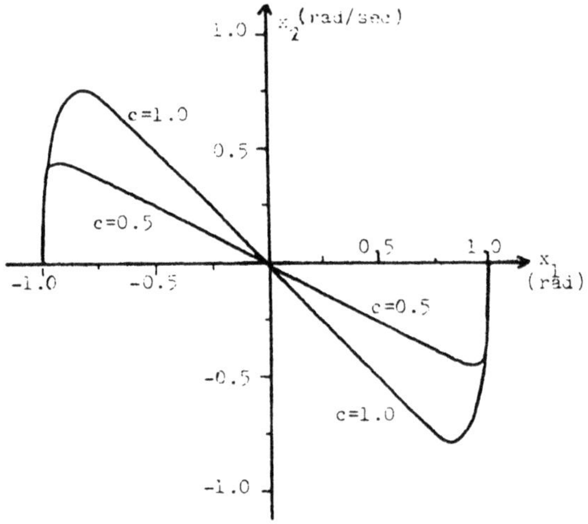

Fig. 12 Phase plane trajectories
$u=\psi_1 x_1 + k_f \mathrm{sgn}\ s.$
$\alpha_1=-\beta_1=1,\ k_f=20$ bits

4. DC DRIVE SYSTEM [10],[11]

An microprocessor-based speed regulating system with a 20 kW, dc motor is illustrated in Fig. 13. A 20 kW dc motor and two sets of three-phase full-wave thyristor converters are connected in antiparallel to form a thyristor Leonard system. The system also includes the following process inputs and outputs (PI/Os) : gate pulse generator, pulse amplifier, filter for motor current detector, A/D converter, discontinuous current state detector, and speed detector. The microprocessor is a 16-bit MOS LSI. A unique feature of this control system is that

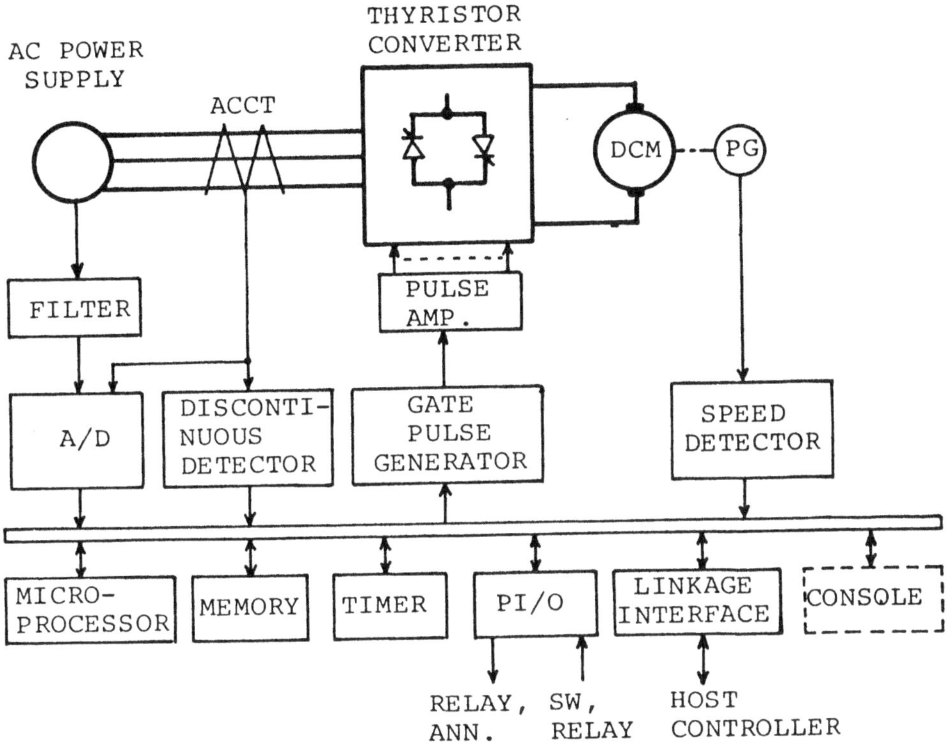

Fig. 13 DC Motor Control System

current-rate control carried out by a feedback mechanism with variable sampling periods synchronized to firing pulses for the converter. The current-rate feedback signal is computed from motor current differences and the sampling time period. Test results exhibit excellent current and current-variation rate regulation and linear compensation capability, demonstrating the method is suited as an internal controller in the microprocessor-based thyristor-drive fast-response speed regulators for motor drives.

Fig. 14 and Fig. 15 show the flowchart of current and current-rate control program and the flowchart of speed control program, respectively. The completely digitalized regulator can permit diagonosis functions, especially in the regulator and gate pulse generator, and bilateral linkage functions with host digital controllers. The proposed method for the current-rate controller was concerned mainly with the application to a fast-response speed regulator for dc motor drives. However, the same method can also be applied to the microprocessor-based adjustable speed control of ac motors fed by thyristor power converters.

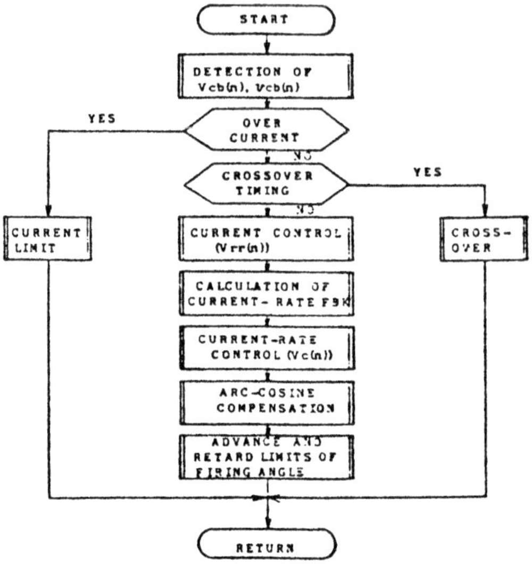

Fig. 14 Flowchart of current and current-rate control program

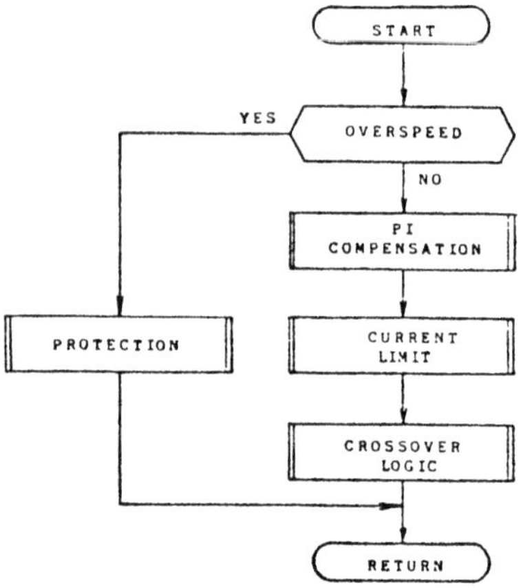

Fig. 15 Flowchart of speed control program

5. INDUCTION MOTOR DRIVE [12],[13]

It is essential to control the vector quantities with their phases to get a quick response with high stability in torque control and speed control of the induction motor (IM) drive. Merely controlling the input voltage amplitude, or input current, or both will not accomplish this. Also, when driving IM by an inverter, the motor input waveform should be as nearly sinusoidal as possible to reduce motor loss and torque pulsation. Pulsewidth modulation (PWM) techniques are useful for this purpose.

The operation of vectors and PWM requires complicated calculation: multiplication, division, power series solutions and trigonometry are all necessary. If this operational circuit is composed of analog and wired logic circuits, the control circuit becomes extremely complex, increasing costs and reducing reliability. Setting up and tuning up this control circuit on the site to achieve desired performance entails some difficult problems.

The microporcessor provides an effective means in solving those problems. There have been a remarkable improvement in microprocessor functions and reduction in its prices. Thus it has become possible to apply the microprocessor readily for motor control as a component used in electronic circuit. An example of the microprocessor systems being applied by vector control based on slip control in IM adjustable speed drive is given here.

Fig. 16 shows the configuration of IM drive system with fixed flux by a current controlled transistor inverter. The computer in this application is composed of an 8-bit microprocessor with 3 MHz clock frequency, a programable interval timer (PIT), 256 bytes RAM, 2K bytes ROM and I/O controllers. Field-oriented control of induction motors can be also realized by cycloconverters or current-source inverters. Microprocessor-based systems of field-oriented controlled converter-fed induction motor have been developed, and they are now in practical uses.

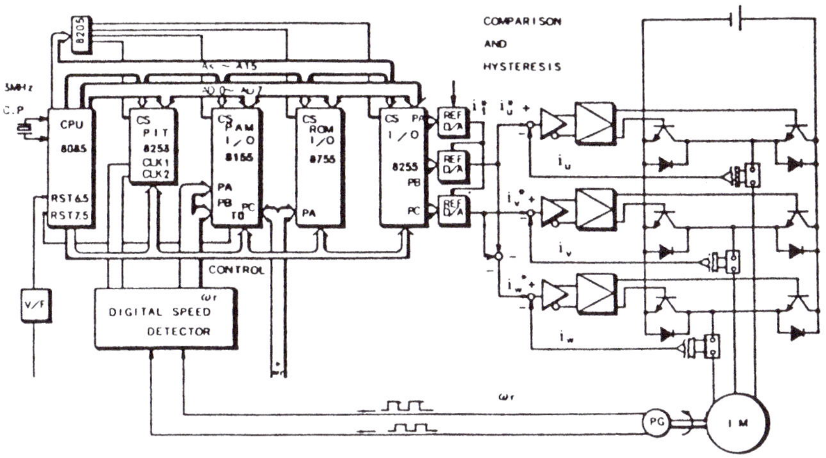

Fig. 16 Microprocessor Application to Vector Control in Current Controlled Transistor Inverter Drive

6. SYNCHRONOUS MOTOR DRIVE [14]

The so-called commutatorless motor, which is already largely applied nowadays in industry as variable speed drive, consists of a synchronous motor fed by a current source inverter, where, for its thyristors triggering, the constant lead-angle has been so far used. This lead-angle has been required to be set in a large value in order to prevent commutation failure, taking into account the armature reaction and the thyristors overlapping period. This large lead-angle, on the other hand, worses the power factor and the motor efficiency. In order to overcome such performance restriction, recently some research has been carried out aiming to maintain the constant power factor angle (effective lead-angle) or to control the commutation margin-angle. The constant margin-angle control is a triggering control scheme applied to an inverter to avoid the commutation failure and to improve the system power factor. Nevertheless, with respect to the constant margin-angle control scheme, there exists intrinsically two different types of instability which have to be well analyzed and elucidated. First, an unstable behavior of a commutatorless motor may occur due to the fact that the voltage drop during the thyristors commutation period acts equivalently as a negative resistance. Since a commutatorless motor presents essentially small existent resistances, that voltage drop effect may become remarkable, making the entire resulting resistance very small or carrying the system to an unstable situation depending on the conditions. Second, there is the instability which may occur in the closed loop usually incorporated to the system to feed back the detected margin-angle (or overlapping-angle). This type of instability takes place while the load is heavy or the margin-angle is set small. Consequently, the advantages of this triggering angle control scheme is upset since the setting of margin-angle in a smallest value, as it is required to achieve the highest power factor operation, is not permitted. In order to avoid instability mentioned above, current feedback loop is effective. Actually, a negative feed back of dc current, as illustrated in Fig.17, can perform this stabilization

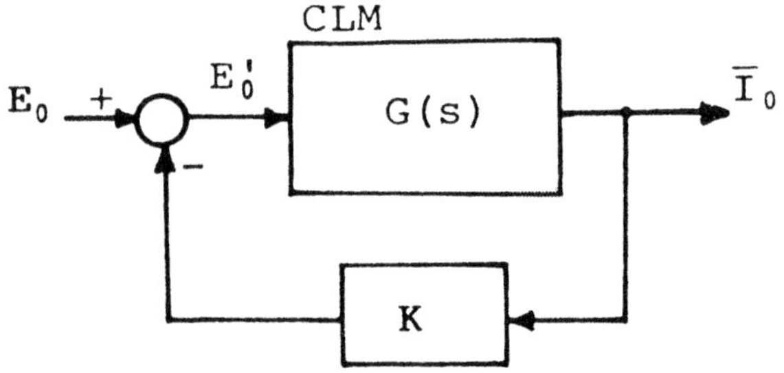

Fig. 17 Commutatorless Motor with Current Feedback Loop

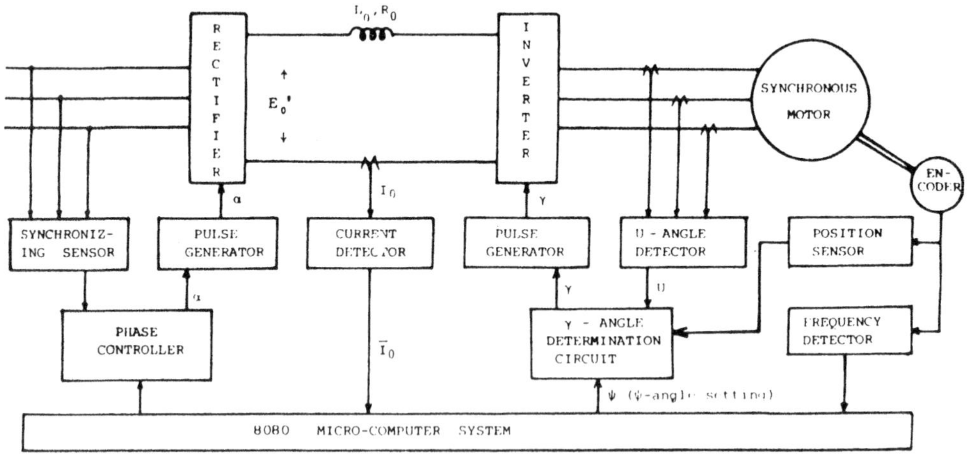

Fig. 18 Functional Diagram of Experimental System

without introducing any additional power loss.

Fig. 18 shows the microprocessor-based control system of a thyristor commutatorless motor with constant margin angle control scheme. On the inverter side, the control circuit provides the triggering angle at each 60° electrical, by adding the set margin-angle and the detected overlapping-angle. The overlapping-angle is detected from the overlap of the armature current. On the rectifier side, by means of the average current detector, the current feedback loop is formed where the loop gain is defined externally and the feedback algorithm is carried out by the microcomputer program.
This system gives a excellent control performance and high energy efficiency over a wide speed range.

7. CONCLUSION [1],[7]

Microprocessors is giving a great impact on modern motor control technology. A tremendous number of discrete and integrated electronic components can be replaced by microprocessor. Control algorithms which previously discarded bacause of their complexity are now becoming economically feasible alternatives. Thus, it is possible to significantly improve the performance of a system at a competitive cost.

The application of microprocessors also makes it feasible to implement modern control strategies including adaptive and learning techniques, stochastic estimation and model-reference following systems. Thus, the widespread use of microprocessors in the control of motor drive systems should result in superior performance, greater reliability and competitive manufacturing costs.

Microprocessor-based control systems of motor drive are already commercially available. In most practical applications, however, general purpose microprocessors are used. In the near future, hardware requirement and control algorithms will be standardized, and custom LSIs specially designed for motor control will be provided.

REFERENCES

[1] Hoft, R.G. : 1981, Proc. U.S. - Japan Seminar on Analysis and Simulation of Power Electronics, Kobe : pp. 237 - 244
[2] Harashima, F., et al: 1981, Systems and Control, 25, pp. 263 - 269 (Japanese)
[3] Harashima, F., and Kondo, S.: 1981, Proc. IVth Power Electronics Conference, Budapest, pp. 227 - 236
[4] Harashima, F., Naitoh H. and Taoka, H.: 1980, IEEE Trans. on IECI, IECI-27, pp. 196 - 201
[5] Harashima F., et al : 1981, IEEE Trans. on IECI, IECI-28, pp 56 - 61
[6] Kitamori T.: 1979, Trans. of SICE of Japan, 15, pp. 131 - 136 (Japanese)

[7] Harashima F. and Kondo, S.: 1982, Proc. 1982 PESC, Boston, pp. 289 - 297
[8] Dote, Y. and Hoft, R.G. : 1980, Proc. IEEE IAS Meeting, pp. 641 - 645
[9] Kaynak, O. and Harashima, F. : 1982, Proc. Int. Conf. on Micro-electronics in Power Electronics and Electrical Drives, pp. 273 - 279
[10] Kaminama, K. et al : 1981, Proc. U.S. - Japan Seminar on Analysis and Simulation of Power Electronics, Kobe, pp. 249 - 258
[11] Konishi, T. et al : 1980, IEEE Trans. on IA, IA-16, pp. 378 - 387
[12] Ohtsuka, K. et al : 1980, Proc. U.S. - Japan Seminar on Analysis and Simulation of Power Electronics, pp. 259 - 268
[13] Nabae, A. et al : 1980, IEEE Trans. on IA, IA-16, pp. 342 - 350
[14] Harashima, F. et al : 1982, Proc. Int. Semiconductor Power Converter Conference, Orlando, pp. 192 - 204

CHAPTER 14

MICROPROCESSOR ALGORITHM FOR THE CONTROL OF SCR MOTOR SYSTEMS

J. B. Plant
Royal Military College of Canada
Kingston, Ontario
Canada

ABSTRACT

This chapter[1] contains a new algorithm for the microprocessor control of three phase SCR controlled DC machines. Two look up tables provide the open loop nominal steady state SCR firing angle and steady state current as a function of the demanded speed. A feedback correction is made to the nominal SCR firing angle according to a linear combination of the sampled deviations of the current, the speed and the sum of the speed deviations from their expected steady state values. Nonuniform sampling is used and sampling occurs at the instant of firing each SCR. The nonlinear discrete-time equations are linearized for the sake of stability analysis in the event of small speed and current deviations. The linearization analysis and digital computer simulations of a three horsepower motor indicate that control can be achieved and constant disturbance torque inputs rejected using only two sets of feedback gains; one set each for the continuous and discontinuous conduction modes of motor operation. The compute delay phenomenon is included in both the analysis and the simulations.

1. INTRODUCTION

The use of a three phase full wave silicon controlled rectifier (SCR) bridge to drive the armature circuit of a separately excited direct current motor is a well established industrial practice. To control such a bridge, the modern technique involves the combined use of both digital and analog components. The state of the art, however, is rapidly nearing the point where the bridge circuit will be controlled with a microprocessor.

This chapter is concerned with a discrete-time mathematical model of the motor plus SCR bridge combination and an algorithm for its control. Both are considered novel.

S. G. Tzafestas (ed.), Microprocessors in Signal Processing, Measurement and Control, 299–321.

The control of a SCR bridge connecting a three phase power supply to the armature circuit of a DC motor as shown in figure 1.1 appears naturally suited to microprocessor control. First, control of the bridge requires the application of logical functions both for SCR protection and armature current limiting. Second, since the SCR's are fired in turn and only if either the current in the preceding SCR in the firing sequence is zero or the voltage across the succeeding one exceeds the voltage across its predecessor, the SCR bridge is in essence a discrete-time device. Third, control is achieved through the selection of SCR firing time or angle. Accurate timing is easily implemented in the microprocessor.

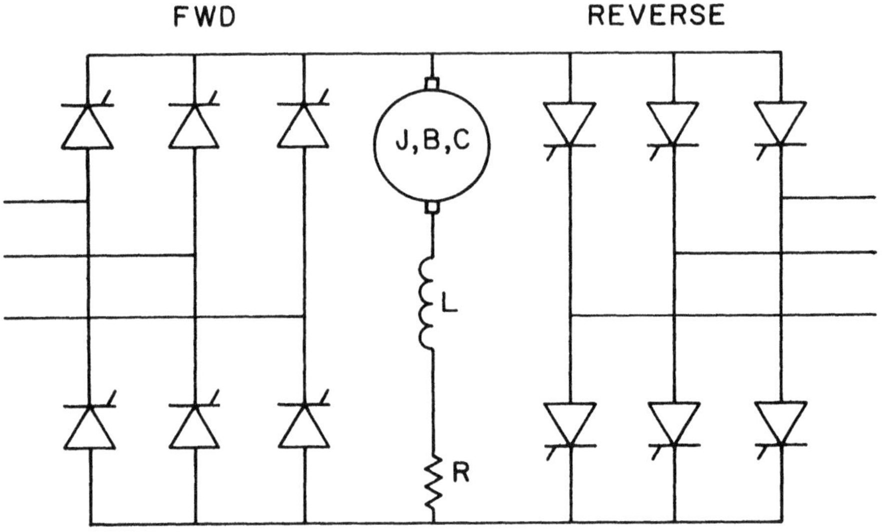

Figure 1.1 SCR Bridge and DC Motor Circuit

The SCR bridge and motor combination is a very nonlinear device. Not surprisingly, there arises a need to generate some nonlinear functions in its control. More precisely, in the algorithm proposed here, it is necessary to express the steady state firing angle and current as functions of the steady state speed. These functions are easily represented in the microprocessor memory as look up tables. And perhaps more practically, these tables can be filled automatically on site by the microprocessor. All that is required is that the firing angle be held constant until steady state is reached and then the speed and current are recorded. Indeed it appears that use of a microprocessor will provide an opportunity to implement a self tuning controller.

In this chapter the problem of speed control is treated in considerable detail, the problem of position control being considered to be solvable by merely placing a good full range speed controlled system in a position feedback loop and readjusting the feedback gains.

The mathematical model includes allowance for both disturbance input torques on the motor shaft and coulomb friction. The discrete-time model is novel in that the integration of the motor differential equations starts at the instant that the SCR is fired. It is assumed that the motor speed and the armature current are sampled immediately prior to firing each SCR and that the firing angle is correctable for every such firing. Thus the system is sampled at the maximum practicable rate which implies a potential for high performance.

Because sampling occurs immediately prior to each SCR firing, the time interval between sampling is nonuniform except in the case of the steady state when it becomes constant and equal to sixty degrees of the supply sinusoid.

The control algorithm is based on linearization about the steady state. A look up table provides a nominal firing angle equal to that which provides the desired speed as $t \to \infty$ if there is no disturbance torque. A second table provides the corresponding steady state armature current. The actual sampled speed and current are compared with the desired speed and the current value provided by the look up table respectively and a correction is made to the firing angle according to a linear combination of the deviations. Both simulation results and analysis of the characteristic polynomials of the small deviation linearized difference equations show that only two sets of gains are required in the linear combination. That is one set of gains each to be used throughout the entire range of both the continuous and the discontinuous conduction ranges of motor operation suffice to give good performance.

It is recognized in both the linearization analysis and the simulation results that a compute delay of one control interval is necessary in implementing the control. In other words, because the microprocessor does not operate instantaneously, the firing angle computed as a result of the measurement of the kth samples is not used until the $k+1^{st}$ control interval.

Disturbance torque inputs are counteracted through the use of an additional adjustment to the firing angle calculated as a linear combination of the sum of the speed deviations. Use of deviation summation feedback is found to provide zero steady state speed errors in the face of either constant disturbance torque inputs or errors in the estimates of the friction coefficients.

This chapter concludes with a description and discussion of the simulation results.

2. MOTOR SPEED CONTROL MODEL

2.1 Continuous Time Model

It is desirable in modeling the motor to include allowance for both linear friction and stiction and also for disturbance (load) torques on the motor shaft or gearing. In the model used herein, the armature circuit is assumed to have constant inductance L and resistance R. The armature circuit equation is;

$$L \frac{di(t)}{dt} + Ri(t) = V(t) - K_v \omega(t) \tag{2.1}$$

where i(t) is the armature current at time t, ω(t) the motor speed, K_v the motor back emf constant, and

$$V(t) = V_0 \sin\left(2\pi ft + (n-1)\frac{\pi}{3}\right), \; n = 1,2,3 \tag{2.2}$$

where n denotes the nth of three phases which are switched "on" in turn by their SCR (see figure 1.1) and f is the frequency of the supply sinusoidal voltage. The supply voltage amplitude V_0 is constant in magnitude, positive for forward speeds, and negative for reverse.

The mechanical circuit has inertia J, linear friction coefficient β, and coulomb friction coefficient C, all of which are constant. The disturbance torque d(t) is variable and K_t is the motor torque constant;

$$J \frac{d\omega(t)}{dt} + \beta\omega(t) + C \operatorname{sgn}\{\omega(t)\}^{2} = K_t i(t) + d(t) \tag{2.3}$$

The disturbance torque may be a random variable as in the wind torque experienced by an antenna drive or piecewise constant as in machine tools when the cutting tool bites into the material to be machined.

Equations (2.1) through (2.3) describe the behaviour of an idealized[3] motor when an SCR is "on" (i.e. conducting) and one of the phases of the three phase supply voltage is applied across the commutator brushes. When all of the SCR's are "off" only equation (2.3)(with i = 0) applies. These equations provide the basis for the development of a discrete-time model appropriate for microprocessor control. This discrete-time model is also found to be useful for the simulation of motor repsonse on a general purpose digital computer and for forecasting controller feasibility and performance.

2.2 Discrete-Time Model

"Discrete-time" or "sampled data" controllers are the subject of many texts such as [1] or [2]. Most texts limit analysis to linear systems and make the assumption that the sample interval is constant (i.e. that the microprocessor controlller samples the system variables every T seconds, where T is a constant).

The SCR controlled DC motor is an essentially nonlinear device. Even though equations (2.1) and (2.3) are linear in i, ω and v, SCR control of the motor is achieved through adjustment of the firing time (or angle) of the SCR's. This amounts to phase switching of the supply voltage and the effect of SCR fire angle variation on speed is thus nonlinear. (SCR firing angle is the angle measured from the zero of the supply sinusoidal phase which is switched by that SCR).

In references [3] and [4] Plant et al. studied the microprocessor control of SCR driven DC motors when the "sample" or "control" interval T was relatively long, that is encompassing many SCR "firings" (where the term firing is synonymous with turning on the SCR). Under these conditions, the time required by the microprocessor to compute the new firing angle was considered negligible as were the effects of changes in the firing angle on the sample interval itself. In this work, quite a different approach is used. Here, the sampling rate is variable becoming constant and equal to 60 degrees of the supply voltage sinuosoid only when the motor is running at constant speed (i.e. in the steady state). Further, the firing angle is altered as necessary each time an SCR is to be fired. With this controller algorithm in use, the microprocessor compute time consumes a significant portion of the sample interval and hence the design includes allowance for compute delay.

More explicitly, the motor speed and armature current are sampled at the instant before each SCR is fired. These sampled values are then used to determine when to fire the SCR for switching the next phase. Thus the sample interval is reduced or increased according to whether the firing angle is reduced or increased respectively. Because the microprocessor requires a non zero amount of time to compute the new firing angle, it is necessary to place a limit on how much the firing angle may be reduced. An arbitrarily chosen limit of $\pm$ 15° gave good results in simulation. Such a limit permits the microprocessor 2.08 milliseconds to complete its tasks assuming a 60 cycle supply.

The discrete-time model is derived by integrating equations (2.1) and (2.3). For this integration, let t = 0 be the firing time of the SCR for phase n and integrate to t_ℓ, the time elapsed until the SCR for phase n+1 is fired. In the steady state $t_\ell = \frac{1}{6f}$ secs (i.e. 60°).

An important consideration in the integation of equation (2.1) and (2.3) concerns whether the motor is in the continuous or the discontinuous conduction mode. In the former, the current remains non zero throughout the sample interval and switching from phase to phase is achieved by virtue of the fact that the voltage of the n+1st phase exceeds that of the nth phase at the time that the latter phase's SCR is fired. In the discontinuous conduction mode, the current subsides to zero before the end of the interval and the SCR of phase n turns off leaving the current at zero until phase n+1 is turned on. In the discontinuous conduction mode, the time t_c of current "cut off" is an important variable which cannot be expressed as an explicit function of the speed and firing angle.

A possible firing sequence is illustrated in figure (2.1) where the firing angle α is shown to increase from α_n to $\alpha_{n+1} = \alpha_n + \Delta\alpha_n$ and from α_{n+1} to $\alpha_{n+2} = \alpha_{n+1} + \Delta\alpha_{n+1}$ where n represents the nth firing of an SCR. The equations which follow are expressed in terms of angle $\rho = 2\pi ft$ rather than in time t. The angle α_n is measured from the zero of the supply sinusoid which is to be switched on by the firing of its SCR.

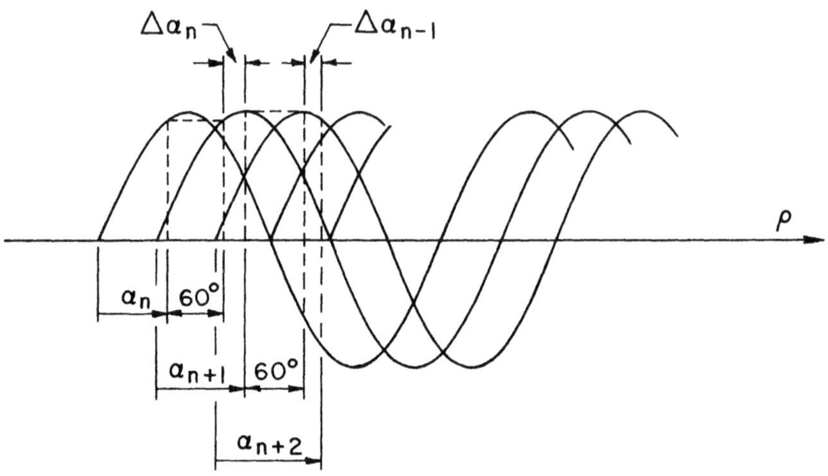

Figure 2.1 Firing Angle Sequence

Since the zero time reference for each sample interval is taken to be the firing time, the phase argument $(n-1)\pi/3$ is dropped from equation (2.2). Let

$$2\pi ft = \rho + \alpha_n, \tag{2.4}$$

$$L' = 2\pi fL, \tag{2.5}$$

and $$J' = 2\pi fJ, \tag{2.6}$$

then equations (2.1) and (2.3) become

$$\frac{di}{d\rho} = -\frac{R}{L'}i - \frac{K_v}{L'}\omega + \frac{V_0}{L'}\sin(\rho+\alpha_n), \tag{2.7}$$

and $$\frac{d\omega}{d\rho} = \frac{K_t}{J'}i - \frac{\beta}{J'}\omega - \frac{C}{J'}\mathrm{sgn}\{\omega\} + \frac{d}{J'} \tag{2.8}$$

with initial conditions $i(0) = i_n$ and $\omega(0) = \omega_n$, the sampled values of current and speed at $\rho = 0$.

2.2.1 Continuous Conduction Mode. While $i \neq 0$ equations (2.7) and (2.8) may be integrated to yield

$$\begin{bmatrix} i(\rho) \\ \omega(\rho) \end{bmatrix} = \phi(\rho) \begin{bmatrix} i_n \\ \omega_n \end{bmatrix} + g(\rho,\alpha_n) + \left(\frac{d_n}{J'} - \frac{C}{J'} \operatorname{sgn}\{\omega_n\}\right) h(\rho), \tag{2.9}$$

$$0 < \rho < \rho_0 + \Delta\alpha_n$$

where $\phi(\rho)$ is a 2×2 real matrix and

$$\phi(\rho) = \pounds^{-1}\{(SI-A)^{-1}\}, \tag{2.10}$$

where

$$A = \begin{bmatrix} -\frac{R}{L'} & -\frac{K_v}{L'} \\ \frac{K_t}{J'} & -\frac{\beta}{J'} \end{bmatrix}, \tag{2.11}$$

I is the 2×2 identity matrix, s is the laplace transform complex variable, $(sI-A)^{-1}$ means the inverse of the matrix sI-A and $\pounds^{-1}\{\ \}$ is the inverse laplace transform operator. If the determinant

$$\left| sI-A \right| = \left(s + \frac{R}{L'}\right)\left(s + \frac{\beta}{J'}\right) + \frac{K_v K_t}{J'L'} \tag{2.12}$$

$$= (s + r_1)(s + r_2) \tag{2.13}$$

has real unequal roots, $-r_1$ and $-r_2$ then it can be shown that

$$\phi(\rho) = \frac{e^{-r_1\rho} - e^{-r_2\rho}}{r_2 - r_1} \begin{bmatrix} \frac{\beta}{J'} & -\frac{K_v}{L'} \\ \frac{K_t}{J'} & \frac{R}{L'} \end{bmatrix} + \frac{r_2 e^{-r_2\rho} - r_1 e^{-r_1\rho}}{r_2 - r_1} \begin{bmatrix} 1 & 0 \\ 0 & 1 \end{bmatrix} \tag{2.14}$$

Also

$$g(\rho,\alpha_n) = \frac{V_0}{L'} \begin{bmatrix} \int_0^\rho \phi_{11}(\rho-\lambda)\sin(\lambda+\alpha_n)d\lambda \\ \int_0^\rho \phi_{21}(\rho-\lambda)\sin(\lambda+\alpha_n)d\lambda \end{bmatrix} \tag{2.15}$$

where the integration may be completed to yield

$$g_1(\rho,\alpha_n) = \frac{V_0}{L'(r_2-r_1)}\left(p(r_1,\rho,\alpha_n) - p(r_2,\rho,\alpha_n)\right) \tag{2.16}$$

where

$$p(r_i,\rho,\alpha_n) = \left(\frac{\beta}{J'} - r_i\right)\left(\frac{1}{1 + r_i^2}\right)\left(e^{-r_i\rho}q(r_i,\alpha_n) - q(r_i,\alpha_n+\rho)\right), \tag{2.17}$$

$$q(r_i,\alpha) = \cos\alpha - r_i \sin\alpha,$$

$$g_2(\rho,\alpha_n) = \frac{V_0\ K_t}{J'L'(r_2-r_1)}\left(\frac{p(r_1,\rho,\alpha_n)}{\frac{\beta}{J'} - r_1} - \frac{p(r_2,\rho,\alpha_n)}{\frac{\beta}{J'} - r_2}\right) \tag{2.19}$$

and $$g(\rho,\alpha_n) = \begin{bmatrix} g_1(\rho,\alpha_n) \\ g_2(\rho,\alpha_n) \end{bmatrix}. \tag{2.20}$$

Finally

$$h(\rho) = \begin{bmatrix} h_1(\rho) \\ \\ h_2(\rho) \end{bmatrix} = \begin{bmatrix} -\frac{K_v}{L'}\left(\frac{1-e^{-r_2\rho}}{r_2} - \frac{1-e^{-r_1\rho}}{r_1}\right) \\ \\ e^{-r_1\rho} - e^{-r_2\rho} + \frac{R}{L'}\left(\frac{1-e^{-r_1\rho}}{r_1} - \frac{1-e^{-r_2\rho}}{r_2}\right) \end{bmatrix} \tag{2.21}$$

To obtain the last term in equation (2.9) it was assumed that $d(t) = d_n$, a constant over the nth sample interval and that $\omega(\rho)$ does not change sign.

In the steady state $\alpha_{n+1} = \alpha_n = \alpha_0$, $\rho = \rho_0 = 60°$, $i_{n+1} = i_n = i_0(\alpha_0)$, and $\omega_{n+1} = \omega_n = \omega_0(\alpha_0)$ and equation (2.9) may be solved to yield

$$\begin{bmatrix} i_0(\alpha_0,d_0) \\ \omega_0(\alpha_0,d_0) \end{bmatrix} = \left(I-\phi(\rho_0)\right)^{-1}\left(g(\rho_0,\alpha_0) + h(\rho_0)\left(\frac{d_0}{J'} - \frac{C}{J'}\,\mathrm{sgn}\{\omega_0\}\right)\right) \tag{2.22}$$

where d_0 is a constant disturbance torque.

In the case of complex roots equation (2.9) is unchanged, however, g, h, ϕ must be recalculated with complex r_1 and r_2. The evaluation of g, h, and ϕ requires a knowledge of $\frac{\beta}{J'}$, $\frac{K_t}{J'}$, $\frac{K_v}{L'}$, $\frac{R}{L'}$, d, and C which is necessary if equations (2.9) through (2.21) are to be used for computer simulation. The functions g, h, ϕ and their derivatives may also be used to design the controller, however, as will be discussed, such a thoroughly analytical approach is not necessary in practice.

2.2.2 Discontinuous Conduction Mode. In the discontinuous conduction mode the armature current reaches zero, "cuts off," at $\rho = \rho_c$ (the cut off angle) before it is time to fire the next SCR (i.e. $0 < \rho_c < \rho_0 + \Delta\alpha_n$). Since $i(\rho_c) = 0$ and assuming that $i_n = 0^4$, equation (2.9) implies that

$$0 = \phi_{12}(\rho_c)\omega_n + g_1(\rho_c,\alpha_n) + \left(\frac{d_n}{J'} - \frac{C}{J'}\,\mathrm{sgn}\{\omega_n\}\right)h_1(\rho_c), \tag{2.23}$$

and $$\omega(\rho_c) = \phi_{22}(\rho_c)\omega_n + g_2(\rho_c,\alpha_n) + \left(\frac{d_n}{J'} - \frac{C}{J'}\,\mathrm{sgn}\{\omega_n\}\right)h_2(\rho_c). \tag{2.24}$$

Now the current remains zero for the remainder of the control interval, thus using equation (2.8) with $i = 0$,

$$\omega(\rho) = e^{-\frac{\beta}{J'}(\rho-\rho_c)}\omega(\rho_c) + \frac{1}{\beta}\left(1-e^{-\frac{\beta}{J'}(\rho-\rho_c)}\right)\left(d_k - C\,\mathrm{sgn}\{\omega_n\}\right), \tag{2.25}$$

$\rho_c < \rho < \rho_0 + \Delta\alpha_n$. Equation (2.23) is implicit in ρ_c an is solved using an iterative technique.

In the steady state $\rho = 60° = \rho_0$, $\alpha_{n+1} = \alpha_n = \alpha_0$ and the solution of equations (2.23) through (2.25) yield two functions

$$\rho_{c0} = \rho_c(\alpha_0, d_0) \tag{2.26}$$

and $$\omega_0 = \omega_0(\alpha_0, d_0) \tag{2.27}$$

replacing equations (2.22) of the continuous conduction mode.

3. DIGITAL SPEED CONTROL

3.1 Speed Control Algorithm

The speed control algorithm is based on the notion of "tracking the steady state."

Let ω_d be a desired steady state speed and let d_0 be a known constant disturbance torque. Assuming the continuous conduction mode, it is found that the second of equations (2.22) may be uniquely solved to find $\alpha_0(\omega_d, d_0)$ the firing angle which provides the desired speed ω_d. If this value of α is substituted into the first of equations (2.22) the necessary steady state current $i_0(\omega_d, d_0)$ is also determined. Similarily, in the discontinuous mode, $i_0 = 0$ and equation (2.27) may provide $\alpha_0(\omega_d, d_0)$ uniquely.

Thus the functions $i_0(\omega_d, d_0)$ and $\alpha_0(\omega_d, d_0)$ may be stored as look up tables in the microprocessor memory and provided an estimate $\hat{d}_n$ of d_n is available, these tables form the basis of an open loop controller. For the most rudimentary form of controller design it is assumed that $d_n = 0$ which reduces the look up tables to the form $i_0(\omega_d)$ and $\alpha_0(\omega_d)$ requiring considerbly less memory. In either case, the memory requirement also depends on precision specifications and this aspect of the research is not yet completed.

It should be noted that the look up tables may be constructed either through direct calculation using the analysis of section 2 or, and perhaps more useful, directly on the machine during a precontrol learning process which can be carried out automatically by the microprocessor. All that is required is for the motor to be sequenced through the range of firing angles while being allowed to reach steady state for each angle. Measurement of $i_0(\alpha_0)$ and $\omega_0(\alpha_0)$ in this manner permits automatic construction of the look uptable during installation of the microprocessor controller.

The proposed control algorithm is one in which the sampled motor speed and armature current are compared with their expected steady state values supplied by the look up tables. The firing angle is then corrected according to a linear combination of the deviations and their sum. The control algorithm is depicted in figure 3.1.

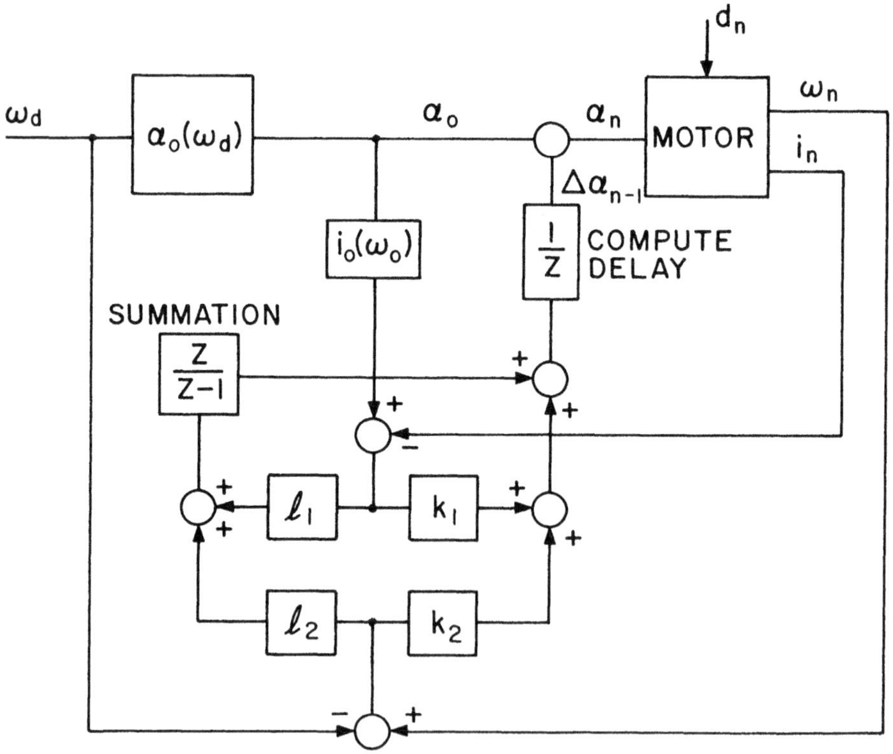

Figure 3.1 Block Diagram - Digital Speed Control Algorithm

The control algorithm of figure 3.1 uses two feedback loops; the first is "proportional" with gains k_1 and k_2 on the current and speed deviations Δi_n and $\Delta\omega_n$ respectively. The second is a "summation" (integral) feedback with gains ℓ_1 and ℓ_2 respectively. Both feedbacks suffer a compute delay of one sample interval represented by the block with the $\frac{1}{z}$ "z" transfer function. The block with z transfer function $\frac{z}{z-1}$ represents the effect of summation of the linear combination $\ell_1\Delta i_n + \ell_2\Delta\omega_n$. The complex variable z represents the z shift operator which is discussed in Appendix A.

The summation loop is used if estimates of d_n are not available and if d_n varies slowly. Indeed if ℓ_1 and ℓ_2 are properly selected, constant non zero disturbance torques will have no effect on the steady state speed.

The four feedback gains k_i and ℓ_i, $i = 1,2$ must be chosen to provide the closed loop system with good reponse characteristics both to changes in ω_d and to errors in speed, current or firing angle.

Optimum gains k and ℓ depend on the desired speed, the disturbance d and on how far the current and speed have deviated from their desired steady state values. These gains may be found approximately through simulation, analysis or tuning of the controller directly on the installed motor. The practicality of the latter will depend on the installation and will probably not be useful on large machines. By analysis is meant "small deviation analysis" wherein the equations are linearized about the steady state and linear system stability concepts are used. Simulation permits tuning without risk of hardware failure. All three approaches involve selection of k and ℓ by trial and error.

In the case of the three horsepower motor which has been simulated by the author, it was found that both the analysis and the direct tuning approaches led reasonably quickly to the selection of useful sets of gains. However, it was found that the combined use of the two techniques was the most satisfactory.

It was also found that despite the fact that optimum feedback gains depend on both ω_d and d, in practice only two sets of feedback gains were required. One set each sufficed to give good performance over the full range of both the continuous conduction and discontinuous conduction modes.

3.2 Small Deviation Analysis (Gain Selection)

The behaviour of the system of figure 3.1 can be analyzed for small values of Δi_n and $\Delta\omega_n$ using a Taylor series expansion of equations (2.9) and (2.23) through (2.25). The expansion is made about the steady state conditions i_0, ω_0, ρ_0, α_0 and ρ_{c0}. To this end let

$$i_n = i_0 + \Delta i_n, \tag{3.1}$$

$$\omega_n = \omega_0 + \Delta\omega_n, \tag{3.2}$$

and $$\alpha_k = \alpha_0 + \Delta\alpha_{k-1}. \tag{3.3}$$

To obtain i_{n+1} and ω_{n+1} from equation (2.9) it is necessary to have the correct value of ρ, the angle measured from $\alpha_0 + \Delta\alpha_{k-1}$ to $\rho_0 + \alpha_0 + \Delta\alpha_k$, thus

$$\rho = \rho_0 + \Delta\alpha_k - \Delta\alpha_{k-1}. \tag{3.4}$$

Equation (3.4) can be deduced from figure 3.2.

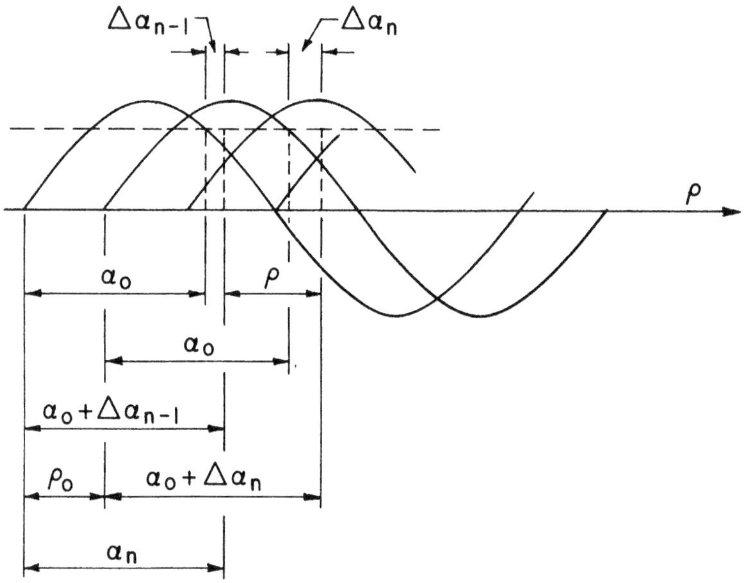

Figure 3.2 "ρ"

We see that $\rho_0 + \alpha_0 + \Delta\alpha_n = \rho + \alpha_0 + \Delta\alpha_{n-1}$ whence equation (3.4). Since the new samples are taken at ρ

$$i(\rho) = i_0 + \Delta i_{n+1}, \tag{3.5}$$

and $$\omega(\rho) + \omega_0 + \Delta\omega_{n+1}. \tag{3.6}$$

<u>3.2.1 Continuous Conduction Mode</u>. Substituting equations (3.1) through (3.6) into equation (2.9) and expanding the functions therein in a Taylor series of first order, after the use of equation (2.22) to cancel constant terms, results in the linear difference equation

$$\begin{bmatrix} \Delta i_{n+1} \\ \Delta \omega_{n+1} \end{bmatrix} = \phi(\rho_0) \begin{bmatrix} \Delta i_n \\ \Delta \omega_n \end{bmatrix} + \phi_\rho(\rho_0) \begin{bmatrix} i_0 \\ \omega_0 \end{bmatrix} + g_\rho(\rho_0,\alpha_0) - \frac{C}{J'} h_\rho(\rho_0)\,\Delta\alpha_n$$

$$- \phi_\rho(\rho_0) \begin{bmatrix} i_0 \\ \omega_0 \end{bmatrix} + g_\rho(\rho_0,\alpha_0) - g_\alpha(\rho_0,\alpha_0) - \frac{C}{J'} h_\rho(\rho_0)\,\Delta\alpha_{n-1} \qquad (3.7)$$

where $\phi_\rho = \frac{\partial\phi}{\partial\rho}$, $g_\rho = \frac{\partial g}{\partial\rho}$, $h_\rho = \frac{\partial h}{\partial\rho}$ and $g_\alpha = \frac{\partial g}{\partial\alpha}$. Equation (3.7) has the form

$$\Delta x_{n+1} = \phi\Delta x_n + \beta_1\Delta\alpha_n + \beta_2\Delta\alpha_{n-1} \qquad (3.8)$$

where Δx is a 2-vector. Equation (3.8) consists of two second order difference equations where both β_1 and β_2 depend on ω_d via their dependence on $\alpha_0(\omega_d)$ and $i_0(\omega_d)$. The compute delay of figure 2.1 is included in the development of equation (3.8) and it was assumed that $d_0 = 0$. Assuming that $\ell = 0$ (i.e. no summation)

$$\Delta\alpha_n = k^T\Delta x_n, \qquad (3.9)$$

where $k^T = (k_1, k_2)$ a row vector.

Substitution of equation (3.9) into (3.8) and use of the z operator (Appendix A) yields

$$\left(z^2 I - (\phi + \beta_1 k^T)z - \beta_2 k^T\right)\Delta x_n = 0. \qquad (3.10)$$

A necessary and sufficient condition for the sequence Δx_n, $n = 1,2,\ldots$, generated by equation (3.10) to converge to $\Delta x_\infty = 0$ is that the determinant

$$Q(z) = \left| z^2 I - (\phi + \beta_1 k^T)z - \beta_2 k^T \right|, \qquad (3.11)$$

which is a polynomial of degree 4 in z, has all of its roots inside the unit circle in the complex plane. It can be easily shown by direct calculation of the determinant that

$$Q(z) = zQ'(z) \qquad (3.12)$$

i.e. $Q(0) = 0$. Thus the problem of selecting k, the feedback gains, is equivalent to suitably placing the roots of $Q'(z)$ in equation (3.12). Since k is a 2-vector and Q' is of order 3, these roots cannot be specified arbitrarily although a state variable feedback controller [1] for equation (3.10) can be devised which would permit such arbitrary root specification. Complete state variable feedback, however, would require at least a third feedback gain element in k and some calculations for observation of the state. Such things complicate the controller increasing the memory requirement and the compute time. They are therefore to be avoided if possible.

Using a polynomial root finder and a determination of ϕ, β_1 and β_2 according to the data obtained from measurements made on a three horsepower motor, it was found that a trial and error selection of k_1 and k_2 quickly produced values which provided suitable stable roots of $Q'(z)$ over the entire range of continuous conduction mode steady state speeds.

3.2.2. Discontinuous Conduction Mode. To derive the linearized equation for the discontinuous conduction mode, it is necessary to expand equations (2.23), (2.24) and (2.25) in a Taylor series. The implicit relationship (equation (2.23)) which provides ρ_c complicates the expansion. Let

$$\rho_c = \rho_{c0} + \Delta\rho_c, \tag{3.13}$$

$$\varepsilon(\rho-\rho_c) = e^{-\frac{\beta}{J'}(\rho-\rho_c)}, \quad \varepsilon^\circ = \varepsilon(\rho_0-\rho_{c0}), \quad \phi^\circ = \phi(\rho_{c0}),$$

$$\xi(\rho,\alpha) = g(\rho,\alpha) - \frac{C}{J'}\,h(\rho), \quad \xi^\circ = \xi(\rho_{c0},\alpha_0), \quad \phi_\rho = \frac{\partial\phi}{\partial\rho}, \quad \xi_\rho = \frac{\partial\xi}{\partial\rho} \text{ and } \xi_\alpha = \frac{\delta\xi}{\delta\alpha}$$

and substitute equation (2.24) into (2.25) to obtain

$$\omega(\rho) = \varepsilon(\rho-\rho_c)\left(\phi_{22}(\rho_c)\omega_n + \xi_2(\rho_c,\alpha_n)\right) - \frac{C}{\beta'}\left(1 - \varepsilon(\rho-\rho_c)\right). \tag{3.14}$$

The Taylor series expansion of equation (2.23) gives

$$0 = \phi^\circ_{12}\Delta\omega_k + (\phi^\circ_{12\rho}\omega_0 + \xi^\circ_{1\rho})\Delta\rho_c + \xi^\circ_{1\alpha}\Delta\alpha_{k-1}. \tag{3.15}$$

The same expansion of equation (3.14) after a further use of (3.14) with $\rho = \rho_0$, $\rho_c = \rho_{c0}$ and $\omega_n = \omega_0$ yields

$$\Delta\omega_{k+1} = \frac{\varepsilon^\circ_\rho}{\varepsilon^\circ}\left(\omega_0 + \frac{C}{\beta'}\right)(\Delta\alpha_k - \Delta\alpha_{k-1} - \Delta\rho_c)$$

$$+ \varepsilon^\circ(\phi^\circ_{12}\Delta\omega_k + (\phi^\circ_{22_\rho}\omega_0 + \xi^\circ_{2\rho})\Delta\rho_c + \xi^\circ_{2\alpha}\Delta\alpha_{k-1}. \tag{3.16}$$

Assuming that equation (3.15) can be solved for $\Delta\rho_c$ the result when substituted into equation (3.16) gives

$$\Delta\omega_{k+1} = \theta\Delta\omega_k + b_1\Delta\alpha_k + b_2\Delta\alpha_{k-1} \tag{3.17}$$

where θ, b_1 and b_2 are scalars. The actual expressions for these coefficients are cumbersome and are not included in the text. In fact, given a simulation program, it appears to be easier to calculate these coefficients using the program and numerical differentiation than it does to evaluate them using equations (3.15) and (3.16).

Equation (3.17) implies that the controlled motor in the discontinuous conduction mode with $\Delta\alpha_k = k_2\Delta\omega_k$ has the characteristic polynomial

$$Q(z) = z^2 - (\theta + b_1 b_2)z - b_2 k_2. \tag{3.18}$$

Thus the problem is to find a value for k_2 which gives Q(z) suitable roots. For the three horsepower motor studied by the author, it was found that a single gain sufficed to give good performance over the entire discontinuous conduction range, however, this gain differed from that found desirable for the continuous conduction mode.

3.2.3 Summation (Integral) Control. In an application where the disturbance torque d_n is significant and where speed precision is required, in spite of the disturbance, it is necessary either to include both an algorithm for estimating d_n and two dimensional look up tables $\alpha_0(\omega_d,d)$ and $i_0(\omega_d,d)$ or a summation feedback control loop with gains ℓ as indicted in figure 3.1. The latter has the property of eliminating the effect of constant disturbance torques in the steady state. This property can be deduced in the small deviation continuous conduction mode case by rewriting equation (3.8) to include the disturbance torque d_n.

$$\Delta x_{n+1} = \phi\Delta x_n + \beta_1\Delta\alpha_n + \beta_2\Delta\alpha_{n-1} + \frac{C}{J'} h(\rho_0)d_n \tag{3.19}$$

If $\ell_1 = 0$ and $\ell_2 \neq 0$, the controller will not permit a constant speed error to be sustained. One has

$$\Delta\alpha_n = k^T\Delta x_n + \ell^T\Sigma_n \tag{3.20}$$

where

$$\Sigma_n = \sum_{i=-\infty}^{n} \Delta x_i \tag{3.21}$$

and

$$\Delta x_i{}^T = (\Delta i_i, \Delta\omega_i). \tag{3.22}$$

Since

$$\Sigma_n = \Sigma_{n-1} + \Delta x_n \tag{3.23}$$

$$\Delta\alpha_n - \Delta\alpha_{n-1} = k^T(\Delta x_n - \Delta x_{n-1}) + \ell^T\Delta x_n. \tag{3.24}$$

Use of the z shift operator in equations (3.24) and (3.19) provides

$$\Big[z^2(z-1)I - \big(z(z-1)\phi + z\beta_1(k^T(z-1) + z\ell^T) + \beta_2(k^T(z-1) + z\ell^T)\big)\Big]\Delta x_n = z(z-1)\frac{C}{J'} hd_n \tag{3.25}$$

which has a characteristic polynomial $Q(z)$ of order six. It may be shown that $n \to \infty$ corresponds to $z \to 1$, a result of which is that $\Delta x_{2,n} \to 0$ and $\Delta x_{1,n} \to$ constant in equation (3.25) provided all the roots of $Q(z)$ lie inside the unit circle in the complex plane.

In the case of the discontinuous conduction mode, equation (3.25) applies with ϕ, β_1, β_2, k, ℓ and Δx_n replaced by the scalars θ, b_1, b_2, k_2 ℓ_2 and $\Delta\omega_n$. One finds that

$$\frac{\Delta\omega_n}{d_n} = \frac{z(z-1)\,\frac{Ch}{J'}}{z^3 - (1+\theta+b_1\ell_3)z^2 + \left(\theta+b_1k_3-b_2(k_3+\ell_3)\right)z + b_2k_3}. \tag{3.26}$$

When the summation feedback loop is used all feedback gains k and ℓ must be readjusted so that $Q(z)$ continues to have all its roots suitably located inside the unit circle in the complex plane. Experiments with the simulation of a three horsepower motor indicate that the summation loop can be added successfully with $\ell_1 = 0$ and ℓ_2 constant over the range of each of the two conduction modes.

4. DIGITAL POSITION CONTROLLER

One way to implement microprocessor position control of the motor is to simply add a position feedback loop with forward gain K_δ around the motor plus digital speed control algorithm. The situation is depicted in figure 4.1 where the box marked M is the computer speed controlled motor, δ_n is the shaft position, $\omega_{d,n} = K_\delta(\delta_{d,n} - \delta_n)$ is the speed demanded of the speed controlled motor and ω_n is the actual speed at the nth sampling.

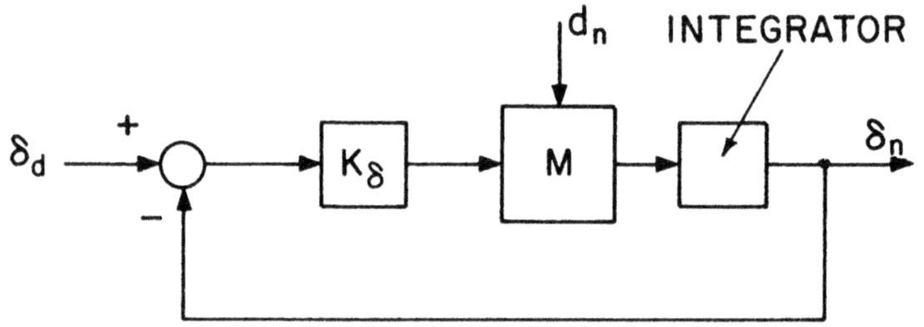

Figure 4.1 Position Controller

The purpose of this section is to show how the structure of figure 4.1 gives position control in the case where the speed controlled motor satisfies a linear difference equation. Research is under way to confirm this approach.

Let

$$Q(z) = z^{\ell} + a_{\ell} z^{\ell+1} + \ldots a_1 \quad (4.1)$$

and $$N(z) = b_{m+1} z^m + b_m z^{m-1} + \ldots b_1, \quad (4.2)$$

be ℓ^{th} and m^{th} order polynomials in z, the shift operator, and let

$$Q(z)\omega_{dn} = N(z)\omega_n. \quad (4.3)$$

Now

$$\delta_n(t) = \int_0^t \omega_n(\tau)d\tau. \quad (4.4)$$

If we approximate equation (4.4) by

$$(z-1)\delta_n = T\omega_n \quad (4.5)$$

which is exact if $\omega(t)$ is constant over the control interval, and apply (4.5) to (4.3) using figure 4.1 we find that

$$\left((z-1)Q(z) + K_\delta TN(z)\right)\delta_n = K_\rho TN(z)\delta_{d_1 n}. \quad (4.6)$$

Equation (4.6) has a DC gain of unity. Thus if $\delta_{d,n}$ is a step input equation (4.6) will produce a sequence δ_n, $n = 0,1,\ldots$ which converges to $\delta_{d,n}$ provided the roots of the polynomial $(z-1)Q(z) + K_\delta TN(z)$ all lie inside the unit circle in the complex plane.

It will be necessary to select new speed and current proportional (i.e. $k^T = (k_1, k_2)$) and integral (i.e. $\ell^T = (\ell_1, \ell_2)$) feedback gains for the computer speed control algorithm as well as K_ρ. This approach to position control is being studied.

5. FIRING ANGLE LIMITATIONS

It is possible to linearize the motor equations for the purpose of understanding and/or selecting the feedback gains when $\Delta\alpha_n$ is small. Equally important, and of significance to the total workload of the microprocessor are the limits which must be placed on α_n in order to protect the motor and the SCR's from burn out (i.e. current limits). These considerations are also important in simulation studies.

5.1 Control Interval Limits

In the control algorithm proposed here, the control interval has a duration of $60° + \Delta\alpha_n$. Since the calculations required to compute $\Delta\alpha_n$ must be carried out by the microprocessor between the times of sampling a lower limit must be placed on $\Delta\alpha_n$. This limit depends on the complexity of the algorithm and the speed of the microprocessor. In the simulations which were used to study performance of the speed controller, both an upper and a lower limit were placed on $\Delta\alpha_n$. An arbitrarily chosen limit of 15° was found to give good results although no attempt has yet been made to optimize this limit. Thus in the case of these simulations the control interval was constrained to lie in the range (45°, 75°).

5.2 SCR Firing Angle Limits

An examination of the three phase sinusoidal voltage waveforms in figure 2. reveals that $\sin(\rho+\theta_{i+1}) > \sin(\rho+\theta_i)$ for $\frac{2\pi}{3} < \rho < \frac{5\pi}{3}$. Assuming that the interval between SCR firings has duration $\pi/3$ and since the SCR being turned on must have a supply voltage which exceeds that of the SCR it is replacing in the firing sequence, it follows that $\frac{\pi}{3} < \alpha_n < \frac{4\pi}{3}$. The lower limit, $\pi/3$, corresponds to maximum forward motor power.

5.3 Armature Current Limiting

In addition to the limits on α_n described in section 5.2, a further limit is necessary to avoid overheating of the armature windings. Depending on the speed of the motor it is necessary to limit α_n so that the maximum safe current is not exceeded.

Neglecting the armature inductance and assuming a control interval of $\pi/3$ radians and a constant motor speed ω_n over the single control interval, the average current is given by

$$\bar{i}_n \simeq \frac{\bar{\nu}_n - K_v\omega_n}{R} \tag{5.1}$$

where

$$\begin{aligned}\bar{\nu}_n &= \frac{3}{\pi}\int_0^{\pi/3} V_0 \sin(\rho+\alpha_n)\,d\rho \\ &= \frac{3V_0}{\pi}\cos(\alpha_n - \pi/3).\end{aligned} \tag{5.2}$$

It follows that a lower limit α_c may be chosen as

$$\alpha_c = \pi/3 + \cos^{-1}\{\frac{\pi}{3V_0}(K_v\omega_n + i_{max}R)\}, \tag{5.3}$$

Where i_{max} is the prescribed current limit. This limit produced good results in the simulation although it sometimes provided a more severe limit on the current than was necessary. The use of this limit requires a $\cos^{-1}$ look up table.

5.4 Motor Reversal

As illustrated in figure 1., the full wave three phase SCR controlled motor drive circuit has a forward and a reverse circuit. Examination of this circuit reveals that an attempt to switch from forward to reverse or back must be carried out only when the armature current $i_n = 0$. To do otherwise will cause the on SCR to be connected across the supply and hence to burn out. Thus in the speed controller, reversal is permitted only in the discontinuous conduction mode. In this mode reversal is made if the new firing angle α_{n+1} exceeds an upper limit α_H. This upper limit is taken to be $\frac{5\pi}{6}$ or $\cos^{-1}(\frac{K_v\omega_n}{V_0}) + \frac{\pi}{2}$ whichever is the least. The second limit is chosen because if α_n exceeds it, no current will flow in the SCR because the back emf of the motor exceeds the supply if it is turned on at that angle. The first limit (i.e. $5\pi/6$) is the value of α which results in zero average voltage being applied to the motor armature circuit in the steady state. These limits gave good results on the simulation.

6. EXPERIMENTAL RESULTS

Equations (2.9),(2.23),(2.24), and (2.25) were used to simulate an idealized DC machine. A "cutting and halving" technique was used to find the zero of equation (2.23) and thus deduce ρ_c given ω_n, i_n and α_n to a precision of $< 60 \times 2^{-23}$ degrees. The same technique was used to detect a transition from the continuous conduction to the discontinuous conduction mode and vice-versa. For example, assuming that $i_n \neq 0$ and knowing $\Delta\alpha_n$, if $i(\rho_0 + \Delta\alpha_n)$ in equation (2.9) has a sign different from that of i_n, the current has obviously gone to zero in the interval and thus the machine has made a transition from continuous to discontinuous conduction mode. In this instance, the simulation program searches the interval $0 < \rho < \rho_0 + \Delta\alpha_n$ to find ρ_c.

Equation (2.22) was used to calculate the steady state speed and current given α_0 a constant firing angle.

Feedback control was implemented according to figure 3.1 and using the limits on α described in section 5.

A three horsepower motor which happened to be available for study was used as a basis for the simulation studies. Measurements made on this machine indicated the following parameter values:

$\frac{\beta}{J} = 0.946 \times 10^{-4}\ \text{sec}^{-1}$ $\frac{K_t}{J} = 0.02186$ newton meters/sec

$\frac{R}{L} = 0.2482$ ohms/henry $L = 0.0045$ henries

$C = 1.383$ newton meters $J - 0.0841$ newton meter sec^2

$\frac{K_v}{L} = 0.4085$ volts/henry sec $V_0 = 208$ volts

For this motor feedback gains of $k_1 = 0.035$, $k_2 = 0.0025$ and $k_1 = 0$, $k_2 = 0.5$ in the continuous and the discontinuous conduction modes respectively, provided good performance over the entire speed range of the motor. The current limiter of section 5 worked satifactorily and settling times for speed errors were less than one or two hundered SCR firings.

When the above gains were used with $\ell_1 = 0$ and $\ell_2 = 0.0001$, the controlled motor maintained constant steady state speed in the face of step changes in disturbance torques on the motor shaft. Optimum values of these gains were not available at the time of writing.

In the simulations conducted thus far it was assumed that the nominal values of firing angle and current provided by the look up tables were exact. In practice, since the memory will be finite, these functions must be quantized which will reduce the precision to which the speed may be controlled. The impact of this fact and methodology for quantization of the functions has not yet been considered in detail.

FOOTNOTES

1. This work was carried out at the Virginia Military Institute, Lexington Va, USA where Dr Plant held the Wachtmeister Visiting Eminent Scholar Chair from 15 Aug to 31 Dec 1982.

2. $\text{sgn}\{x\} = -1,\ x < 0;\ 0,\ x = 0;\ +1,\ x > 0$

3. i.e. magnetic circuit saturation, current dependent armature resistance and windage etc. effects have been neglected.

4. The "transition" from (to) continuous conduction to (from) discontinuous conduction is not considered.

REFERENCES

[1] James A. Cadzou and Hinrich R. Martens, "Discrete-Time and Computer Control Systems," Prentice-Hall Electrical Engineering Series, 1970, Prentice-Hall Inc., Englewood Cliffs, New Jersey.

[2] Paul Katz, "Digital Control Using Microprocessor," Prentice-Hall International Inc., 1981.

[3] J.B. Plant, S.J. Jorna and Y.T. Chan, "Microprocessor Control of Position or Speed of an SCR-DC Motor Drive," IEEE Trans. Indust. Electron. Contr. Instrum, pp. 228-234 Vol. IECI-27, No 3, August 1980.

[4] Y.T. Chan, A.J. Chmiel, J.B. Plant, "A Microprocessor-Based Current Controller for SCR-DC Motor Drives," IEEE Trans. Indust. Electron. Contr. Instrum pp. 169-176, Vol. IECI-27, No 3, August 1980.

APPENDIX A

THE Z OPERATOR AND DIFFERENCE EQUATIONS

Given a sequence x_n, $n = 1,2,\ldots$ let

$$zx_n = x_{n+1}, \tag{A.1}$$

and $$x_n = z^{-1}x_{n+1}, \tag{A.2}$$

and $$z^k x_n = x_{n+k}, \text{ k an integer.} \tag{A.3}$$

z^k is a k-shift operator.

Now consider the difference equation

$$y_{n+k} + a_k y_{n+k-1} + \ldots a_1 y_n = b_{m+1}u_{n+m} + \ldots b_1 u_n. \tag{A.4}$$

Using the k-shift operator z^k we can rewrite (A.4) as

$$(z^k + a_k z^{k-1} + \ldots a_1)y_n = (b_{m+1}z^m + \ldots b_1)u_n \tag{A.5}$$

and then

$$\frac{y_n}{u_n} = \frac{b_{m+1}z^m + \ldots b_1}{z^k + a_k z^{k-1} + \ldots a_1} \tag{A.6}$$

is the z-transfer function of (A.5). It is well known [1] that equation (A.5) is asymptotically stable if, and only if, the roots of the polynomial $z^k + a_k z^{k-1} + a_{k-1}z^{k-1} + \ldots a_1$ lie inside the unit circle in the complex plane.

Consider the sum

$$y_{k+1} = \sum_{\ell=-\infty}^{k+1} u_\ell \tag{A.7}$$

then

$$y_k = \sum_{\ell=-\infty}^{k} u_\ell \tag{A.8}$$

therefore

$$y_{k+1} - y_k = u_{k+1} \tag{A.9}$$

and $$(z-1)y_k = z\, u_k \tag{A.10}$$

and thus the summation operation in (A.7) has the z-transform $\frac{z}{z-1}$.

APPENDIX B

SMALL DEVIATION LINEARIZATION DATA

B.1 Continuous Conduction Mode

$$\Delta x_{n+1} = \phi \Delta x_n + \beta_1 \Delta \alpha_n + \beta_2 \Delta \alpha_{n-1} \tag{B.1}$$

$$\beta_1 = \phi_\rho(\rho_0) \begin{bmatrix} i_0 \\ \omega_0 \end{bmatrix} + g_\rho(\rho_0, \alpha_0) - \frac{C}{J'} h_\rho(\rho_0) \tag{B.2}$$

$$\beta_2 = - \beta_1 + g_\alpha(\rho_0, \alpha_0). \tag{B.3}$$

ϕ is given in equation (2.14).

$$\phi_\rho(\rho_0) = \frac{r_2 e^{-r_2\rho_0} - r_1 e^{-r_1\rho_0}}{r_2 - r_1} \begin{bmatrix} \frac{\beta}{J'} & - \frac{K_v}{L'} \\ \frac{K_t}{J'} & \frac{R}{L'} \end{bmatrix} + \frac{r_1^2 e^{-r_1\rho_0} - r_2^2 e^{-r_2\rho_0}}{r_2 - r_1} \begin{bmatrix} 1 & 0 \\ 0 & 1 \end{bmatrix}. \tag{B.4}$$

$$h_\rho(\rho_0) = \begin{bmatrix} - \frac{K_v}{L'} (e^{-r_2\rho_0} - e^{-r_1\rho_0}) \\ r_2 e^{-r_2\rho_0} - r_1 e^{-r_1\rho_0} + \frac{R}{L'} (e^{-r_1\rho_0} - e^{-r_2\rho_0}) \end{bmatrix} \tag{B.5}$$

$$g_\rho(\rho_0, \alpha_0) = \frac{V_0}{L'(r_2 - r_1)} \begin{bmatrix} \frac{\partial p}{\partial \rho}(r_1, \rho_0, \alpha_0) - \frac{\partial p}{\partial \rho}(r_2, \rho_0, \alpha_0) \\ \frac{K_t}{J}\left(\frac{1}{(\frac{\beta}{J} - r_1)} \frac{\partial p}{\partial \rho}(r_1, \rho_0, \alpha_0) - \frac{1}{(\frac{\beta}{J} - r_2)} \frac{\partial p}{\partial \rho}(r_2, \rho_0, \alpha_0) \right) \end{bmatrix} \tag{B.6}$$

where

$$\frac{\partial p}{\partial \rho} = \left(\frac{\beta}{J} - r_i\right)\left(\frac{1}{1 + r_i^2}\right)\Big(-r_i e^{-r_i\rho_0}(\cos\alpha_0 - r_i \sin\alpha_0) + \sin(\alpha_0 + \rho_0) + r_i \cos(\alpha_0 + \rho_0)\Big). \tag{B.7}$$

To find $g_\alpha(\rho_0, \alpha_0)$ replace $\partial\rho$ by $\partial\alpha$ in equations (B.6) and (B.7).

CHAPTER 15

MICROPROCESSOR CONTROL OF PHASE-CONTROLLED CONVERTERS AND CYCLOCONVERTERS

S. K. Tso
University of Hong Kong,Hong Kong

1. INTRODUCTION

Phase-controlled thyristor converters [1, 2] provide AC/DC controlled power conversion. Depending on the circuit arrangement power can flow in either direction, the converters thereby acting respectively as controlled rectifiers or synchronous inverters. Apart from low-power systems the AC supply used is normally three-phase, or multi-phase as derived from suitable transformer connections or otherwise. The thyristors in the converter are sequentially triggered into conduction and in the normal process of switching, the device(s) previously conducting will experience a reverse bias, presented by the supply voltage, and turn off after a short period of time. This mechanism is known as "natural commutation", because no additional or external means is utilised for device turn-off or commutation purposes.

Phase-controlled cycloconverters, [2, 3] on the other hand, provide direct AC/AC controlled power conversion with the ability to change smoothly both the frequency and the amplitude of the output. Only the naturally commutated types, which share a similar principle of operation with phase-controlled converters, are considered in this Chapter.

Phase-controlled converters and cycloconverters are collectively treated because of their many similarities particularly in terms of the control features. Their very extensive industrial applications justify the development of special digital microprocessor-based controllers, hereinafter called "micro-controllers", intended for multi-phase thyristor phase control.

Phase-controlled naturally commutated converters find the following important applications, among others:

- adjustable speed DC drives of the single, two or four quadrant type,
- commutatorless motors, and high-voltage DC transmission converters, both having a DC link,

S. G. Tzafestas (ed.), Microprocessors in Signal Processing, Measurement and Control, 323–343.

. synchronous inverters for facilitating slip-power recovery in slip-ring motors and connecting local energy storage/generation sources to the power network,
. DC supplies for electrochemical processes and battery charging, and as the primary stage of uninterruptible power supplies,
. DC supplies for rotating machine excitation, having special importance to large alternators, and for electromagnetic clutches and eddy-current couplings,
. general-purpose supplies for various electronic equipment and apparatuses, often including a switched mode regulator stage; isolated supplies, specifically for transistor or valve amplifiers and inverters, and also for thyristor inverters (oscillator-driven or otherwise driven) as used in AC motor drives and induction heater coils (tuned).

Phase-controlled naturally commutated cycloconverters are useful in the following areas:

. variable speed four-quadrant induction motor drives, in particular very large, low-speed or multi-motor drives, or drives fed from a high frequency source,
. variable-speed constant-frequency power supply systems, particularly for aircraft,
. commutatorless motor drives fed directly from AC supply,
. subsynchronous and supersynchronous doubly-fed induction motor drives,
. asynchronous ties between two different AC networks,
. static reactive compensators (or static var generators), either in the double-ended arrangement or using a high frequency base.

The last two areas represent interesting developments [4] which have not yet reached the stage of extensive commercial use, but applications in the other areas are already long-standing and ever widening. Actually the phase-controlled converters and cycloconverters have established themselves as highly demanded power apparatuses and quite indispensable, like the omnipresent workhorse - the electric motor. Any attempts which directly exploit the rapid advances in VLSI technology to reduce the cost of converters or cycloconverters, and improve their performance level and capabilities are readily considered to be both natural and worthwhile. Examples of such attempts are by no means hard to find in the recent literature. [5-9]

2. REVIEW OF BASIC PRINCIPLES

Before discussing the digital microprocessor-based methods for controlling phase-controlled converters or cycloconverters, it is important to review briefly their basic operating principle, and the problems involved in terms of their control.

To start with consider the simplest three-phase converter circuit shown in Fig. 1a. Connected to the three supply phases are three thyristors, each one controlling the passage of current from the respective supply phase to the load.

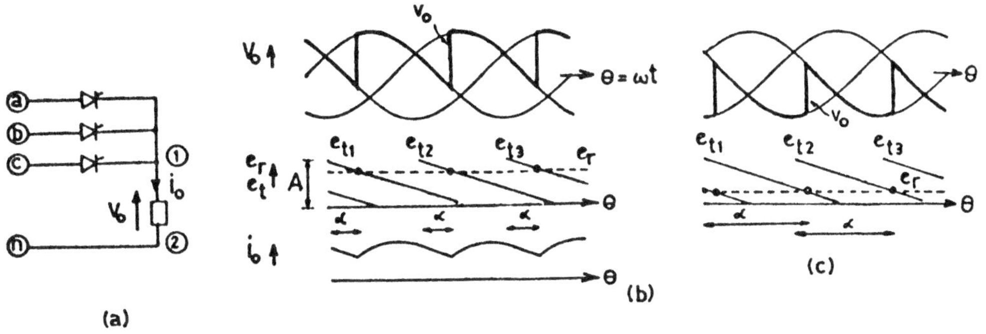

Fig. 1 Phase-controlled 3-ph converter (positive current converter)
(a) Power circuit diagram
(b) Output voltage, DC reference signal, timing signals and output current waveforms (positive mean output voltage; $\alpha < \pi/2$ rad)
(c) Output voltage, DC reference signal and timing signals waveforms (negative mean output voltage; $\alpha > \pi/2$ rad)

Let the phase voltages be

$$\left.\begin{aligned} v_a &= \sqrt{2}\,V \sin(\omega t + \frac{\pi}{6}) \\ v_b &= \sqrt{2}\,V \sin(\omega t - \frac{3\pi}{6}) \\ v_c &= \sqrt{2}\,V \sin(\omega t - \frac{7\pi}{6}) \end{aligned}\right\} \quad (1)$$

where V is the RMS value of the supply voltage, and ω is the angular supply frequency in rad/s.

On the assumption of regular firing from one phase to the next and a continuous flow of output current i_o being maintained, the output voltage v_o will appear as shown in Fig. 1b or 1c, depending on the firing angle α. It is easy to see that the mean output voltage $\bar{v}_o$ is a cosine function of the firing angle:

$$\bar{v}_o = \left(\frac{3\sqrt{6}\,V}{2\pi} \right) \cos \alpha \tag{2}$$

Eqn. (2) implies that, as long as continuous conduction prevails, the mean output voltage is positive for $\alpha < \frac{\pi}{2}$ (Fig. 1b) and negative for $\alpha > \frac{\pi}{2}$ (Fig. 1c). It is assumed in Fig. 1 that a set of linearly falling timing waves e_t are used, synchronised to the supply waves, and the firing points for the thyristors are obtained by comparing a DC reference signal e_r with the respective timing waves. From Fig. 1 (setting $\theta = \omega t$ for convenience), one can write generally

$$e_{tj} = A[1 - \frac{\theta - 2\pi(j - 1)/3}{\pi}], \; j = 1, 2, 3, \ldots . \tag{3}$$

and at the intersection $e_{tj} = e_r = A(1 - \frac{\alpha}{\pi})$ (4)

where both e_{tj} and e_r are arbitrarily assumed to be able to vary between 0 and A. In fact, with the timing waves synchronised to the supply waves, the starting point of each timing wave shown in Fig. 1 corresponds to the datum for measuring α for thyristor firing.

It follows that for the above synchronous firing control, the mean output voltage is given by

$$\bar{v}_o = \left(\frac{3\sqrt{6}\,V}{2\pi} \right) \cos \left[1 - \frac{e_r}{A} \right) \pi], \tag{5}$$

and hence it is not related to the input reference e_r in a linear fashion. This relationship is of course the result of the particular (linearly falling) timing waves used. Other types of timing waves may be employed, reflecting various degrees of circuit complexity and yielding different results in the transfer relationship. One can observe that the choice of falling ramps for the timing waves ensures that, as e_r increases, $\bar{v}_o$ becomes larger. The opposite would be true, for example, if linearly rising ramps were used.

The circuit of Fig. 1a leads to what is called "two-quadrant operation," since only the mean output voltage (not the current) can be reversed in polarity. Because the output waveform reveals 3 pulses per supply cycle, the circuit is known as a "3-pulse circuit".

By interchanging the anode and cathode positions of the thyristors in Fig. 1, the circuit appears as shown in Fig. 2, where the corresponding waveforms are also indicated. This is of course the dual of the previous circuit, since the output current can only be negative, and under continuous conduction conditions the mean output voltage is

$$\bar{v}_o = -\left(\frac{3\sqrt{6}\,V}{2\pi}\right)\cos\alpha \qquad (6)$$

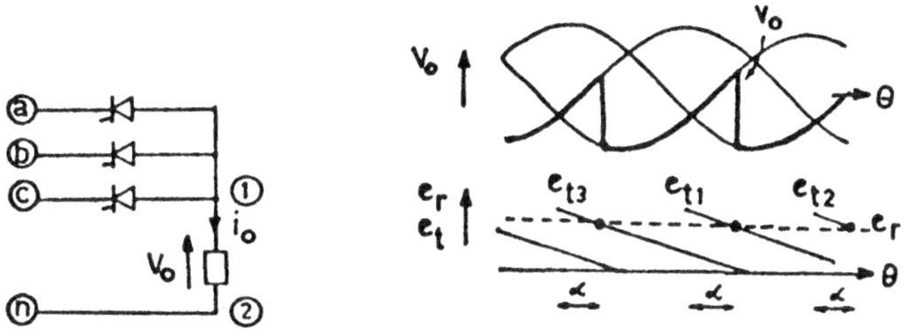

Fig. 2 Phase-controlled 3-pulse converter (negative current converter)

The two preceding circuits provide the basis for discussing the multiphase operation in the present Chapter, because they are sufficiently general, and yet not so complicated as to tend to obscure the essential issues involved. The basic principles are not at all affected by increasing the pulse number from 3 to any larger number. The two circuits are seldom used in isolation with direct connections to the supply system (except perhaps for very low ratings) because DC would otherwise be injected into the system. They are usually made practical with the aid of properly connected input transformers, not shown in the diagrams. However, when the two circuits are combined in series, the most popular "three-phase bridge converter" — illustrated in Fig. 3 — is obtained.

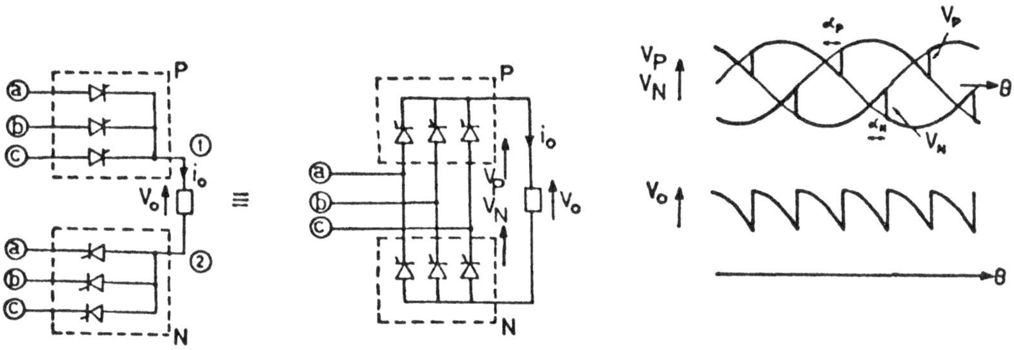

Fig. 3 Phase-controlled 6-pulse converter (3-ph bridge)

It is noted that the neutral is no longer needed. One group of the thyristors, connected to the upper rail ① of the output, is called the positive or P group; and the other group connected to the lower rail ② is the negative or N group. It is easy to see that for the same input conditions and firing angle the bridge (which is a series connected circuit) provides an output voltage twice as that obtained with any one group of thyristors as in Fig. 1 or Fig. 2. Under regular firing the output voltage gives rise to 6 pulses per supply cycle owing to the phase staggering effect. The 6-pulse converter has clearly a higher ripple frequency and lower ripple level. Other circuit arrangements can be constructed giving higher pulse numbers with monotonic improvement in performance, but since the basic operating principle remains unchanged these are not separately discussed here, for economy of space.

It can be readily deduced that in the steady state, under regular firing conditions, the interval Δ_o between consecutive firing instants is given by

$$\Delta_o = \frac{2\pi}{p} \text{ (rad)} \tag{7}$$

where p is the converter pulse number.

All circuits in Fig. 1-3 are of the two-quadrant type. When a P group and an N group are combined in parallel across the load, as illustrated in Fig. 4a for p = 3, a "four-quadrant converter" is formed.

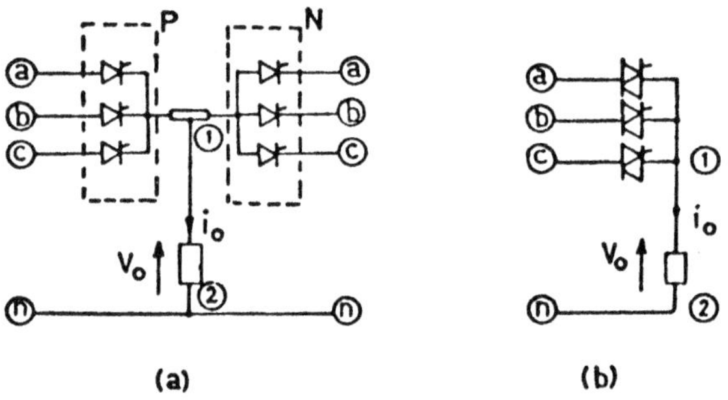

Fig. 4 3-pulse 4-quadrant converter or cycloconverter circuits
(a) Using thyristors,
(b) Using triacs

Both groups in Fig. 4a can produce either a positive or a negative mean output voltage, but each group can facilitate current flow in only one direction. The current in the load can therefore acquire either polarity. The output is shown to be obtained via a centre-tap reactor with a view to limiting any current that may circulate from one group to the other. It can be dispensed with, if the thyristors are controlled such that the two groups can never conduct simultaneously. Four-quadrant operation can be obtained using 3 triacs (Fig. 4b) instead of the 6 thyristors. In this case, although each triac acts virtually as two thyristors belonging to two groups connected in inverse parallel, there is no means for inserting the centre-tap reactor physically, and the converter must necessarily operate without permitting any circulating current to flow, in order to avoid shorting the supply lines.

Finally, by appropriately modulating the relative firing angles of the two groups of devices (actual or virtual) of the four-quadrant converter as a function of time, an alternating output voltage and current can be established in the load circuit, and, because the modulating action need not be related to the supply frequency, the output frequency can be made quite arbitrary. However, as the output frequency increases, fewer voltage segments become available for synthesising each cycle of the output wave. Waveform distortion, therefore, puts an effective limit on the upper frequency achievable in practice. The output voltage and hence current level can however be adjusted, independently of the output frequency, by proper control of the absolute values of the firing angles. In general the closer the angles are to $\pi/2$ rad, the smaller the output amplitude becomes.

With reference to Eqn. (2) corresponding to p = 3, the pseudo-instantaneous output voltage of the P group under continuous conduction conditions may be approximated by

$$\tilde{v}_p(t) = \left(\frac{3\sqrt{6}\,V}{2\pi} \right) \cos\left[\alpha_p(t)\right] \tag{8}$$

Fig. 5 shows one possible way of waveform synthesis, in which synchronised timing is obtained using again linearly falling ramps for simplicity, and applying a time-varying reference signal $e_r(t)$ to produce a sine-like variation in output voltage.

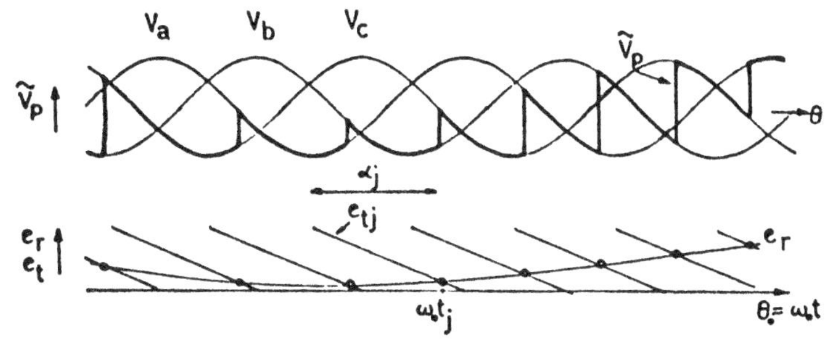

Fig. 5 3-pulse cycloconverter waveforms (for P group only)

In the case of linearly falling ramps, the jth intersection occurs at

$$e_{tj} = (1 - \frac{\alpha_{pj}}{\pi}), \; j = 1, 2, 3, \ldots\ldots \tag{9}$$

Then, selecting $c_r(t)$ as

$$e_r = 1 - \frac{1}{\pi} \cos^{-1} [r \sin(\omega_o t + \rho)] \tag{10}$$

where both e_{tj} and e_r are assumed to vary between 0 and 1 for simplicity, ω_o is the desired angular frequency of the output in rad/s, and r is an amplitude ratio varying between 0 and 1, it is clear that at the jth intersection the firing angle is

$$\alpha_{pj} = \cos^{-1} [r \sin(\omega_o t_j + \rho)] \tag{11}$$

From Eqn. (8) it follows that the pseudo-instantaneous output voltage of the P group becomes

$$\tilde{v}_P(t_j) = \sqrt{2}\, r\, V_o \sin(\omega_o t_j + \rho) \tag{12}$$

where $V_o = (\frac{3\sqrt{3}\, V}{2\pi})$ for p = 3. The amplitude V_o can in fact be changed by altering the pulse number or the supply voltage. Of course the actual output voltage is not smooth but consists of segments with a predominant sinusoidal component at a frequency ω_o as illustrated in Fig. 5.

Provided that the firing angles are by design equal to

$$\alpha_{Nj} = \cos^{-1} [-r \sin (\omega_o t_j + \rho)], \tag{13}$$

it can similarly be established that the pseudo-instantaneous output voltage of the N group follows again a sinusoidal variation:

$$\tilde{v}_N(t_j) = \sqrt{2}\, r\, V_o \sin (\omega_o t_j + \rho) \tag{14}$$

If circulating current is allowed to flow, the output voltage is the mean of v_P and v_N. Whenever this is prohibited, the output voltage is either v_P or v_N depending on the output current polarity. This is the basis of cycloconverter operation.

Reference will be made to Eqns. (2), (6), (10), (11) and (13) when considering software implementation using microprocessors.

3. MOTIVATIONS FOR MICROPROCESSOR CONTROL

The trend towards the digital implementation of thyristor phase control is rooted in the basic limitations of the rather complex analogue circuits for phase control and the need to improve qualitatively and quantitatively the functions to be incorporated in the controller. The rapid development of the LSI and VLSI technology in making the micro-controller more practically realisable and cost-effective has contributed much in accelerating this trend.

At the component level, it is generally difficult with analogue components to maintain a high degree of symmetry, consistency or stability in multi-phase and multi-channel operation. The digital approach

- reduces the need for high precision and costly components,
- is much less affected by drift and aging problems,
- minimises circuit adjustments during manufacturing and/or after it, and
- provides greater ease to achieve the required degree of accuracy or resolution.

At the converter level the move towards compact and powerful LSI and VLSI circuits leads to enhanced reliability, availability and maintainability despite the intrinsic functional complexity of the converter. This is so because their introduction

- leads to component and wiring reduction in the entire converter circuits,
- makes realistic the incorporation of self-diagnostic and testing facilities for the various circuits or sub-systems involved, and
- tends to either reduce noise susceptibility or simplify somewhat the measures needed to overcome the problem.

At the system level which, in the long run, may turn out to be an even more decisive factor than the other levels in considering its higher-volume production potential the phase control micro-controller is by far more versatile, flexible and capable than its conventional counterpart, particularly when provided with a proper bus structure. Specifically the microcontroller

- offers greater opportunities for product standardization and modular expansion/replacement,
- can be integrated much more easily with other digital controllers at the same or other hierarchical levels and/or other interfacing equipment,
- makes automatic production testing and site testing easier,

- promotes the realization of advanced and optimal control strategies for performance enhancement, hardly practicable with analogue circuits, and
- facilitates functional changes and on-site adjustment with minimum effort.

Most of the advantages described above at all levels have direct or indirect cost implications. Each one advantage or a combination of them may provide sufficient justification for the use of the micro-controller in a given particular phase control application.

4. POSSIBLE SCOPE FOR DIGITAL PHASE CONTROL

There are many aspects of multi-phase thyristor firing control, based on the phase control principle, which may be subject to digital or software implementation. Usually and desirably several aspects are simultaneously coped within a microprocessor scheme, but for the sake of clarity they are here separately described. The description is not intended to be exhaustive. It is rather hoped to shed some light mainly on the possible approaches which may be usefully adopted to fulfil some of the basic and less sophisticated objectives.

4.1 Static processing of DC reference inputs

According to Eqn. (2) or (6), under the assumption of synchronous firing control and synchronized timing waves, the mean output voltage takes the form

$$\bar{v}_o = V_m \cos \alpha \tag{15}$$

With linear ramps used for timing, it is seen that a non-linear transfer relationship results between the reference input signal e_r and the output $\bar{v}_o$, as given by Eqn. (5). This would create difficulties under feedback control situations since the system gain becomes load-dependent. In conventional analogue circuits, it is customary to employ the less easily derived cosine timing waves represented by

$$e_{tj} = A \cos \left[\theta - \frac{2\pi(j - 1)}{3}\right], \; j = 1, 2, 3, \ldots. \tag{16}$$

Then at the intersection (Fig. 6) we have

$$e_r = A \cos \alpha \tag{17}$$

Here e_r and e_{tj} are usually allowed to vary between -A and A for convenience of signal generation.

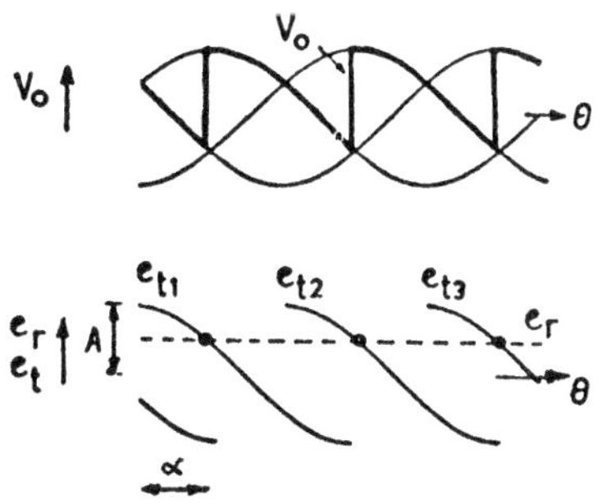

Fig. 6 Firing control using cosine timing waves

Accordingly

$$\bar{v}_o = (\frac{V_m}{A})\, e_r \tag{18}$$

and so a constant gain results. With digital control, however, it is no longer essential to generate cosine timing waves since the non-linearity can be compensated in another way. Assuming again that linearly falling ramps are used as timing waves, the input command signal ε is first transformed to an appropriate reference signal e_r (used for comparison with the timing waves) according to

$$e_r = A[1 - \frac{1}{\pi} \cos^{-1} (\frac{\varepsilon}{A})] \tag{19}$$

Here e_r is assumed to be able to vary between 0 and A.

Consequently at the intersection, with e_r given by

$$e_r = A(1 - \frac{\alpha}{\pi}) \tag{20}$$

the resulting average output voltage is found to be

$$\bar{v}_o = V_m \cos [(1 - \frac{e_r}{A})\pi]$$

$$= (\frac{V_m}{A})\, \varepsilon \tag{21}$$

which is now a linear function of the actual input command ε, as desired. This is clearly an acceptable way of digitally processing the input signal for a desirable objective. Because of the speed limitation of present-day microprocessors, the transformation based on Eqn. (19) is not normally executed by direct trigonometric calculation, but is obtained by looking up a pre-stored table. The question of actually determining the firing angle α will be discussed in Section 4.3.

Although it has so far been implicitly assumed that timing is arranged synchronously with respect to the supply voltages, firing control may also be based on adjusting the firing interval Δ in a closed loop manner. According to Eqn. (7) the steady-state interval, under regular firing conditions, is $\Delta_o = 2\pi/p$ rad. In order to change the output voltage or current, it is possible to effect only a momentary change in Δ. If Δ is transiently reduced, the firing angle is advanced and a progressively increasing output would tend to result. Likewise a reduction in output would follow from momentarily widening Δ. In this approach, sometimes called 'asynchronous firing control', the expansion or contraction of Δ must be actuated by an error related to the output voltage or current, or to some other dependent variable. Since the voltage error is expected to be of the form

$$\Delta\bar{v}_o = -V_m \sin\alpha\,\Delta\alpha, \qquad (22)$$

in order to obtain a linear gain for the control system (assuming that no other nonlinearities are present) one needs to introduce the following compensation:

$$\Delta\alpha = -\left(\frac{1}{\sin\alpha}\right)\varepsilon \qquad (23)$$

where ε is the incremental input command signal used to drive $\Delta\bar{v}_o$ towards zero. Again, the micro-controller can accommodate the trigonometric conversion by means of the table-lookup technique. In response to any non-zero signal ε, the value of cosec α obtained from the stored table is multiplied by ε, to determine the required incremental change in the firing interval Δ. The angle α may be based on the latest stored value, measured or estimated.

If there exist other known nonlinearities in the control system, they can also be compensated either in the same stroke or separately.

4.2 Real-time generation of time-varying reference inputs

The generation of time-varying signals by analogue means may call for rather complex and costly circuits. Various types of oscillators are used for sinusoidal references, but generation of multi-phase variable-frequency and variable-amplitude signals of high purity, accuracy and stability is clearly not a trivial task. Of this type are of course the reference signals needed by cycloconverters based on cosine timing waves. It would be even less conceivable to generate functions, such as described by Eqn. (10), using purely analogue circuits. On the other hand, there is relatively little difficulty in performing this task with the micro-controller. In this case two look-up tables need be created, notably a sine table and an arccosine table, for computing the function $\cos^{-1}(r \sin \omega_o t)$.

Eqn. (10) does not represent the only form of reference signals suitable for use. Since the ultimate objective is to synthesise the 'optimum' output waveform, other variations of the basic scheme or different algorithmic approaches can be well developed depending on the criterion of optimality. In any case the micro-controller provides a practical means for their generation. It would not be possible to generate them on a truly continuous basis, since it must take a finite time to evaluate each point of the time function. The way in which this problem is tackled is discussed in the next section.

4.3 Firing point determination for synchronous firing control

By synchronous firing we mean the ability to keep track of the datum for defining α from one firing point to another, and hence some means of monitoring the supply voltages. As far as the micro-controller is concerned, continuous measurement of the voltages is not essential for this purpose. To minimize its overhead requirement it would be generally preferable to inform it (by means of interrupts or otherwise) at the precise instants of only the datum points.

Given the datum or synchronisation points, the timing waves can be generated, if required according to their specified form, for instance as linear ramps. In practice, once the reference signal and timing waves are explicitly known, the points of intersection between them, and hence the actual firing instants for the respective devices, can be determined. A few alternative approaches, however, are possible for their determination.

For example, both the reference and timing waves may be physically generated by the micro-controller. Because it takes a finite time to obtain a value of either function, they are produced in the stair-case form as shown in Fig. 7.

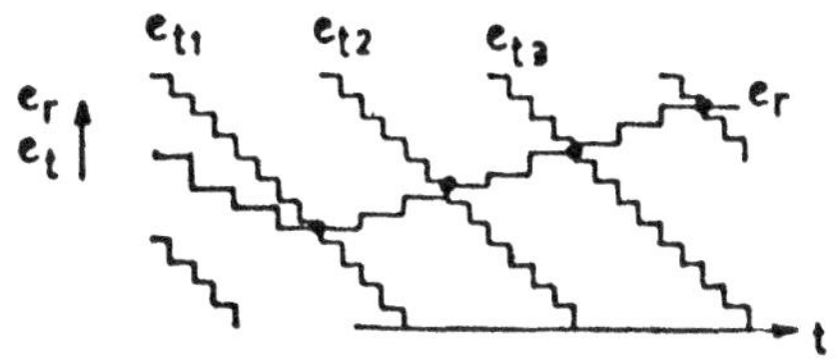

Fig. 7 Digital firing control based on natural sampling

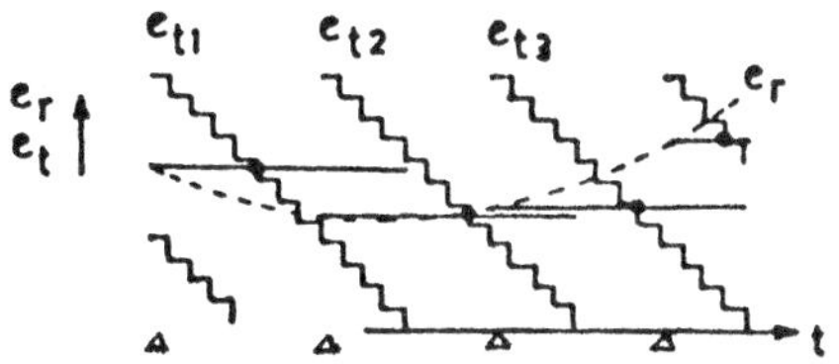

Fig. 8 Digital firing control based on regular sampling
Δ - sampling instant

Linearly falling ramps are illustrated, which start from the synchronization points. A synchronized clock may be employed for timing the width of the steps of the timing waves, which, because of the extremely simple computation involved in their generation, can be very small. There is no need for the steps of e_r to be of equal width; indeed the step width for e_r may well be much larger. By means of a digital comparator the inter-section may be readily determined in real time. This approach is seen to be a direct digital translation of the conventional analogue method based on natural sampling, and for the cycloconverter this corresponds to a physical means of seeking the satisfaction of the mathematical relationships embodied in Eqn. (11) or (13) to obtain the firing instants.

An alternative realization based on regular sampling is illustrated in Fig. 8.

Determination of the firing instants is no longer based on the natural crossing of the two waves. Instead, the reference signal values are determined at only the synchronisation points (i.e. at regular intervals) and these are used for comparison with the digitized timing ramps. Regular sampling is known to result in a reduction of the sub-harmonic amplitudes of the synthesized output waves, and may therefore be preferred to natural sampling, particularly since it is no more difficult to be implemented digitally.

It may have been thought that physical generation of the reference and timing waves is not advisable. In principle the natural intersection points can be established numerically, since both functions are explicitly known. Indeed this has been attempted, but the iterative operations, if not simplified in some ways, are much too lengthy for present-day micro-controllers.

It is only with the exploitation of the regular sampling concept, together with the pre-processing of the reference signal, where necessary, that the possibility of completely determining the firing points by software has become a practical proposition. Indeed, in this approach the reference signal values are periodically obtained or computed at only the synchronization points, and, based on the known slopes of the linear ramps, the intersection points or delay angles can be directly deduced by very simple calculation. Hence neither the timing waves nor the reference wave need be generated in real time.

The pure software approach of firing point determination is not only simple but also highly efficient and consistent with multi-phase operation. This is so because it is no longer concerned with generating a set of timing waves having the exact phase relationship with respect to each other, the correct waveshape (e.g. cosinusoidal), and the precise amplitude.

4.4 Firing execution and channel discrimination

As already explained, thyristor firing should take place at the relevant intersection between the reference signal and timing signal. When both waves are physically generated, the firing command should be issued immediately after the intersection is detected by the comparator.

The problem is of course different if the firing point is determined at some instant prior to the actual firing. In this case, the value of the delay angle for firing must be computed, usually defined with reference to the synchronization point. This delay value is then converted to a time delay counted in real time from a suitable datum, the synchronisation point normally being chosen once again for convenience. The counting can be performed by programmable timers.

In individual timer control, one timer is needed for each firing channel, and the firing angle is determined once for each channel. When the firing angle value for each particular channel becomes available, it is loaded to a dedicated timer at the selected instant. From the point of view of execution, the interval between the instant at which the value is ready and the loading instant is quite immaterial. However if this interval is too large, the converter is responding too slowly to the time-varying input commands; a fast-response scheme must therefore somehow seek to reduce or compensate this additional dead time introduced by the control methods. As long as the timers are dedicated to the firing channels, discrimination is effected when the firing delay information is transferred to the timer. This method is simple to implement, but has to commit a large number of timers especially when the pulse number used is large.

In shared timer control, the firing angle is computed independently of the channels. Once determined, its value is loaded to a timer at regular intervals, and the firing channel accepting the timer output must be selected in a cyclic manner. This method would work, provided that there would never exist any timing overlap. A glance at Figs. 1, 2, 7 or 8 will readily reveal that under these normal circumstances of timing, overlap will indeed occur. One way of circumventing this difficulty is to subdivide each supply cycle into timing belts. For a 6-pulse circuit, for example, belts of $\frac{\pi}{3}$ rad intervals are formed, as shown in Fig. 9.

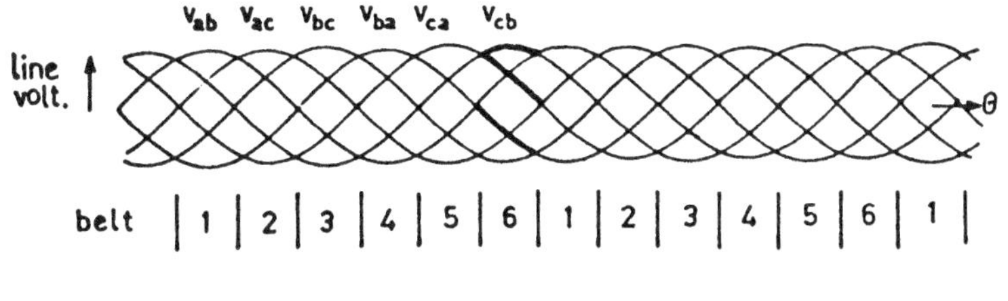

Fig. 9 Firing channel discrimination based on timing belts and firing angle value

It is recognized that, ignoring commutation overlap, at any one time only one pair of thyristors are conducting presenting at the output a voltage equal to a line voltage. Fig. 9 shows the timing belts with reference to all possible line voltages. Only one of three possible pairs can be fired, within any specified belt because the firing angle must lie between 0 and π rad. The correct pair to be selected is revealed precisely by the absolute value of the prevailing firing angle itself. For instance, in belt 6, the output can only 'see' either v_{ba}, v_{ca} or v_{cb}. If the true firing angle for this timing belt, already determined, is smaller than π/3 rad, the thyristors to be fired should result in v_{cb} being presented to the output; if the angle lies between π/3 and 2π/3 rad, v_{ca} should be presented after firing; and if the angle lies between 2π/3 and π rad, v_{ba} should be presented. The corresponding thyristors to be fired can thus be identified. A table can be formed and stored in the micro-controller to facilitate the logical decision and device discrimination, according to the timing belt reached at each time as well as the absolute value of the prevailing firing angle. A useful consequence of this approach is that the inherent dead time due to phase control can normally be contained to within one-sixth of the cycle period (since decision is taken every one-sixth of the supply cycle).

Although there can appear timing overlap in synchronous firing control, there is always a distinct interval between any two consecutive firing. This immediately suggests that the overlap problem can be simply by-passed, if asynchronous firing control based on adjustment of the firing interval Δ is adopted. After the firing interval value Δ is absolutely determined, the timer is loaded with this value at the firing point. At the end of the timing, the device(s) in the next channel must be fired while the new Δ value (already computed) should also be transferred simultaneously to the timer. Clearly only one timer is needed. As in all forms of shared timer control, the micro-controller can be interfaced to all the output channels, appropriately addressed, or alternately the single firing command can be transmitted via some form of hardwired distributor or ring-counter.

4.5 Firing signal generation

There are a few ways in which the firing signals can be produced. A long pulse of any desired width can be generated by the micro-countroller software, or more simply if required to be of constant width, by the use of a monostable multivibrator. If pulse trains are required for firing, the long pulse just described may be gated with the output of an auxiliary clock or master-oscillator. Whenever only a limited number of short pulses are desired at each firing, for example 5 pulses, these may be generated by software without reflecting a substantial burden on the micro-controller.

4.6 Imposition of prescribed limits

The introduction of nonlinearities, particularly in the form of bounds, can be easily accommodated by software. For instance it has been implicitly assumed, in the execution of the firing angles that the firing angle, determined, always falls within prescribed bounds. It is in fact normal and easy to incorporate checking procedures within the micro-controller to safeguard against inadvertent excursion beyond these bounds. The upper practical bound, for reliable inversion operation, must be somewhat below π rad. The bound is imposed basically because of the finite time required for thyristor recovery before forward bias is reapplied, but the acceptable bound is dependent on the prevailing DC level owing to commutation overlap. The situation for a 3-phase converter is illustrated in Fig. 10. The relationship between the recovery angle δ, angle of advance for firing β, the commutating inductance per phase L, and the load current $\bar{i}_o$ is governed by the equation

$$\cos\beta = \cos\delta - \left(\frac{2\omega L}{\sqrt{6}\,V}\right)\bar{i}_o$$

$$\text{or} \quad \beta = \cos^{-1}\left[\cos\delta - \left(\frac{2\ L}{\sqrt{6}\,V}\right)\bar{i}_o\right] \qquad (24)$$

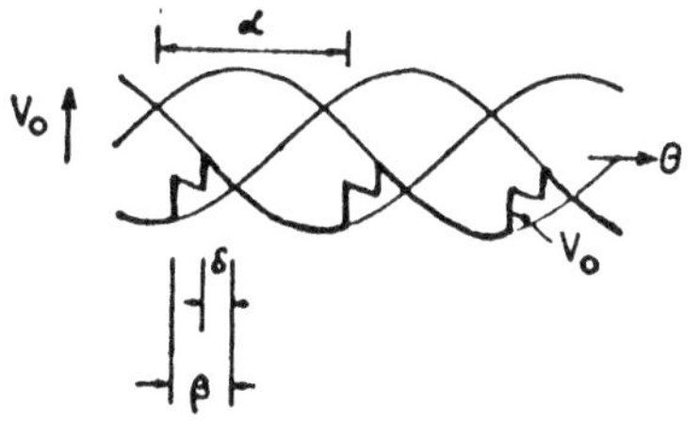

Fig. 10 Synchronous inversion waveform with commutation overlap
β = angle of advance for firing $(\pi - \alpha)$, δ = recovery angle provided for thyristor, $(\beta - \delta)$ = overlap angle

In Fig. 10 the commutation overlap angle is given by $(\beta - \delta)$. Provided that the DC $\bar{i}_o$ is available as a transduced signal or otherwise estimated, the micro-controller can be used to evaluate dynamically the limiting value for β as the load current changes, in order to keep the recovery angle δ always at an acceptable level. For each evaluation only one multiplication and subtraction step and one table-lookup (arccosine) step are involved. This clearly would not entail too much computational burden for the micro-controller. In the absence of the output current information or of a suitable controller, the conventional approach is to impose simply a constant limit β_c that can safely meet the expected full-load or worst condition, a limit which thus becomes pessimistic at all other loads.

4.7 Other functions

There are many other functions related to the control of phase-controlled converters and cycloconverters which can be efficiently performed by the micro-controller. For example, in the four-quadrant converters or cycloconverters, if the circulating current mode is prohibited, the requirement to properly service the current crossover for safe, effective and efficient operation exists. The problem is often complicated by the possible occurrence of multiple current zeros. Converter blanking arrangments which would tend to minimize waveform distortion are generally desirable.

In closed-loop control systems the problem of dynamic response and stability always arises. It is already noted that in asynchronous firing control the closed loop inevitably exists. For all but the simplest applications, the converter normally includes an output current loop to speed up response and to facilitate the incorporation of a protective limiting condition on this variable. All these imply the need for dynamic compensation. The micro-controller can readily accommodate a PID-type or a more sophisticated type of control algorithm to optimize performance. In view of the finite and variable dead time introduced by the inherent phase control mechanism and by the particular digital control methods adopted, as well as other possible uncertainties present, the converter assumes a variable structure that would defy the efforts of non-adaptive controllers. The micro-controller can be used for a variety of signal or parameter estimation operations to facilitate advanced control.

The above discussion has been deliberately confined to regular symmetrical firing, without other features such as controlled freewheeling, etc. However it is obvious that the micro-controller would be more necessary for the more complicated schemes. The flexible nature of software control should make the required modifications and extensions much less problematic.

In Section 3 we listed many of the worthy objectives of microprocessor-based control lying at the converter and system level of implementation. Although the present treatment is not intended to go much beyond the basic problems of phase control, it should be mentioned that these objectives will influence the practical development of firing control including its reliability and performance aspects.

4.8 A micro-controller example

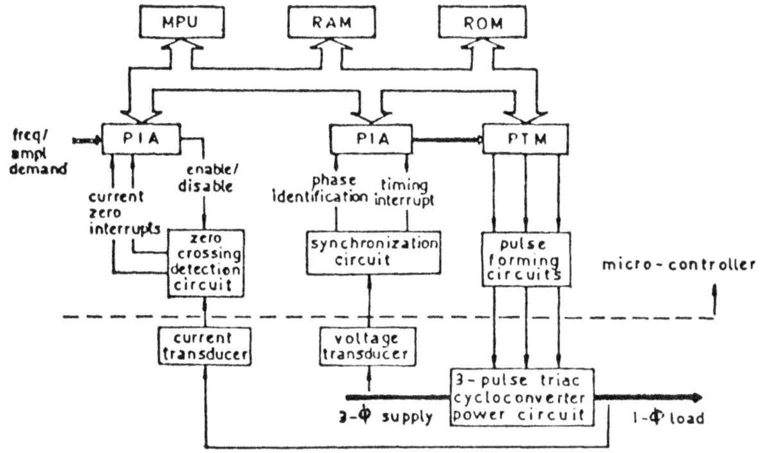

Fig. 11 Micro-controller for a triac cycloconverter

Fig. 11 illustrates the arrangement of a micro-controller based on Motorola products controlling a 3-pulse triac cycloconverter. For the sake of brevity, the circuit requirements for only one output phase are shown, although 3-ph output schemes are typically implemented. The micro-controller includes a microprocessor (MPU), memory (ROM and RAM), and peripheral interface adapters (PIA). A programmable timer module (PTM) is used for firing delay execution. There are three independent timers in the module, each responsible for a triac connected to one of the input phases. They are periodically loaded with the binary words representing the computed delay values. The firing pulses are obtained from three monostable multivibrators triggered respectively by the timers. Both the PIA's and the timers are connected to the micro-processor bus and addressed as memory. There are two other important interfacing circuits. The synchronisation circuit provides both the regular interrupts for initiating timing control and a means of identifying the respective input phases. The zero crossing detection circuit generates an interrupt whenever the single-phase output current crosses zero from either polarity. Two sets of pulses are produced corresponding to the alternate half cycles. In order to minimise the risk of transmitting false zero crossing information into the micro-controller, this detection circuit is enabled only in the region of an expected cross-over and is otherwise disabled.

Not shown explicitly are some buffers, decoders and counter latches added to the interfacing circuits. The frequency/amplitude command for the cycloconverter is entered as shown through a PIA as a binary word into the micro-controller, assuming open-loop control.

It is clear that the MPU is interrupt driven. The program contains several interrupt routines and as soon as an interrupt is received and checked, the appropriate routine will be executed. In this example, the software tasks include the real-time generation of a time-varying reference input, firing point determination for synchronous firing control based on regular sampling, firing execution based on individual timer control, imposition of firing angle limits, P and N converter group selection, blanking control, and also pre-processing of the input command for soft changeover conditions. The software repertoire may of course be further expanded, especially when feedback control is introduced in the cycloconverter system.

An indication of the performance obtainable with a typical microcontroller applied to a 3-phase triac cycloconverter is given in Fig. 12. The performance depends, of course, on the actual algorithms implemented, which are really beyond the scope of the present discussion.

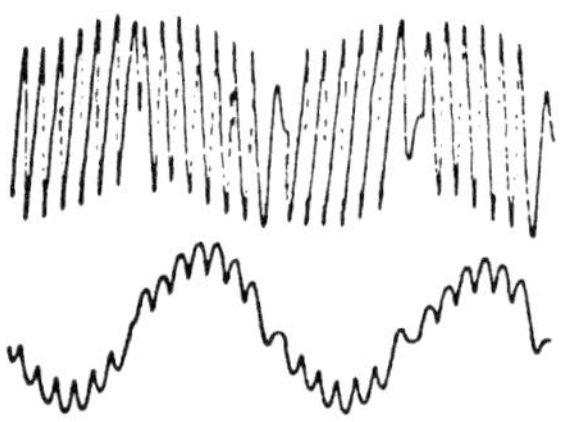

Fig. 12 Output voltage and current waveforms of microprocessor-controlled 3-pulse triac cycloconverter

5. CONCLUDING COMMENTS

An attempt has been made in this Chapter to review the principles of phase control as applied to converters and cycloconverters, and to provide some insight into their operation which makes digital and microprocessor control favourable. Development in this direction has been greatly motivated by the wide range of converter and cycloconverter applications, as well as the rapid advances in microprocessor technology.

Actually, there is a relatively high degree of permanency in the general concepts of phase control and the possible solution methods, although there may be future revisions, refinements or extensions. Therefore the phase control techniques described above may be regarded as the basis for practical experimentation leading to large-scale commercial applications. In contrast, the microprocessor/microcomputer technology is rapidly changing. It is usually difficult, for example, to draw the line between a microcomputer and its interfacing circuits. This is why there is much less emphasis given in this Chapter on the hardware implementation aspects, examples of which may be found in the recent literature.

6. REFERENCES

1. Schaefer, J.: 'Rectifier Circuits, Theory and Design' (Wiley, 1965).

2. Pelly, B.R.: 'Thyristor phase-controlled converters and cycloconverters' (Wiley-Interscience, 1971).

3. Mc Murray, W.: 'The theory and design of cycloconverters' (MIT Press, 1972).

4. Gyugyi, L., and Pelly. B.R.: 'Static power frequency changers' (Wiley-Interscience, 1976).

5. Chen, H.H.: 'A microprocessor control of a three-pulse cycloconverter', IEEE Trans., 1977, IECI-24, pp. 226-230.

6. Singh, D., and Hoft, R.G.: 'Microcomputer-controlled single-phase cycloconverter', IEEE Trans., 1978, IECI-25, pp. 233-238.

7. Tso, S.K., Spooner, E.D., and Cosgrove, J.: 'Efficient microprocessor-based cycloconvertor control', IEE Proc. B, 1980, 127, pp. 190-196.

8. Tso, S.K., and Ho, P.T.: 'Dedicated-microprocessor scheme for thyristor phase control of multiphase convertors', IEE Proc. B, 1981, pp. 101-108.

9. Tang, P.C., Lu, S.S., and Wu, Y.C.: 'Microprocessor-based design of a firing circuit for three-phase full-wave thyristor dual converter', IEEE Trans., 1982, IE-29, pp. 67-73.

CHAPTER 16

INDUSTRIAL ROBOTS AND THEIR MICROCOMPUTER CONTROLS

M. Bertino
SOFT POWER, TORINO
Torino, Italy

The development and expansion of industrial robots is directly related to the progress reached in electronics, and particularly to the availability of powerful microprocessors and sophisticated sensory systems. Robots are characterized by their mechanical structure (type of motion, degrees of freedom), by the actuator's type (pneumatic, hydraulic or electric motor), and by the hardware and software architecture of the control unit. The presently most versatile structure seems to be the anthropomorphic one. The control of motion for a robot of this type has a high degree of complexity, since it is necessary to correlate the movement of the joints to obtain a rectilinear path for the end-effector. The solution of the equations of motion requests a powerful microcomputer, in terms of computation capability and instruction execution time. The control unit must, also, have considerable Input / Output real time capabilities to connect the robot with the external environment. The problems related to the man-machine interaction and to the sensory perception are not yet satisfactorily solved; the development in this field will actually bring a new generation of more powerful robots.

1. TYPES OF INDUSTRIAL ROBOTS: MECHANICAL STRUCTURE, DEGREES OF FREEDOM, HARDWARE AND SOFTWARE CHARACTERISTICS, AREA OF USE.

Industrial robots made their first appearance in 1960, in the United States. These machines were developed not to imitate man's movements or behaviour, but to solve specific automation problems related to production processes.

At the beginning, these machines had limited capability and poor reliability.

Since then electronics made tremendous bounds forward and the LSI technology, together with the development of microprogramming

S. G. Tzafestas (ed.), Microprocessors in Signal Processing, Measurement and Control, 345–365.

techniques, made it possible to build the "microprocessor". Microprocessors together with fast and high capacity semiconductor memories set up the basic nucleus of microcomputers whose characteristics are comparable to those of more expensive and cumbersome devices. Among others microcomputers have found useful applications in the industrial field, due to such features as low cost, reduced power consumption, high reliability and realtime Input/Output capabilities. From the very beginning, robots are constantly improved following the electronics progress, so as not only to improve the existing devices but also to allow the construction of new generations machines.

Industrial robotics was stimulated (and actually is) to the realization of a variety of automatic machines, from the simple to the more complex ones, by the automation demands of different production processes.

This caused a fast and tremendous growth and also a lack of meaning for the term robot: sofar it doesn't exist a universally accepted definition stating precisely the necessary properties for an automatic machine to belong to the robot class. Actually, not all presently produced devices,which are defined as robots, are really robots.

The Robot Institute of America defines a robot as a "reprogrammable multifunctional manipulator designed to move material, parts, tools, or specialized devices through variable programmed motions for the performance of a variety of tasks". The Japanese Industrial Robot Association defines as a robot any machine possessing the characteristics stated in the previous definition, and also includes manual manipulators performing fixed or preset sequences. This less restrictive definition is not generally accepted by European Association, wich usually agree with the American Institute.

Thus the fundamental characteristic of the robot seems to be the working cycle reprogrammability and the capability to change the trajectory and end points. This presupposes the availability of some teaching method for the working task, and the existence of a memory to store the related information. It is also presupposed a more or less precise perception of the outside world through the use of sensors (ON/OFF or proportional), so as to correlate the working cycle with the external events.

Here by the term robot we will refer to devices with the above characteristics, not considering manipulators or loading arms with limited, repetitive, not intelligent movements, or remote controlled manipulators having a human factor in the control loop (for example those used for the treatment of radioactive materials). Note however that the aforementioned restriction set a lower boundary between"robot" and "not-robot" only approximately.

A first robot classification can be made on the basis of their geometrical and mechanical characteristics.

The three most common structures of a robot are the polar (or cylindrical), the cartesian and the anthropomorphic configuration (fig. 1). Each one has its own advantages and is adopted on the basis of the

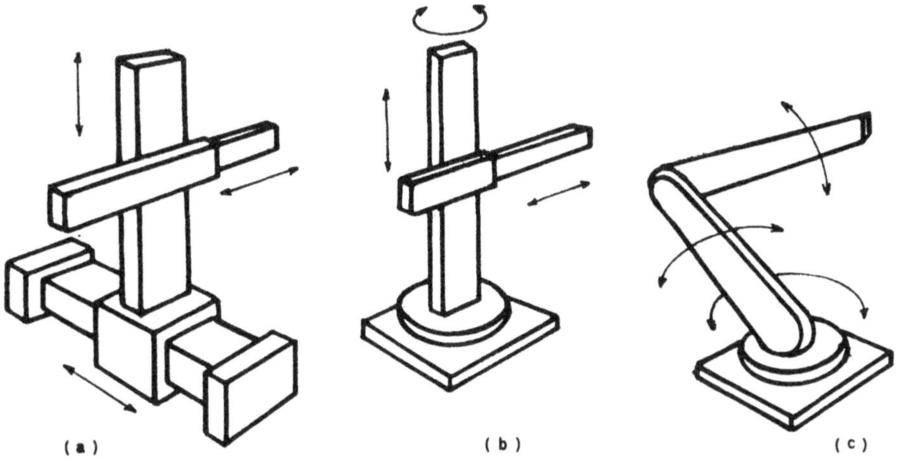

fig. 1 - Robot's mechanical structure (3 degrees of freedom)
Cartesian (a) - Cylindrical (b) - Anthropomorphic (c)

specific tasks for which the machine is intended. The anthropomorphic structure possesses probably the largest working area, related to the robot dimensions, but it is more critical, regarding pay-load and positioning accuracy. It also requests a more complex control strategy, if the end-effector is to be moved along straight spatial path. The cartesian (rectilinear) structure does not exhibit the above control algorithm difficulties. It is however quite cumbersome; its use is generally restricted to the area of automated assembly and part mating, where position accuracy and speed are of primary importance. The polar structure can ben considered to lay between the previous two ones. It is generally adopted for devices of medium complexity featuring a good pay load capability and positioning accuracy.

Having defined the geometric structure, we can classify a robot considering their available degrees of freedom.

It is well known that to identify an object in the space, three coordinates are necessary to define the point, and three more coordinates to define the object orientation to that point. Therefore a robot should have at least 6 degrees of freedom, that is 6 independent movement possibilities. More degrees of freedom are necessary if we require the end-effector to reach a point and orientation in the space following different paths, that are determined by external constraints.

Clearly the more the axes of a robot are, the more the mechanical structure and the control unit will cost.

Fortunately, an analysis of the current applications in industrial automation shows that rarely robots necessitate more than 6 degrees of freedom. The majority of the presently available manipulators have a number of independent movements between 3 and 6. Machines with more than 6 degrees of freedom, featuring straight path movements and tool attitude control for the end-effector, reach the maximum complexity obtainable with the current technology, as far as commercial feasibility is concerned. Spray painting robots can be an exception, regarding the degrees of freedom, but in this case the control strategy is simpler and the mechanical structure is less subject to the constraints of repeatability and pay-load.

To drive the articulated joints of a robot devices of different type can be utilized: namely pneumatic, hydraulic or electric actuators. Normally only one type is selected for axes movement of a particular robot; so it is justified to characterize the robot as pneumatic, hydraulic or electric one.

Pneumatic actuators are less expensive, but they allow limited control capabilities and poor pay load capacity. They are usually employed in less complex machines, performing simple repeatitive work-cycles without trajectory control between starting and ending points. Such manipulators are currently named 'pick-and-place', and as a consequence of their low cost are largely employed in medium complexity and highly repeatitive automated processes; at the present time the majority of industrial robot installations concerns the pick-and-place type.

Hydraulic actuators allow a more precise control of the articulated structure, thanks to the availability of transducers (servo-valves) delivering an oil flow-rate proportional to the control current. A system with such regulating capabilities can advantageously adopt the feed-back concept with the aid of transducers (position, velocity, torque), so as to realize, through the control unit, servo-loop for position and velocity. Hydraulic robot can reach the highest degree of complexity along with the largest versatility that present tecnology may achieve. They show two main disadvantages. To achieve good system dynamic qualities, high oil pressures are requested (about 110 bars) and as a consequence cumbersome hydraulic power units. High fluid pressure implies in general the risk of leaks, of clogging and, above all, of component wear.

Electric servo-motors possess comparable or even superior characteristics over the hydraulic ones. The development in the area of numerically controlled machine-tools and, later, in industrial robots stimulated the research and development of D.C. servo-motors particularly suited for these types of applications. Very good dynamic responses (a few tens of milliseconds to a step change) are common features, in accordance with the driving capabilities of the control unit. Less power consumption, more torque and better reliability make more and more preferable the choice of D.C. servo-motors for industrial robots, from the simpler to the more sophisticated ones. The only disadvantage is the weight and the volume of these devices, wich is higher as compared to the hydraulic actuators. As a consequence,it is often not easy (or not possible) to mount the electric motors near the joints of the structure by increasing the number of gears for the transmission of the movement.

The characteristics of the electronic control units settle further differences between robots, regarding performance and flexibility of application. They define essentially two classes: servo-controlled and non servo controlled robots. The presence of feed-back of any type is strictly dependent on the availability of a transducer system, sending to the central unit information related to the controlled parameters. The greater complexity of a servo controlled system, is justified by the better performance obtainable. In fact a non-servo loop control system can not perceive the effect of the driving commands it sends; so it is not guaranteed that movement follows the theoretical behaviour. This is particularly true when in the working cycle there are time varying conditions (i.e. the pay-load) which can affect spatial repeatability.

The different capabilities of a control system set the boundary between the classes of point-to-point and continuous path robots.For point-to-point manipulators there is no control between starting and ending point of each trajectory, and the axes movements are not correlated to each other. Continuous path robots belong to a superior class, allowing a complete control of the tool center point. Given the starting and ending point of a spatial trajectory, it is possible to correlate articulation velocities so as to move the end effector along a predetermined path with desired orientation. Since the control is continuous,it is also possible to fit the velocity to the forces exerted on the tool and, consequently, modify the motion. The aforementioned characteristics give a high degree of versatility, allowing satisfactory results in many different tasks.

Versatility is strongly related to the programming methods for an industrial robot; i.e. to the selected procedure to teach a robot the assigned task. Since the teaching process interacts with human factors (the operator programming the machine), it should be easy to understand and to operate, even in complex working cycles. The type of programming method is also strictly conditioned by the specific task for which the robot is intended.

There are three primary programming methods for a robot: walk through, where the robot is physically manipulated through the desired motions; lead through, in which the operator controls the robot through a remote teach box; and plug-in , where the robot operates via prere - corded off-line programs. The first method is almost exclusively adopted for spray-painting robots, where the programmed motions cannot be easily characterized, but rather subjected to the empirical experience of the operator. The end effector, carrying the painting tool, is physically trailed along the whole path. During this phase, the control unit sam - ples the position transducer signals and stores the corresponding data; they are later read to play-back the work cycle taught. The sophistica- tion requested to the control unit is poor, because it is not necessary to employ complex elaboration techniques nor advanced sensory systems. On the other side the data storing capacity must be very high, given the amount of points to be memorized: to do this magnetic supports such as floppy disks are normally employed.

The two following teaching methods show various levels of complexity which depend on the application, and, sometimes, can be ad - vantageously combined to split the programming process in two subsequent phases. Through the teaching box, which can be portable or fixed at the main panel, the operator moves, under the central unit control, the axes of the robot, and records only the ending point of each segment of the working cycle. A set of additional information is associated to these coordinates;such as parameters defining the trajectory characteristics (velocity, orientation, etc.), the input/output behaviour and the de - cision-making situations. Off-line programming, extensively used for numerically controlled machine-tools, shows special characteristics when applied to robots.It is used in a fairly rudimentary form in low complexi ty manipulators (i.e. point-to-point), where the amount of stored infor mation is minimum and the low cost of the machine do not justify on-line programming facilities. It is used however in high technology, multitask robots for which on line programming by the operator could become diffi cult.

The characteristics and the relations existing among the three main components of a robot, namely mechanics, electronics, software, establish whether a given machine belongs to a definite area of use. Although robots are spread among many different industrial processes,we may find four main sectors of application, i.e. spray painting, handling, assembling and measuring robot. Each sector requests well defined featu res by the manipulator, apart from their sophistication level. A spray- painting robot must have a relatively light and highly articulated me- chanical structure, so as to allow the operator to teach the machine working movements very similar to those performed by a human being. An assembly robot must be fast to reduce the operating time, precise to mate parts, and equipped with a sensory system to control the assembly process. A handling robot must be flexible, so as to cope with the associated highly varying process; it should have a good position repeatability, a large working area and a good pay-load capability. A measuring robot must attain very high positioning accuracy within the limits requested by the measurement; it must be fast and capable to adapt to the positioning variability of the object.

Given a certain robot class, it is possible to compare performances on the basis of the aforementioned characteristics, even if it is not always easy to interpret the available technical data. As stated at the beginning of the paragraph, there is a great indetermination in the use of terms and in the characterization procedure, due to the lack of standards. A practical example may clarify the problem.

A common parameter is "positioning accuracy"; but it is meaningless if there is not a standard definition on terminology and characterization. Referred to a robot' positioning accuracy' may signify: positioning repeatability at the end point of each trajectory segment; degree of agreement between theoretical and actual position; difference between teaching and playback position. The associated numerical value may be a maximum, a medium or a standard deviation.

Although a heavy standardization process should not be carried out, as it can hinder the rapidly developing field of the industrial robot, efforts should be made to set standards for some important items. This will be advantageous not only for the user but also for the manufacturer himself, who will thus be able to characterize more appropriately the features of the machine.

2. PATH CONTROL FOR SERIAL LINK (ANTHROPOMORPHIC) MANIPULATORS.

Although the joint coordinates of this type of manipulators fully specify the position and orientation of the end-effector, they are not suitable as a working coordinate system because they are not orthogonal and do not separate position from orientation. We can think of a continuous arc welding or grinding process: the workpiece is usually referred by the operator to a cartesian three-dimensional space. The operator will define the path followed by the robot during the working cycle in terms of lines belonging to this orthogonal frame. Therefore it is necessary for the tool to move making reference to cartesian coordinates, even if the joints forming the kinematic chain do not accomplish cartesian movements. It follows that the correct movement of the end effector is obtained by combining the joints movements.

From a general point of view, the principles establishing the manipulator control philosophy are well understood. Although other information can be used for control strategy, we shall restrict our discussion to closed-loop controls, based on measurements of the manipulator state (position and/or velocity) in joint space.

Closed-loop controls are able to continuously compare theoretical computations performed on suitable system function to measured quantities relating the actual state of the system, and output the necessary correction commands. The more frequent this comparison, the more narrow the manipulator deviation from theoretical behaviour is. It is the necessity to continuously repeat a set of computations and measurements in short time that limits the complexity of the system function and requests the use of more or less justifiable approximations. For this reason the development of control algorithms for serial link robots is still a subject of research. Note, furthermore, that such developments are partially subject to the availability of fast central units having a very high computation rate, at a justifiable commercial cost.

To deal with the problem, the proposed solutions can be grouped according to two different strategies: the first one utilizes the system's dynamic properties to derive equations based on corresponding quantities, the second analyses the kinematic relations of the structure. The requirements of the problem are in any case the same; given path and orientation of the end-effector in cartesian coordinates, it is necessary to determine path and velocity of the joints in a suitable coordinate reference system and drive the corresponding actuators properly.

The dynamic control strategy must take into account the high degree of non-linearity for the related parameters. To drive a robot carrying variable or unknown loads along a predetermined trajectory, one must repeatedly compute the required forces and torques which drive all the joints appropriately, at an adequate sampling frequency. Each joint is driven by an actuator (i.e. D.C. motor) and each motor axis is coupled with a position transducer (encoder, resolver, etc.) and a velocity transducer (tachogenerator). To describe the motion of the manipulator, let $\vec{q}(t)$ be a n-dimensional vector representing the actual displacements of the n joints. From the robot structure and the movement characteristics we find the dynamic quantities involved in the equation of motion. In general this equation can be written as:

$$\vec{J}(q)\cdot\ddot{q}+\vec{V}\cdot\dot{q}+\vec{F}(\dot{q}_i\cdot\dot{q}_j,q)+\vec{G}(q)=\vec{f}$$

where:

$\vec{J}(q)$ = inertia matrix (nxn)
$\vec{V}$ = viscous friction matrix (nxn)
$\vec{F}$ = vector defining Coriolis and centrifugal terms (nx1)
$\vec{G}$ = vector defining the gravity terms (nx1)
$\vec{f}$ = vector of input generalized forces (nx1)

The inertia of the robot as well as the gravitational force are dependent at each instant on the joint position, and so are dependent on $\vec{q}$. The Coriolis and centrifugal forces, howewer, depend on the product $\dot{q}_i \cdot \dot{q}_j$ (i,j = 1...n) of the joint velocities, besides their dependence on position. The equation of motion is mathematically exact, provided that we know the parameter values entering the equation itself. The computation starts from the measured values q and $\dot{q}$ and ends up at the determination of $\vec{f}$, i.e. at the amount of the command for the joint actuators.

The computation must be performed "on-line" with a suitable repetition frequency to guarantee the convergency of the control algorythm. Consequently the formulation of the equation of motion, which affects the speed of computing the output forces, determines the feasibility of implementing the control schemes.

Until now many formulations of the equation of motion are proposed, along with different levels of system approximation. In particular, for serial link manipulators, the method of Lagrangian generalized coordinates leads to a systematic representation. An alternative procedure for deriving the equation is through the method of virtual work, but this approach offers only a different sequence of steps needed for the derivation. Both procedures suffer from the fact that they need a very large quantity of calculations, unless it is possible to make severe approximations, at the expense of solution general validity. A third formulation is based on the Newton-Euler method, wich yields a set of recursive equations that applies sequentially to all joints. Due to this property it is possible to share the computation process between several, parallel working, elaboration units. Problems however may arise, concerning the coordinate systems. Complex coordinate transformations are actually needed to express the results in a cartesian system.

The kinematic approach seems to have greater possiblity of success, even if the necessary computations tend to be cumbersome, when the robot structure is particularly complex and the number of links is higher than 6. A serial link manipulator consists of a sequence of mechanical links connected together by joints (fig. 2). This structure constitutes a kinematic chain and can be analyzed by mathematical methods. This analysis yields the equations expressing manipulator end-effector cartesian position and orientation in terms of the joint coordinates. These equations can always be obtained, independently of the number of links or degrees of freedom. It may be possible that a unique solution does not exist; i.e. given a certain spatial position of the end-effector, more than a single set of joint coordinates may exist, satisfying this condition. In practice only one solution is desired which can be obtained by imposing further constraints on the joint movements.

It is possible to formulate the kinematic equations by using essentially two methods.

The first is the so called "trigonometric" method in which a reference cartesian frame is assigned coincident with the robot base. The n-joint coordinates are derived with reference to this frame. The method is called trigonometric because the derivation of the joint coordinates is made with the aid of angles and segment projections and their trigonometric relations. From a mathematical point of view, the procedure does not present particular conceptual difficulties. But if the implementation in a control loop is concerned, considerable problems may arise when the number of joints is high: the equation relating joint positions to the end-effector in the reference cartesian system becomes complex and requests the computation of a very high number of trigonometric functions.

The second method is based upon the assignement of a coordinate frame to each robot joint. This leads to the definition of a distance d_n and angle ϑ_n between two links connected with the same joint (fig. 2). In the case of revolute joint ϑ_n is the joint variable, whereas

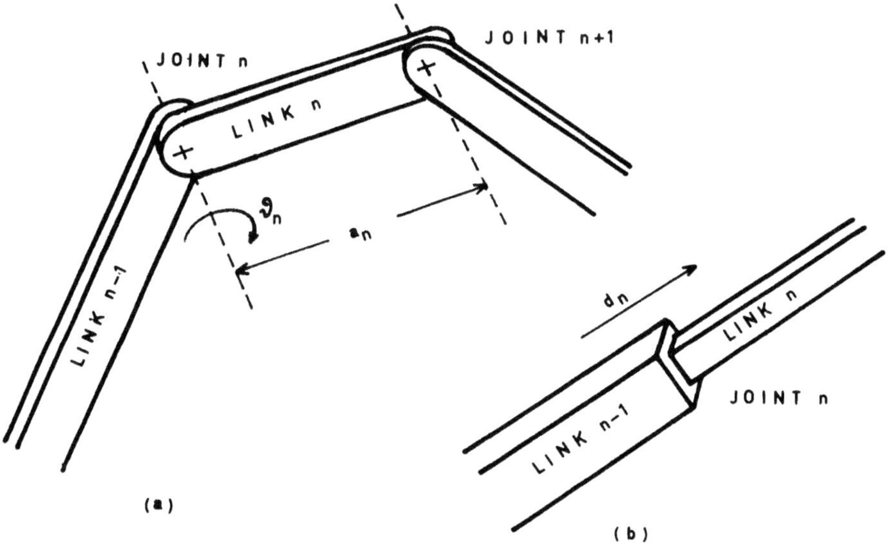

Fig. 2 - Revolute joints (a) and prismatic joint (b)

in prismatic joint the joint variable is d_n. The relationship between successive frames n, n+1 can generally be obtained with rotation and traslation. Therefore coordinate frame of link n+1 related to the coordinate frame of link n may be expressed as the product of four homogeneous transformations represented by a matrix A_{n+1}. So it is possible to express position and orientation of the final link, with respect to the base, in terms of the variables ϑ_n and d_n

$$T = A_1 \cdot A_2 \dots A_m$$

As one can see, the formulation is straightforward and the method is applicable to complex structures. In practice difficulties may arise, due to the necessity to compute matrices containing, among others, trigonometric functions.

Obviously, due to the requirement to solve only linear equations, the kinematic approach is not subject to the approximation errors which occurred in the dynamic approach. The solution is mathematically exact, subject solely to rounding and truncation errors during the process of the computation.

The success of a control strategy depends not only on the correct formulation of the equation of motion, but also on the capability to solve these equations a large number of times during the trajectory. This computation frequency can be derived with good approximation.

The computation of joint coordinates, allowing the output to joint actuators, is a discrete process lead at a fixed sampling period t_o. This period is a function of the maximum rate of change of manipulator set point and determines the tracking accuracy of the preprogrammed trajectory. The maximum rate of change of the set point of a manipulator can be approximately related to the resonant frequency and damping factor of the structure. In order to control such a structure without oscillation problems two conditions must be met:

1. The frequency spectrum of the driving function (the output to the actuators) should be limited to about 1/8 of the resonant frequency. A higher frequency will only excite structural resonances.
2. A sample rate of approximately 8 times the highest frequency component is needed to describe such a function accurately. Thus

$$t_o \leqslant \frac{1}{f_{resonant}}$$

Considering that serial link manipulators have typical values of $f_{resonant}$ ranging between 10 and 30 Hz, and that it is a good practice to sample with a period $t_o/2$, to ensure a good stability margin for the control loop, it follows that the function's computation must be performed at least 60 times per second.

3. HARDWARE STRUCTURE OF A CONTROL UNIT FOR A SERIAL-LINK MANIPULATOR WITH SIX DEGREES OF FREEDOM.

When the hardware structure of a robot control unit is considered, the first question arising is whether we are concerned with a multi-micro or single-micro architecture. We may find however this difference only in highly complex manipulators, like the one under consideration. For medium and low level manipulators the single microcomputer architecture is generally adopted. In a multi-micro control unit a separate microcomputer is dedicated to each movement axis, together with the related interface modules for the position transducer signals, and the actuator's drive command. Normally an additional microcomputer exists, co-ordinating and supervising the entire process and carrying out all common tasks which cannot be distributed. A single-micro architecture employs only one computer to control the robot. Actually several arguments can be given against or in favour of either solution, but the fundamental points are involved in the following discussion.

The distributed intelligence solution allows the subdivision of tasks between microcomputers (at a loss of simplicity of the hardware structure), so as to avoid saturation of the elaboration capability of the central unit. In fact, as mentioned, the software for a serial-link manipulator with five or six degrees of freedom is complex and the control algorithm of the trajectory is subjected to severe time limits. If we consider however the enormous improvements attained in the semiconductor field, which offers 16 and 32 bit microprocessors with powerful instruction set and small cycle time, the distributed architecture philosophy loses very much its necessity. Today, with the microprocessors of the last generation, valuable devices are available allowing the centralized control of a robot. In general we may state that the multi-micro and single-micro options do not originate from different philosophical statements, but rather from the availability of less or more powerful devices.

Whatever approach is adopted, the necessity of a modular structure is always emphasized. This is because there is such a high complexity that only through modularity the engineering design, the industrial feasibility and the maintainability of the control unit are guaranteed to be satisfactory. Modularity must possess the following features:

- Functional modularity, which reflects the tendency of a system to remain constant over many possible system sizes. This concept involves both hardware and software because only a modular hardware structure allows the configuration of a functional modular software. In practice this implies, for example, that if we wish to add a further movement axis we do not have to reconfigure the whole system, but only to integrate a set of new functions (hardware and software) in the existing one.

- Physical modularity which concerns more directly the hardware architecture. This must be structured in such a way that one or more (similar) physical units (cards) correspond to each functional unit. This organization gives undoubtely many advantages regarding the maintainability and the overall reliability of the system.

In the following we describe a single-microcomputer control unit for an anthropomorphic manipulator with six degrees of freedom (fig. 3). The proposed architecture is as much as possible in agreement with the above aspects of modularity.

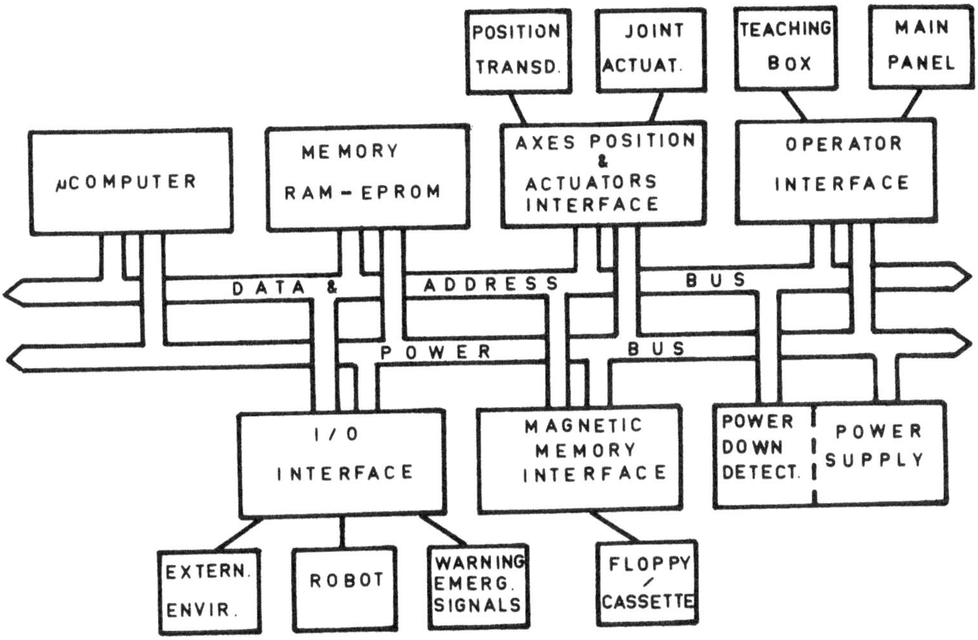

fig. 3 - Robot's control unit block diagram.

The principal unit is the microcomputer with the related memory banks: this constitutes the intelligent element of the system. The microprocessor choice and, as a consequence, the architecture of the above unit determines the control unit power and flexibility. On the ground of the preceding discussion, the microcomputer's characteristics must be very high; this is translated in a series of requirements, as for example:

- arithmetic and logic unit (ALU) capable of 16 bit data manipulation and complex arithmetic operations, such as multiplication and division.
- powerful real-time Input/Output and interrupts management.

- wide instruction set with many addressing modes.
- short instruction cycle-time.

Some commercially available devices possess characteristics which are very close to the above requirements. Examples are the 9000 series of Texas Instruments microprocessors, the Motorola 68000 , the Intel 8086 and so on. For a quantification of the aforementioned qualitative features, the corresponding data sheets may be examined. The memory is composed of a read-only type (PROM, EPROM or ROM), storing the general control program of the robot, and of a random access type (RAM) used as a working area for program execution and for the storage of the variable working cycle taught to the robot. RAM may be in turn subdivided into volatile memory (it loses all information during power-down), and non-volatile memory (usually CMOS with battery back-up) used to save information that the control unit can not loose even during power-off.

From a physical point of view,the microcomputer is normally subdivided into a set of "cards" containing respectively the computing unit and the memory banks. Concerning the memory we mention that generally a magnetic support is present (cassette or floppy disk), storing the working programs of the robot; the one selected is transferred to the main memory and then executed. This magnetic memory is considered as a peripheral unit, rather than as an extension of the main memory, and, therefore is considered not to belong to the microcomputer.

The microcomputer interacts through a signal bus with a set of interface modules, providing the necessary information,and receiving and traslating the commands.

The unit used for the acquisition of the position data and for the output to the joint actuators constitutes the interface for the robot's movement. A position (incremental) transducer (optical encoder, resolver, etc.) is connected to each robot's joint, through which it is possible to know the position of the articulations at every instant. The interface converts the transducer signals to a convenient numerical information, continuously updated, which is read by the microcomputer and utilized for the movement command's computation. These commands are converted to appropriate signals, normally of the analog type, which drive the specific actuators employed.

A typical closed loop for the robot position control, based on a kinematic algorithm , is realized in the following manner.

At constant time intervals the central unit computes the six joint coordinates necessary to move the robot's end-effector of the quantity corresponding to the above time interval, along the working trajectory and with the necessary orientation (on the basis of the requested travelling speed). These coordinates are compared with those received from the transducers. From this comparison, error terms are derived, which are passed through a Proportional-Integrative-Derivative (PID) function with coefficients dependent on the system characteristics.The six error functions are sent to the driving actuator interface. As stated in the previous section, the whole process must be carried-out at high frequency, to guarantee feedback loop stability and minimize path deviation from theoretical behaviour.

The third functional module of the control unit can be called "operator interface". This is a very important unit, because the selected architecture determines largely the robot's versatility and ease of operation. Normally the operator interacts with the control unit through two different terminals: the main panel and the teaching box. The main panel is always in an active state and supports all the push-buttons, the selectors, and the signaling devices necessary to operate the robot. The teaching box is physically apart from the control unit and connected to it through a bidirectional line (serial line). It is used by the operator during the teaching phase and must guarantee some degree of portability, since the operator must be close to the robot joints in order to position the machine correctly along the working points. The physical realization of the operator interface is highly variable, and depends on the tasks the robot must perform as well as on the versatility and complexity level. Present efforts are largely directed towards the design of architectures which allow a more powerful and easy man-machine interaction. The principal characteristics needed are the following:

- the commands and operator actions must be closer to human logic, and not to the machine logic.
- the operator must be guided during the various stages of the process, especially during the teaching phase.
- a powerful editing capability must exist to allow the correction and modification of the working cycle tought to the robot.

To achieve these features one needs a complex software structure and an extensive use of alphanumeric display systems, as well as a highly interactive communication link for the teaching box itself.

The last functional module is the Input/Output interface, physically composed by a set of hardware units handling various ON/OFF signals (D.C., A.C.) and analog signals coming from transducers or sent to proportional actuators.

The robot is surrounded by a working environment which,in some way, determines and conditions the cycle behaviour. Therefore an exchange of information needs to take place between the manipulator and the external environment. This is done by the Input/Output module which has the ability to receive and interpret the information concerning the process (e.g. the presence of the working object) and to transmit the commands specifying the sequence of operations (e.g. the command to a conveyor -

belt).The logical sequence and the proper selection of the I/O lines are defined during the teaching phase. The I/O module also handles some devices mounted on the robot itself, like axes end-of-stroke, gripper or specialized end-effector tools.

Lastly the control unit is equipped with a series of hardware and sotfware devices protecting the robot and the environment from da - mage caused by malfunctioning devices, and speeding-up the maintenance and repair times. Some of this type are, for exemple, thermic and over-voltage protection devices, circuits detecting a broken connection between the control unit and the manipulator, extra-torque detectors of the joint actuators, test and diagnostic programs, etc.

Having examined the problem of the path control and the hard-ware architecture of the control unit, we will consider how this unit works. The proposed example is referred to a system with a single mi - crocomputer, controlling an anthropomorphic robot with six degress of freedom, and using a kinematic equation of motion.

Each robot function is controlled by a program stored in a non-volatile type memory (EPROM-PROM-ROM). This program is composed of four main modules:

- the movement control module
- the I/O management module
- the supervisor module
- the interrupt handling module.

The movement control routine is activated by a real time clock interrupt with a suited frequency (50÷100 Hz). This routine handles a set of function directly related to the motion of the axes; namely the joint coordinates computation from the position transducer data; the computation in cartesian coordinate of the next point to be reached at the next sampling period; the transformation from cartesian coordina - tes to joint coordinates; the evaluation of the position error and the output to the motor drives.

The I/O module (clock driven) is normally activated at the end of each block execution (each programmed segment of trajectory). It performs a set of functions related to the synchronization between the robot and the external environment. Some of this type are:

- Timer. A programmed time interval between the end of the actual block and the start of the following block.
- Output. The output signals that must be set or reset to drive external actuators.
- Interlock. The external input signals that must become true before the movement execution can resume.
- On Input Go To. The execution of an alternative sequence of the working program if a specified input is true.

The I/O module also handles the robot signals specifying warning or emergency conditions, like axes end-of-stroke, overtorque of the motors, etc.

The supervisor module, asynchronous with respect to the clock, enters the teaching or the working mode, selected by a switch on the operator panel. The teaching routine interprets the commands from the box and creates the blocks of program to be later executed by the robot. The operator moves the axes along a selected trajectory and then records the starting and ending point through the activation of a 'Memory' key. He then completes the current block of program with the introduction of the necessary end-of-block parameters. The teaching routine is also responsable of all editing and verifying facilities, so as to allow an easy modification and correction of the program by the operator. The working routine starts the robot's working program and executes the commands sent by the operator through the main panel. These can be the 'Stop' at the end of the cycle, the temporary 'Hold' of the movement, the modification of the programmed velocity, etc.

The fourth software module handles several types of interrupts coming from the whole system; namely the main panel pushbuttons and selectors, the programming box communication link, the magnetic mass memory, the interface for the acquisition of the axes position, the power-down detector. These interrupt signals possess different levels of priority, according to their importance; so for example the operator panel and the programming box have the lowest level of priority,whereas the power-down has the highest.

The performances of this system are very good; typical figures are a resolution of 1/100 mm in the computation of the trajectory,a linear velocity for the end-effector of 1 m/sec with the above precision, acceleration and deceleration ramps with constant slope at each starting and ending point,an optimal dynamic control through the adjustement of the coefficients of the PID function.

4. APPLICATIONS AND FUTURE DEVELOPMENTS OF INDUSTRIAL ROBOTS.

Forecasting robotic growth is difficult because the number of installed robots is still quite small, their future prices are quite uncertain, and the majority of future applications have not been analyzed with sufficient accuracy to know when and if they will be cost-effective.

It is nevertheless sure that the robot expansion will follow the development and improvement of available sensory systems. The replacement of a human operator by an automatic machine, in any process type, presupposes the correct reaction of this machine to a set of events, perceived before, consciously or unconsciously,by a man. The more complex the process is, the more numerous are the sensory elements involved and their inter-correlations. The robot should take into account these ele -

ments and, often, improve the perceptive capacity so as to turn the human qualitative evaluations into quantitative ones.

To better clarify the importance of the 'sensory' aspect, we will examine the problems encountered in the use of robots in continuous arc welding and assembly processes.

The use of arc welding is increasing because of the superior welding quality, reliability, and higher economy obtained. The use of welding robots also increases, because the welding process is a repetitive process which is dangerous for some aspects. However, there are some factors against the use of robots, such as the variation in work accuracy and the difficulty (or impossibility) to operate on geometrically complex work pieces. A careful analysis of these factors involves all robot components; the mechanical structure, the control unit, and, above all, the sensory apparatus.

From the mechanical point of view, a welding robot should have a highly articulated structure, so as to feature the largest work area. Actually the structure better fulfilling this requirement is the anthropomorphic (or pseudo-polar) one; in fact the most advanced arc welding robots (Asea, Cincinnati-Milacron, Kuka, Yaskawa) employ this type of mechanics. Some unsolved problems remain for the wrist articulation. This must possess a high degree of freedom to allow the electrode to follow complex contours in the space. Above all it must allow a constant electrode orientation with respect to the working surface (tool attitude). These features should be obtained without any loss of overall mechanical dimension, in order not to compromise the accessibility to difficult areas of the workpiece.

The electrode positioning and orientation not only enhance the mechanical problems, but also the control system ones. The teaching phase is not easy, because of the complicated spatial trajectories, and the control of the movement involves the computation of geometrically complex path. Moreover the possibility of non-univocal trajectories adds many severe problems, not yet completely solved even theoretically.

The power and input/output capabilities of the control unit are directly correlated to the factors concerning the sensory apparatus. The examination of this problem involves two distinct aspects; the first is concerned with the availability of transducers accomplishing a satisfactory connection between the robot and the external world, and the second is concerned with the control unit capability to correctly analyze the sensory parameters. Both of these aspects undergo a great development, but they are still far from having reached satisfactory results.

An ideal "robotized" welding process should be carried out having in the central unit memory the experimental data concerning a set of parameters like: working object shape, arc luminous energy and arc noise. The comparison between stored data and those collected during the process determines and adjusts the values of the parameters affecting the quality of the results: voltage and current of the electrode, travelling speed, relative distance and orientation between electrode and working surface. It involves a complex sensory system, needing a T.V. camera for shape analysis, welding path and arc intensity inspection; one or more microphones for sound detection; a set of sensors to locate exactly the working piece and its distance from the electrode. Without this apparatus, the results are satisfactory only if randomly varying parameters are maintained within narrow limits in comparison with the standard configuration used to teach the robot. This is a condition not easily met in current industrial processes.

Similar but more complex problems, not yet well defined, are encountered in assembly robots.

From a schematic point of view, assembly can be categorized into two principal activities; part mating or what occurs when two or more parts interact during the assembly process, and the aggregation of this process activity into the assembly of an entire product. If the "assembly system" is considered, then more factors must be taken into account, namely: the feeding of the parts and the inspection steps (quality or conformity control). With the present industrial robots, even the more sophisticated ones, part mating with strict tolerance is possible, but the probability of success can not be stated. This is mainly due to the effect of friction and misalignment, which are, at least, time and temperature dependent. The examination of these problems led to the analysis of a new aspect of robot movements namely: the joint position is not only determined a priori by geometric or kinematic relations, but also by the constraints of the working objects (compliant motion). The necessity of such behaviour by an assembly robot is evident if one thinks of the task of round peg-hole insertion, which is by far the most frequent task in the assembly of metal products. A little misalignment or a high friction coefficient is sufficient to cause very high insertion forces, often behind the robot capabilities, if the movement takes place only by path considerations.

Compliant motion can be realized with passive mechanical devices or with active ones, equipped with force sensor array. The first ones are quite simple, cheap and, at present, faster and more reliable even if they do not give satisfactory results in all cases. Active devices offer more capabilities and repeatability, although they need greater developments in the sensory system and in the programming technique.

The mechanical structure of an assembly robot do not request a high degree of complexity. Studies carried out on different assembly processes showed that less then six degrees of freddom are necessary and that the pay-load rarely exceeds 2 Kg.

The development of a highly sophisticated sensory system seems to be absolutely necessary. The importance of force transducers for the compliant motion was established. The availability of a very versatile vision system is of equal importance. The feeding of workpieces from disarrayed bins and the proper orientation and inspection of the objects request real time vision analysis, in order for the process not to slow down. This means to carry out the analysis of a very high amount of data, coming from the T.V. camera, in less than 1/10 sec. Untill now such a system is not practically available. Camera use makes it possible for a robot to see; it is however, very difficult for a machine to analyze what it has seen.

Regarding the control and programming technique, developments are directed towards the definition of high level languages, simplifying the teaching phase of the working cycle. The major current research issue in robot teaching involves implicit programming techniques. It requires much more computer power and assumes that the robot and work comprise a surveyed environment which has been stored in the computer as a "world model". The desired robot actions are no more expressed in terms of explicit point in space and of a sequence of logical actions, but, more generally, locations and orientations are referred to by name, and displacements are tracked automatically. Similar approaches request the use of hierarchical language systems involving both explicit languages, describing robot joints and position, and implicit languages describing the task to be performed rather than the spatial points through which the robot will pass.

- R E F E R E N C E S -

Paragraph 1

- Allan, R.: Electronic Design, April 29, 1982, pp.87-100
- Robotics at the Regie Renault; The Industrial Robot, March 1980, pp. 13-23
- Inagaki,S.: The Industrial Robot, March 1980, pp.46-49
- Inagaki,S.: 9th I.S.I.R., pp. 679-690
- Froehlich, L.: Datamation, January 1981, pp. 85-96
- Ozaki, S.: Metalworking, September 1981, pp. 82-87
- Nonaka, H.: Metalworking, September 1981, pp. 94-99
- Nevins, J., Whitney, D.: Scientific American, February 1978 pp.62-74
- Saveriano, J.W.: Robotics Age, Summer 1980, pp. 4-17.

Paragraph 2

- Paul, R.P.,Shimano,B.,Mayer,G.E.: IEEE Trans System, Man and Cyber. 6,

June, 1981, pp. 449-455
- Taylor, R.H.: IBM Research Rep. RC 6657, July 1977
- Paul, R: IEEE Trans. System, Man and Cyber. 11, November 1979, pp. 702-711
- Luh, J.Y., Walker, M.W., Paul, R.P.: Journal of Dynamic Sys., June 1980, pp. 69-76
- Raibert, M.H., Horn, B.K.: The Industrial Robot, June 1978, pp.69-73
- Gruver, W.A., Canady, R.M.: Technical Report MSR 80-13, 1980
- Gruver, W.A., Sachs, E.W.: Algorithmic Methods in Optimal Control, Pitman Publ., London, 1980.
- Bertino, M., Furxchi, M.G., Gola, M.: Int.Conf. on Cyber. Society, Oct. 1980
- Hollerbach, J.M.: IEEE Trans. on Syst.Man, Cybern., 11, Nov. 1980
- Horn, B.K., Raibert, M.H.: The Industrial Robot, June 1978, pp.69-73.

Paragraph 3
- Gruver, W.A., Canady, R.M.: Technical Report MSR 80-13, 1980
- Snyder, W.E., Gruver, W.A.: Proc.of Ist.Internat.Conf. on Mini and Microcomp. in Control, San Diego, CA, Jan. 1979.
- Bertino, M., Furxchi, M.G., Gola,M.: Int.Conf. on Cyber. Society, Oct. 1980
- ASEA:IRBT Operator Manual .
- SIEMENS: SIROBOT Operator Manual.

Paragraph 4.
- Froehlich, L.: Datamation, Jan. 1981, pp.85-96
- Allan, R.: Electronic Design, April 29, 1982, pp. 87-100
- Sugarman, R.: IEEE Spectrum, Sept. 1980, pp.53-57
- Nevins,J.L., Whitney, D.E.: The Industrial Robot, March 1980, pp.27-43
- Nonaka,H.: Metalworking, Sept. 1981, pp.94-99
- Mason, M.T.: IEEE Trans.on Sys., Man and Cyber., 6, June 1981, pp. 418-432
- Artificial Intelligence, vol. 14, 1980, pp. 79-107
- Cook, G.E.: IEEE Trans. on Ind. Applications, 17, Nov./Dec. 1981, pp. 619-625

CHAPTER 17

MICROCOMPUTER APPLICATIONS IN POWER SYSTEM SUBSTATION AUTOMATION

C.T. Nguyen, R. Grondin, A.St-Jacques and H.H. Le
Institut de Recherche d'Hydro-Québec (IREQ)
Varennes, Québec, Canada

ABSTRACT

The object of this chapter is to describe power system applications of microcomputers, particularly in the area of power substation automation.The major functions required in substation automation are reviewed and two applications described. The first one is a digital system for transmission line protection. This application requires a high processing power not generally available with low-cost microprocessor. The design described in this chapter gives the necessary power by implementing a parallel processing scheme. The second application is a voltage regulation system which controls tap changing power transformers. A microprocessor-based system was designed to virtually accommodate any substation configuration.

1. INTRODUCTION

A need for advanced instrumentation and control equipment has resulted from the rapidly advancing technology and the increased concern as power systems continue to grow in a context of environmental and economic pressure. In particular, the increased capability of low-cost micro-computers has made economically and technically attractive the utilization of such devices for power system substations automation.

This chapter is divided as follows: A general description of the functions required by substation automation is presented in section 2. A brief historic review of computer systems penetration in the power system industry is also covered in this section. Section 3 describes the power substation environnement and discusses the basic design considerations for the implementation of computer hardware. In section 4, two examples of microcomputer applications are described. These systems have been designed for the Hydro-Quebec power system. The first application is related to the transmission line protection function (distance protection), the second to the voltage regulation by transformer tap changes. Finally, in the conclusion, general remarks about the future

S. G. Tzafestas (ed.), Microprocessors in Signal Processing, Measurement and Control, 367–386.

trends in the design of microprocessor based equipment in power system are noted.

2. BACKGROUND: POWER SYSTEM SUBSTATION AUTOMATION

2.1 Historical note

The first applications of computer technology in the power system industry have been in the area of load flow and power stability calculations for planning purposes. Analog simulators had been the main computational tool used by the power industry. They were rendered obsolete when, in the late 50's, digital computers and efficient algorithms were becoming available to the industry (1). As computer technology developed, applications were extended to the power dispatch center for load frequency control and economic dispatch (2).

The advent of integrated circuits and minicomputers in the late 60's has made possible the application of sophisticated control and analysis equipment for dedicated functions. Minicomputers have been used for such applications as generating station start-up and control (3), and supervisory control and data acquisition (SCADA) in power substation (4).

The next step of technology advances was taken in the mid 70's, when large scale integration produced the microprocessor. The early processor, the 8080 or the 6800, was a relatively slow device with a limited word length, and has not found many applications in power system. The reliability of the new device was looked upon with suspicions, and the vendor did not always follow through for the software support and maintenance. These facts and the fast pace of development in the microprocessor industry, where new models are introduced continuously, have made difficult a firm commitment from the power authorities to develop the applications for microcomputers.

However the potential of a low-cost computer was fully realized within the industry, as the decrease in hardware costs made possible the dedication of a specific function to each microprocessor and the consequent simplification of software. Now that models with increased capability are being introduced (the 8086 with arithmetic accelerator 8087, the 68000, etc.), and more experience has been gained from actual implementations in many utilities and research establishments (5), the microcomputer is going to play an increasingly major role in the design of power system control and monitoring functions.

2.2 Substation automation

A non exhaustive list of substation automation functions is given below.

- Line fault protection
- Line fault location

- Bus bar protection
- Transformers protection
- Voltage regulation and reactive power control
- Load shedding
- Oscillography
- Data logging (analog metering)
- Sequence of events recording
- Transformer tap settings
- Breaker status
- Breaker tripping
- etc.

These functions may be conveniently classified into the following classes (6):

Data acquisition

This class of functions includes data acquisition for local or remote manual control of the substation. It involves the acquisition of voltage and current data, power flows, frequency and the status of switching devices (circuit breaker, etc.). Data acquisition functions are also necessary for post-incident analysis, including events recording (in which the sequence of operation of protective devices, circuit breakers, etc. is time-tagged and recorded) and oscilloperturbography (in which the instantaneously varying voltage and current waveforms are recorded).

Substation operation and control

These functions are those required for the normal local or remote manual surveillance and control of the substation. It includes the following functions: transformer tap-change automation, transfer of circuit from one bus bar to another, the restoration or taking out of service of any designated equipment for maintenance or other purposes, reactive compensation by shunt reactor switching, etc.

Protection

This class of automation function is particularly demanding in term of operation speed and reliability/security criteria. Faults on transmission line or bus bar must be cleared by tripping the faulted portion in order to preserve the system integrity. Other protection functions include transformer protection and back-up protection against any failure of the primary protection (e.g. circuit breaker failure to trip a faulted line).

Automatic post-fault switching

This class of function is concerned with automatic actions designed to return healthy circuits and equipment as fast as possible after a power system fault. It includes such functions as automatic reclosing, system restoration, load shedding and generation control, etc.

All of these functions can be performed by a single computer (centralized approach) or by a distributed system of computers (decentralized and integrated approaches). In the present state of the art, no installations have been made to date implementing all of the functions listed.Most have concentrated on the data acquisition and monitoring functions.Protection functions are still very much in the experimental phase.

2.3 General requirements for substation automation equipment

It is required that substation automation must be performed with:

a) high security
b) high reliability
c) high availability

A system is reliable when it always performs its intended function when required to do so. It is secure when it performs its function only when called upon to do so.

Security and reliability are important objectives (particularly for the protection functions) and are closely related to each other. For instance, a failure to trip a line when a fault has occurred on that line is an event that is generally dealt with by duplicating hardware and providing back-up procedures. However, a false tripping on a healthy line is potentially more damaging to the power system integrity than a failure to trip, and usually an increase of the system reliability makes more probable a spurious operation.

The availability of an equipment is measured by its failure rate and the mean time required for repair. For functions which are not continuously active, such as events recording or protection, a fail condition must be recognized as soon as possible, preferably at the instant of failure, to provide the required levels of security and reliability. This ability is made possible with intelligent equipment capable of self or cross-diagnosis.

The satisfaction of all these requirements involves a compromise with regard to costs, maintenance requirements and facilities. Digital computers present an opportunity to alleviate some of the difficulties associated to these general requirements, but introduce also their own problems that must be dealt with at the design stage. Some of the problems are the commercial availability of the hardware, the cost involved with the software design and the specialized knowledge for hardware and software maintenance.

3. BASIC CONSIDERATIONS FOR HARDWARE DESIGN

3.1 Implementation approaches

As far as the hardware implementations and concerned, three major trends can be identified (7). The first is the centralized approach, based on a single computer of sufficient power to accommodate all or most of the substation automation functions. The second approach is the decentralized one, where each function is implemented on a specific microprocessor. The third approach is the local area network approach (for integrated approach) where the functions are distributed to a number of microprocessors, which are then connected in an integrated system.

The centralized approach relies on a powerful computer to achieve the desired degree of flexibility and speed. Although technically feasible, this solution is not cost competitive and furthermore does not satisfy the reliability requirement, specially for protection functions.

In the decentralized approach, each function is implemented on a microprocessor with the inherent advantage of reliability since hardware duplication can be considered. Tailor fitting the microprocessor to the function is also possible. The over-all picture that emerges is not so different than the conventional, analog-based automation systems presently used. The notable difference being that advanced algorithm may be developed to improve the performances of the equipment.

The integrated scheme is the most recent approach that has been proposed (8,9). In this scheme, all functions are distributed to a number of processors which are interconnected in an integrated system. The connection is generally a tight communication subsystem based on a high-speed digital pathway. This approach offers some unique features not shared by the previous two. Extensive data acquisition and monitoring is possible resulting in an increase of the levels of reliability, maintainability and availability. Useless duplication of hardware may be eliminated by optimizing the system design, a feature not possible with the decentralized approach.

3.2 Environmental Constraints

The hardware design of a microprocessor-based system is strongly related to the environment in which it has to operate. The power substation being a severe environment for this type of equipment, it is worth presenting the main constraints related to it and the techniques used to deal with it. Three types of constraints are considered: electrical, chemical and climatic.

Electrical constraints are related to the surge disturbances which may have three origins: lightning, switching surges and ground potential rises. This produces a wide spectrum of transient phenomena that have been studied and reported in the litterature (10,11). These studies have led to the formulation of standards on transient overvoltages. Two of these are widely used for the power substation environment (12,13). They define characteristic waveforms to be systematically applied to the device inputs and outputs. Under any surge, the device should not fail nor produce erroneous outputs. Typical failure levels are presented in (14) for integrated circuits. Mitigation techniques are divided in two broad categories: attenuation techniques and diversion techniques. The first category includes such procedures as shielding (electrostatic and electromagnetic), insulating (magnetic, optic), fusing, bonding and grounding (15,16). In the latter category, surges are diverted from the device with surge suppressing (17) and filtering (18) techniques. Usually, for a particular device, a combination of all these techniques are used and field experience has shown that they do provide adequate protection against electrical disturbances (19,20).

Chemical constraints refer to the corrosive atmosphere that is found in power substation buildings. It is due to acid vapors produced by accumulators and to humid atmosphere. The connectors are the most corrosion sensitive elements in this kind of equipment and it is imperative to use chemically neutral metals or alloys for every contacting surfaces.

Equipment operating in a power substation generally has to meet stringent environmental conditions. Temperature range of 0°C to 55°C, with a rate of change of 20°C/hour and relative humidity of 10% to 95% without condensation are considered (21). This environment influences the choice of power supplies (since their efficiency is function of temperature), electronic arrangement within enclosures, ventilation techniques (for hot point reduction) and humidity protection (fungus proof electronic might be required). The actual trend is to use low-power consumption components (MOS, Low-power Schottky, ...) when possible and to reduce therefore heat losses. Natural convection is thus used as ventilation technique which offers higher reliability than air forced alternatives. In fact, no fan maintenance is needed and air filtering elements (required with substations dusty environments) are being replaced less often.

3.3 Man-Machine Interfaces

Man-machine interfacing is of main concern in substation automation. It requires combined flexibility and simplicity. Flexibility is needed since multiple functions are implemented under this interface (22). On the other hand, simplicity is necessary to provide an efficient man-machine dialogue.

This subsystem is required to interface with power apparatus and computer equipment. It is particularly easy to provide flexible man-machine interface to the power apparatus since all data pertinent to them are made available in a digital format through the data-acquisition and control automation functions. Relevant informations can be preprocessed and presented to the operator at the central control desk through computer-driven terminal, graphic displays, etc..

Man-machine interface also means interaction with the computer itself to monitor its operating status. Status informations are easily available due to the inherent self-checking capability of this type of equipment. The security of operation is directly proportional to the completeness and repetition rate of diagnostic tasks. The repetition rate is particularly important for short time response devices such as protection relaying, continuous VAR or voltage regulation etc. These diagnostic tasks, together with an effective man-machine interface, provide the tools for reducing the Mean-Time-To-Repair and thus for increasing the system availability.

4. MICROCOMPUTER APPLICATIONS

4.1 Digital relay for line protection

4.1.1 Distance relay

Distance relays are commonly used to protect power system transmission lines. The operation principle is fairly simple: Line circuit-breakers tripping decisions are based on the apparent line impedance. A simplified diagram of a transmission line is given in figure (1).

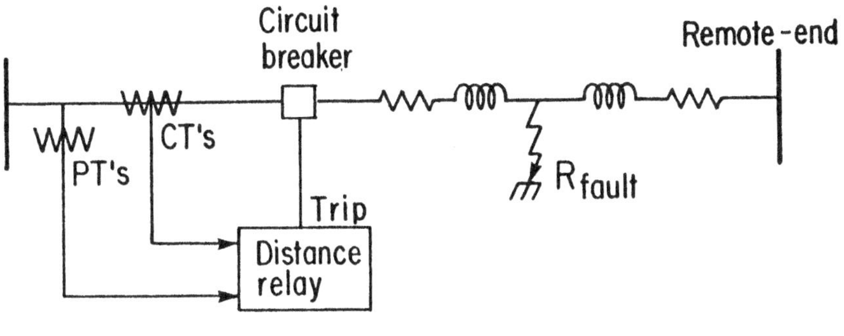

Fig. 1 Transmission Line Relaying.

Assume a solid fault (zero impedance fault) at distance x from the substation-end of the line. Then the impedance evaluated from the measurements of voltage and current can be used to determine the location of the fault relative to a predetermined zone. Line is tripped if within the zone.

The actual situation is more complicated: Fault may not be solid and currents are injected from the remote end of the line. Fault may also be occurring on or between any of the three phases of the transmission line. In order to cover all eventualities, 6 impedances (3 phase to ground, 3 phase to phase) must be considered simultaneously. Furthermore, the decision to trip a line must be taken in a very short time to avoid cascading damages to the power system and eventual complete shutdown. Typically a trip output must be sent to the circuit-breaker within one cycle (16.6 ms for a 60 Hz system) from the instant of fault occurrence.

Over the past decade, many digital protection schemes have been advocated (7) to replace or complement conventional electromechanical or electronic relays. The advantages of digital relays may be summarized as follows:

a) Arithmetic and logic capabilities. Improved protection schemes are possible.
b) Adaptability. Operating parameters, such as zone characteristics, pick-up values, etc. may be easily modified according to the present or foreseeable operating conditions.
c) Coordination and data interchange. Data can be easily interchanged between a number of digital systems to build-up a coordinated protection scheme and to enhance the reliability of the the protection functions.

4.1.2 Digital relaying algorithms

The major approaches used in digital transmission line relaying algorithms are the following:

1) Algorithms based on estimating steady-state line currents and voltages.
2) Algorithms based on the differential equation relating voltage to current waveforms.
3) Algorithms based on travelling waves.

The first type of algorithm extracts the fault information at the power system fundamental frequency, with various means to minimize the transient component in the current waveforms. From the steady-state values of voltage and current, impedances are computed. In the differential equation approach, the transient component is directly taken into account. However, since a very simple R-L line model is assumed, it is necessary to reject harmonics. The last approach relies on the particular characteristics of the forward and reflected travelling waves. In the case of faulted conditions, the two waves are time shifted by a delay equal to twice the propagation time from the relaying point to the fault.

The choice of a particular method is generally a matter of system protection practice and philosophy within a power utility and the availability of processing power and measurement interfaces. In the system described herein, a Fourier-like algorithm is used to determine the steady-state values of the currents and voltages. For the voltage wave forms, the complex phasor $\vec{V}n$, at instant n, is evaluated iteratively by:

$$\vec{V}_n = \vec{V}_{n-1} + (v_n - v_{n-N}) \cdot \exp(j2\pi n/N)$$

where v_n is the value of the voltage sampled at instant n, and N is the number of samples per period.

The transient component in the current waveforms is a decaying exponential which can be minimized by numerical differentiation. The current phasors $\vec{I}_n$ will then be evaluated in an iterative procedure as follows:

$$\vec{I}_n = \vec{I}_{n-1} + (i'_n - i'_{n-N}) \cdot M \cdot \exp j((2\pi n/N)+\phi)$$

where $i'_n = i_n - i_{n-1}$

$$M = \sqrt{2(1-\cos(2\phi/N))}$$
$$\phi = \pi/2 - \pi/N$$

M and ϕ are the module and phase of the phasor $M \cdot \exp(j\phi)$ used to compensate for the numerical differentiation in the time domain.

Impedances are continuously evaluated using the following relations. For phase to phase impedances, such as $\vec{Z}_{ab}$:

$$\vec{Z}_{ab} = (\vec{V}_a - \vec{V}_b)/(\vec{I}_a - \vec{I}_b).$$

For phase to ground impedances, such as $\vec{Z}_{ag}$:

$$\vec{Z}_{ag} = \vec{V}_a/(\vec{I}_a + \vec{K}.\vec{I}_o)$$

where $\vec{K}$ is a complex compensating factor taking into account the earth return impedance, and I_o the zero-sequence current.

4.1.3 Description of the system

The general system configuration in shown in figure (2). The outputs of the power system transducers are low-level analog voltage waveforms corresponding to the transmission line currents and voltages. The signals are first processed by the Insulation and Signal Conditioning interface which consist of step-down voltage and current sensors and low-pass anti-aliasing filters. This interface also provides the necessary electrical insulation from the power system. The coordination and tripping signals I/O interface allows the relay to communicate with the remote line-end counterpart for protection coordination and to output a trip order to the appropriate circuit-breaker. Communication to a central computer for permanent data storage and post-fault evaluation is also provided through this interface.

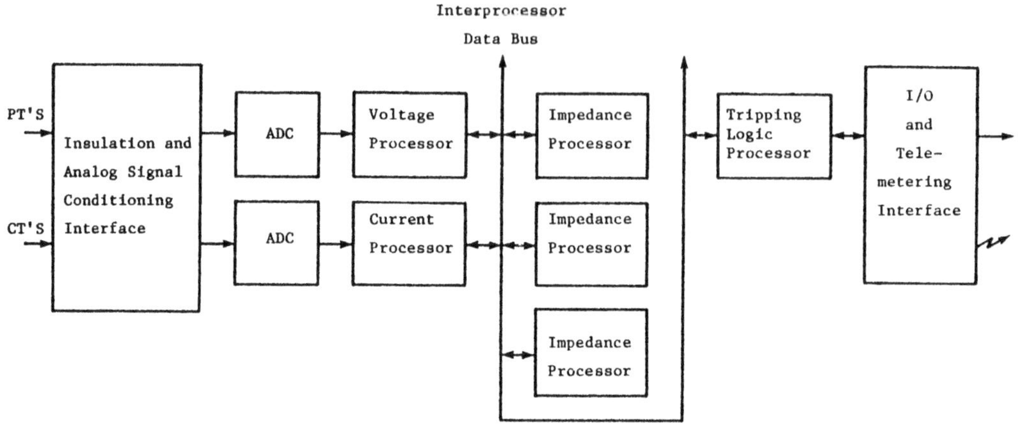

Fig. 2 System Configuration of the Distance Relay.

The determination of the voltage and current phasors, impedance calculations and the trip logic processing are performed simultaneously by different processors. Each of these processors is pipe-lined into an overall system operation as illustrated in figure (3).

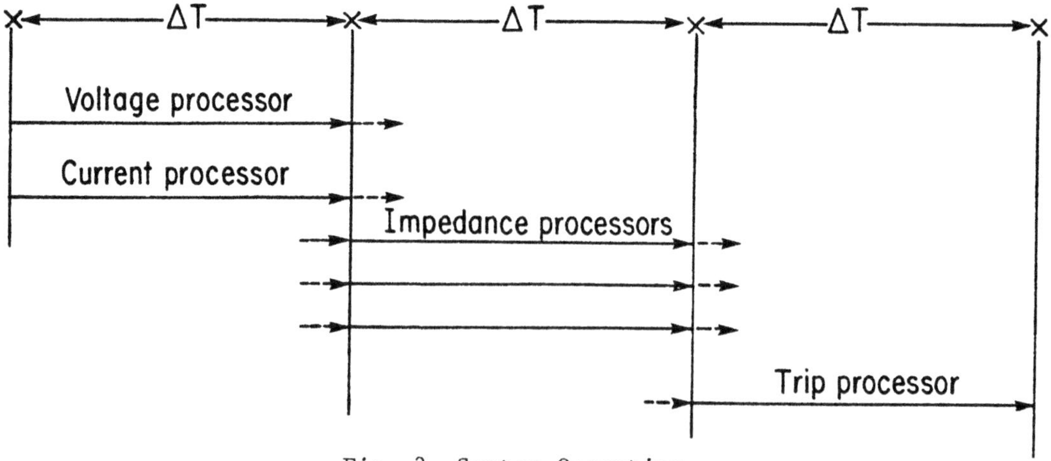

Fig. 3 System Operation.

The processing time ΔT is maintained uniform by balancing the computational load between the processors. With a sampling interval of ΔT, the system will be able to achieve a data processing rate of $1/\Delta T$. A computation delay (between 2 to 3 ΔT) will be required to fill the pipe. Processing time of 1 ms is achieved with off-the-shelf low-cost microprocessor, thus giving a sampling rate of 16 samples per cycle for a 60 Hz power system The interprocessor communication system is shown in figure (4).

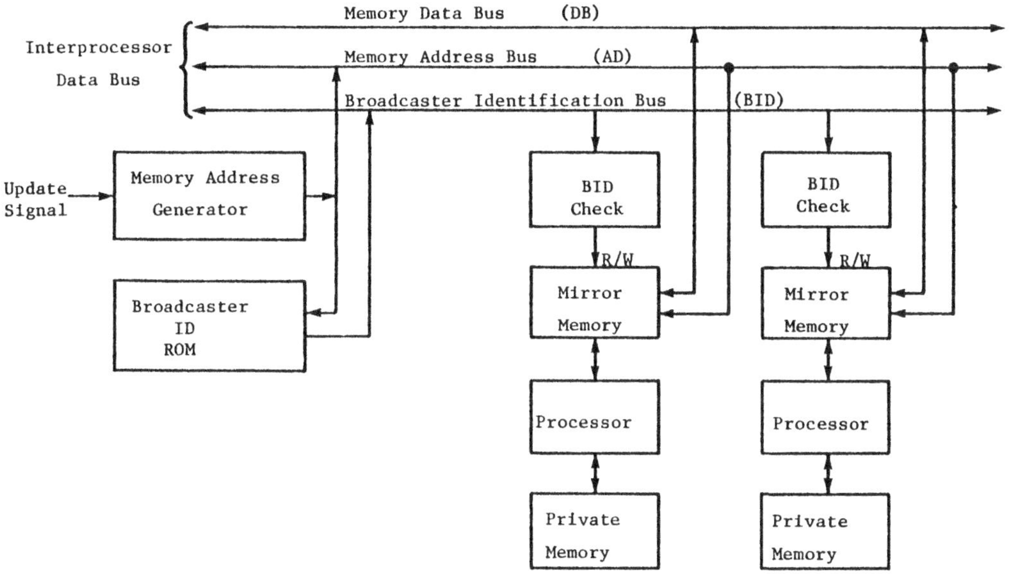

Fig. 4 Interprocessor Communication System.

It consists in a data bus (DB), an address bus (AB) and a Broadcaster Identification bus (BID). Data are exchanged between processors via the mirror memories. Each memory is identified by a code, the BID number, which is stored in the Broadcaster ID ROM. When the system is activated (by an update signal, for instance), the Memory Address Generator sequentially addresses all mirror memory locations. The content of the ROM location at the address specified by the AD bus is the BID code identifying the Broadcaster memory. This code is loaded in the BID bus, and is checked by each mirror memory. If a mirror memory recognizes its own BID code, a Read signal is generated, and the content at location AD is loaded into the Data bus. If not, a Write signal is generated and the data on the Data bus is loaded into the memory.

More details concerning the design and the field testing may be found in references (23,24).

4.2 Substation Voltage Regulation System

4.2.1 Description of the voltage regulation function:

Many power substation transformers are provided with on-load tap changers. As part of a voltage regulation loop, they are used to compensate for the variations of voltage and loading on the network by modifying the voltage transformation ratios. The regulation process may be described as follows: Transformer voltages and currents are measured along with the status of the substation circuit-breakers and switches to compute the effective voltage at a regulated bus. If voltage corrections are needed, "raise" and "lower" commands are sent to the tap changer device. Response time of the regulation process is in the order of 60 seconds, since only slow variations of voltage and loading are considered in this application.

In the reality, many substation configurations and regulation schemes are operated by power utilities (25). Typical configurations are shown in figure (5) and a list of regulation schemes is given below:

1) Regulation of radial operating transformers
2) Regulation of parallel operating transformers by reduction of the reactive power flow or the active power flow between the transformers
3) Regulation of coupled twin-secondary transformers
4) Alternate regulation on high and low voltage buses depending on their relative stiffness
5) Load-drop compensation for feeders in distribution substations, with correction for shunt capacitor banks when applicable.

One advantage of computer-based regulation system over conventional, hard-wired control loop is that the many regulation schemes and the various substation configurations are easily taken into account by soft-

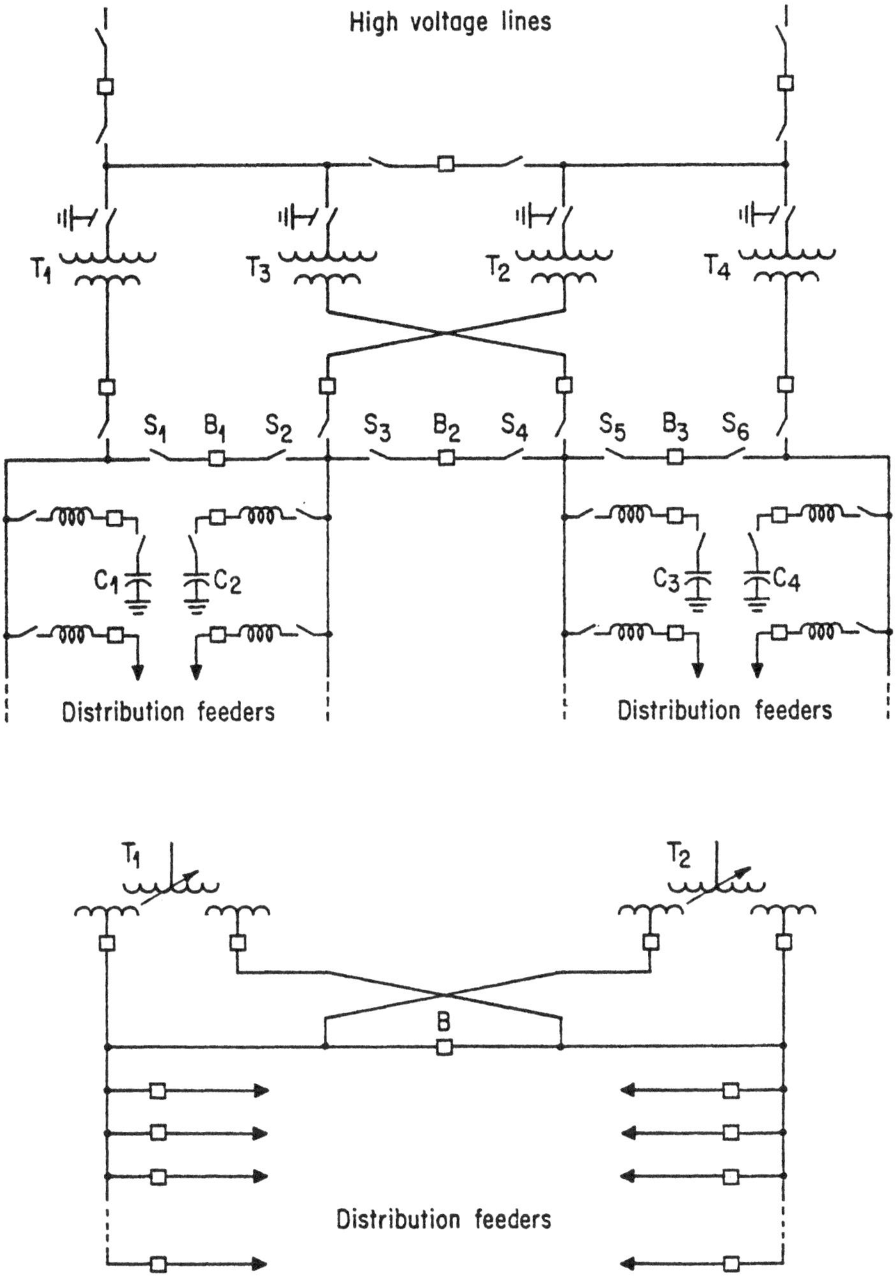

Fig. 5 Typical Substation Configurations.

ware. Furthermore, voltage settings may be remotely adjusted or periodically changed by the computer itself for an improved power system operation flexibility.

4.2.2 Voltage Regulation Algorithm

With the simplest configuration (single transformer with no feeder compensation), the modulus of the transformer voltage is compared to minimum and maximum settings and any deviation from this range starts a timer. A tap change will be initiated at the end of it. The elapsed time is function of the error magnitude, such that a large error will produce a fast tap change response. With the load-drop compensation scheme, the voltage drop along a feeder is computed and subtracted from the measured voltage:

$$\vec{V}_t - (R\cos\theta + X\sin\theta) \cdot (\vec{I}_t - \vec{I}_c) = \vec{V}_{user}$$

$\vec{V}_{user}$ is the voltage that has to be regulated, $\vec{I}_c$ the current flowing into a capacitor bank. R and X make up the feeder image impedance, while θ is the lagging angle between vectors $\vec{I}_t - \vec{I}_c$ and $\vec{V}_t$. In most substations, the capacitor currents $\vec{I}_c$ has to be removed from transformer current $\vec{I}_t$, to obtain the actual feeder current. An equivalent circuit is shown in figure (6).

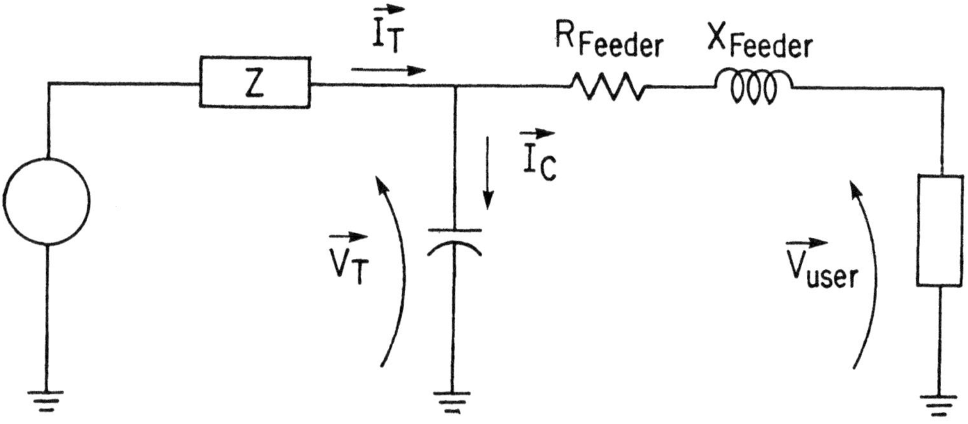

Fig. 6 Equivalent Circuit for Feeder Compensation.

When transformers are operated in parallel, VAR or Watt losses in transformers are considered. In either cases, a third term, proportional to the losses, is included in the voltage computation. This term is such that it reduces the losses in the transformers to a minimum. Derivation of these equations is given in (26).

Coupled twin-secondary transformers and alternate HV/LV regulation configurations introduce additional computations. In fact, with the former configuration (usually 2 coupled transformers per substation, as shown in figure 5), the light and heavy buses must be determined with active power computation in order to perform the regulation on the heavy bus only. On the other hand, a voltage maximum limit must be fixed for the light bus since it is then always operated over its nominal setting. With the latter configuration, the stiffness of high voltage and low voltage buses must be determined every time a tap change is performed. This is done with a sensitivity analysis $\Delta V/V$ on both buses. The stiffest bus has the lower sensitivity and is considered to be the generation side. The other bus is considered to be the load side and is consequently regulated.

4.2.3 Description of the system

An overall block diagram of the equipment is presented in figure (7). The system is based on a 16-bit microprocessor with a 64K words adressing capability. The memory build-up for this application consists in 14K of ROM, 4K of RAM and 1K of EAROM (Electrically Alterable ROM). The EAROM memory is used to store parameters that are substation dependent, and is very useful in the actual implementation since the power supply is not uninterruptible. Data acquisition is performed by memory mapped I/O technique: each external device is addressed by the system as memory location via the memory bus. This permits a faster data throughput and relieves the normal I/O bus for communication with the man-machine interfaces (front panel, printer, terminal, etc..).

Voltage and current signals from the substation PTs and CTs are first processed by analog boards. These consist in a fully surge protected differential input with an accurate resistive divider feeding an insulation amplifier (for the voltage) and a precision shunt with insulation amplifier for the current. These circuits connect to signal conditioning boards which first filter the signal to obtain the 60 Hz fundamental component. Thereafter, the signal is sampled and holded every 1 ms. The processing mentioned above is performed simultaneously on each channel (voltage and current) and all waveforms are synchronously sampled. The data-acquisition is made with dual multiplexer-A/D converter arrangement. Status data (such as tap position, control and telecontrol inputs, confirmation signal, alarms, etc.) are processed through surge-protected, insulated and debounced inputs. Each input becomes a bit of a 16-bit board and produces an interrupt when complemented. The output boards are similarly designed (surge protected and insulated 16-bit board) and are completely based on semiconductors.

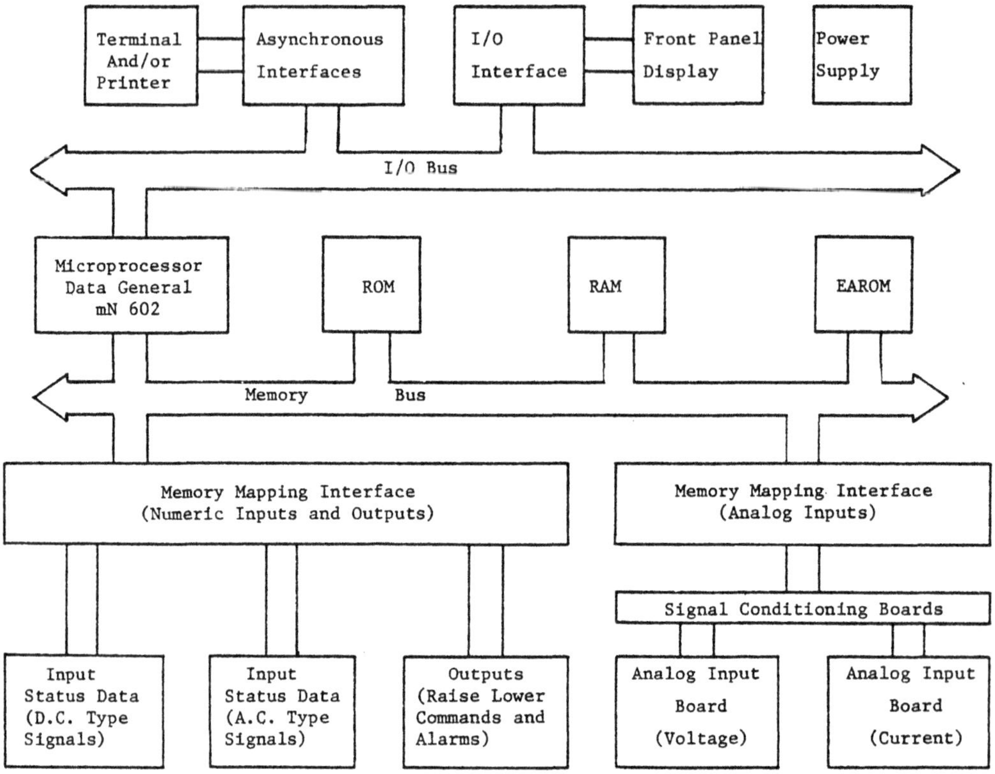

Fig. 7 Block Diagram of the Voltage Regulation System.

Furthermore, periodic calibration of the system is performed without manual intervention. Analog inputs are temporarily disconnected and 60 Hz reference signals are injected sequentially into the voltage and current inputs. Eventual offset and gain errors are automatically corrected.

The front panel is used for the following functions:

- Alarm displaying
- Voltage setting modifications
- Feeder impedance modifications
- Tap position displaying
- Manual tap changes

A small printer is permanently connected (as opposed to the terminal) to the system for alarm event recording while the terminal is used for initialization sequences and for diagnostic purposes. The tasks implemented in the system are presented below:

1) Initialization: substation parameter entry and processing
2) Analog data acquisition: 16 samples/channel/cycle every 100 ms
3) Regulation related computations: solution of equations discussed above
4) Self-diagnosis: exhaustive self-test of the system run on background. Generates alarm messages on the printer. A complete description of the method employed is presented in (27)
5) Self-calibration: periodic calibration of the analog inputs
6) Interrupt handling: servicing of the 7 system interrupting facilities
7) Control, telecontrol interfacing: processing of the related sequences
8) Status input processing: tap position, output confirmations, breaker/switch status
9) Real time clock functions: timer refreshing and task initiation
10) Real time scheduler: appropriate activation of the nine tasks presented above.

Tasks are developed on a modular basis to facilitate software evolution. The real time monitor is designed "in-house" to insure maximum simplicity and efficiency for this kind of applications. Tasks are scheduled according to several external and internal conditions that are recognized and updated by the monitor. The whole program requires less than 14k words.

5. CONCLUSIONS

Microprocessor technology is going to play a major role in the power system data acquisition and control applications in the near future. The approach for integrated architecture appears to be very efficient since resources are distributed and complementary. At the same time, the microprocessor is perfectly designated to fill in most of the tasks within this context and will be more and more attractive in the next years with "off-the-shelf" equipment offered at decreasing cost. The present developments are focusing on performance, reliability and security improvements of already implemented functions and on the integration of the new technology in more conventional equipments.

Performance is a matter of both software and hardware. Microprocessor manufacturers tend to increase the computation power and memory capacity of their microcomputer systems and new generation of "super-micros" is emerging. Implementation in high level language is being eased by the support capabilities of these micros, and is of considerable advantage as far as maintainability and legibility of software are concerned. Reliability and security improvements may be obtained with hardware redundancy and with self-checking and cross-checking methods. Self-checking is usual with single processor system. An efficient scheme that may be implemented for this purpose is described in (27).

Cross-checking is more appropriate with multiprocessor systems and is implemented during the processors idle phases (28). A defective processor may be replaced by another with no down-time penalty for the system. Multiprocessing structures will probably be more and more in use since their computing power may be matched to virtually any application.

6. REFERENCES

(1) St-Clair H.P., Stagg G.W., "Digital Computer Solution of Power System problems", Proc. Am. Power Conf., Vol. 20, pp. 614-629, 1958.

(2) Cohn N. et al. "On-line Computer Applications in the Electric Power Industry", Proc. IEEE, January 1970.

(3) Tawa R.H., Nguyen C.T. & Lalonde P., "Hydrogenerating Station Integrated Computer Control", IEEE Trans. IECI, Vol. 26, pp. 6-11, Feb. 1979.

(4) Staff Report, "SCADA-An Exploding Technology", Elec. World, p. 27-38, January 1976.

(5) Nilsson S.L., "Epri R & D of New Substation Control and Protection Equipment", Proc. of the IEE Conf. on Developments in Power System Protection, IEE Publ. No 185, pp. 88-92, 1980.

(6) Naylor J.H., "Trend in Automation in HV and EHV Substations", Electra., No. 52, pp. 33-52, May 1977.

(7) Kezunovic M., "Digital Protective Relaying Algorithms and Systems - An Overview", ELect. Power Syst. Research, Vol. 4, pp. 167-180, 1980.

(8) Ibrahim et al, "The Integrated Substation Concept - Opportunities for the Future", Proc. Am. Power Conf., April 1979.

(9) Nguyen C.T. et al. "A Microprocessor-Based Parallel Processing System for the Relaying of Power Transmission Lines", Proc. IEEE-IECI, pp. 175-180, March 1980.

(10) Nilsson J.H., "Digital Techniques for Control & Protection of Transmission Class Substation", Epri Report WS-79-184, April 1980.

(11) Hicks R.L., Jones D.E., "Transient Voltages in Power Station Wiring", IEEE Transactions on Power Apparatus and Systems, Vol. PAS-90, No. 1, Jan-Feb 1971.

(12) ANSI/IEEE C 37.90--1978, "IEEE Standard Relays and Relay Systems Associated with Electric Power Apparatus".

(13) IEEE Std 587-1980 "IEEE Guide for Surge Voltages in Low-Voltage AC Power Circuits".

(14) Keuren E.V., "Effects of EMP Induced Transients on Integrated Circuits" IEEE Electromagnetic Compatibility Symposium Record 75 CH1002-5 EMC, 1975.

(15) IEEE Std 518-1977 "IEEE Guide for the Installation of Electrical Equipment to Minimize Electrical Noise Inputs to Controllers from Electrical Sources".

(16) Sutton H.J., "Transients Induced in Control Cables Located in EHV Substation" IEEE Transactions on Power Apparatus and Systems, Vol. PAS-89, No. 6, July/August 1970.

(17) Clark O.M., "Electrostatic Discharge Protection Using Silicon Transient Suppressors" Electrostatic Discharge Symposium, Reliability Analysis Center, 1979.

(18) Martzlov F.D., "Coordination of Surge Protectors in Low Voltage AC Power Circuits" IEEE PES Summer Meeting F 79-635-4, Vancouver, Canada, July 1979.

(19) Grondin R., Girard P., Gauthier P., "Expérience d'Automates Programmables dans un Poste de Répartition" IEEE Conférence Canadienne sur les Communications et l'Energie, CH1583-4/80/0000-0, Montréal, Canada, Oct. 1980.

(20) Nguyen C.T. et al, "Un Système Intégré d'Analyse et de Localisation de Défauts: Théorie et Expériences en Réseau", Proc. of the IEEE Comm. & Power Conf., Montreal, Oct. 1982.

(21) ANSI/IEEE C 37.1-1979 "IEEE Standard Definition, Specification and Analysis of Manual, Automatic and Supervisory Station Control and Data Acquisition".

(22) Westinghouse Electric Corporation, "Substation Control and Protection Project" EPRI EL-1813 Project 1359-1, Interim Report, April 1981.

(23) Nguyen C.T. et al, "A Computer Based System for EHV Line Monitoring and Control", Proc. IEEE-PICA, May 1979.

(24) St-Jacques A., Santerre G., "A Multiprocessor-Based Distance Relay: Design Features and Field Tests Results", Submitted to IEEE-PICA conf., 1983.

(25) Grondin R., "SD-130 Microprocessor Voltage-Regulation Device" Canadian Electrical Association Report, January 1981.

(26) Grondin R., "Computer Dedicated Voltage Regulation Method for Distribution Substations", IEEE Transactions on Power Apparatus and Systems Vol. PAS-100, No. 5, May 1981.

(27) Grondin R., "Fault Detection and Location Tools for Microprocessor Based Systems" Paper to be presented to IEEE IECON 1982 Conference in Palo Alto, California, Nov. 1982.

(28) Siminoncini L., Saheban F., Friedman A.D., "Design of Self-Diagnosable Multiprocessor Systems with Concurrent Computation and Diagnosis" IEEE Transactions on Computers, Vol. C-29, No. 6, June 1980.

CHAPTER 18

MICROCOMPUTERS IN THE PULP AND PAPER INDUSTRY

Paavo Uronen
University of Oulu
Oulu, Finland

1. INTRODUCTION

The first digital computer control applications in the pulp and paper industry date from the early 1960's, the very first ones being the paper machine control systems. The first application in the pulp mill area was the development of Kamyr digester control system. These first control systems installed at that early date have since times been replaced by systems of the next generation.

At the time of these first installations enthusiasm was great and very ambitious plans were presented for putting the whole pulp and paper mill under the control of one large computer. Some of the first projects of this kind have unfortunately failed and therefore pessimism and a stagnation of progress in this area resulted.

The reasons for this seem now to be quite clear. The most important of them are:

1. An underestimation of the problems and difficulties to be encountered, i.e., too few personnel, too tight a time schedule, etc.
2. The reliability, speed and capacity of the computers available at that time were not sufficient for the tasks at hand.
3. A lack of the necessary instrumentation and plant process mathematical models.
4. A shortage of competent and specially trained personnel for the development groups.

After a delay of several years in the middle 1960's, the problem of computerization in pulp and paper mills was attacked from another direction, i.e., one defined smaller, well-limited potential targets from among the different subprocesses in the pulp and paper mill. One also put emphasis on the required instrumentation and the necessary process models, economical benefits, etc. In addition, these new projects were initiated with a class of smaller computers i.e. minicomputers instead of the earlier, business type, "big" computers. The result

S. G. Tzafestas (ed.), Microprocessors in Signal Processing, Measurement and Control, 387–409.

has been that many successful, so called, dedicated, computer system packages for these purposes were developed. For example, the package systems for paper machine control, for kraft batch digester control, etc. These were successful because:

1. They were developed through cooperation between computer specialists and user engineers to solve well-defined problems.
2. They had the necessary instrumentation and process models as background.
3. The computers used were already reliable enough and capable enough at that time to handle these smaller problems.
4. The economic results could be clearly verified in a reasonable period of time.
5. The problems involved with the man-machine interface were noted and solved using the technology then available.

All of this took place during the years 1967-1969. Since then the developments in this area have been quite fast (Gee, J.W., and Chamberlain, R.E.). Today we can estimate that there are about 2600 digital computer control systems now in operation in pulp and paper mills throughout the whole world. Figure 1 presents this development.

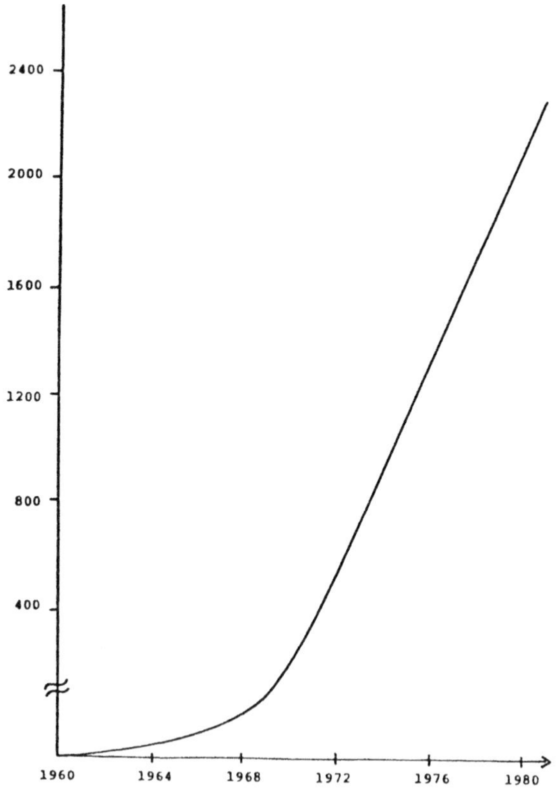

Figure 1. Development of computer control installations in pulp and paper industry.

According to the TAPPI questionnaire (Keyes, M.A.: 1975, Tappi 58, No 6) there were about 500-600 systems in operation in the pulp and paper industry in the USA in 1974 and an additional 681 systems were being planned for insallation during the next four years (1975-1979). Most of these plans have since been fulfilled even if the poor marketing and economic situation of the paper industry in 1977-1978 has postponed many of these installations. According to the same questionnaire, most of the existing systems (about 82 %) were working in paper mills in on-machine and related applications. Further, the next most rapidly increasing areas seem to be those of the digester, bleaching and energy management systems. Most systems at that time were leased. The average return on investment was about 59 % net after taxes for all types of systems.

A very interesting point is the distribution between these so-called packaged systems and "do it yourself" systems. According to the above mentioned study of Keyes, 69.5 % of all pulp and paper systems in the U.S.A. are packaged systems, 21 % are in-house systems and the rest, 9.5 %, are user modifiable packaged systems. The situation in Finland and in the rest of Scandinavia is quite similar. However, in Sweden the relative number of pure packaged systems is somewhat lower (about 50 %)(Eriksson, L.: 1975, Svensk Papperstidning 78, pp. 81, Uronen, P.: 1977, Paper presented at Scandinavian Meeting of Paper Engineers Kuopio, Finland). The use of such tightly controlled (i.e., proprietary) packages means that the user has no allowance to modify any of the computer programs, etc., without the special permission of (and additional payment to) the vendor.

Concurrent with this development in the process control area, there has also been similar remarkable progress and change in the use of "administrative" or "business" computers. They have become more powerful and now have more efficient operating systems. Also the relative prices of these machines have decreased. These machines are now widely used in the pulp and paper industries in such administrative jobs as payroll, inventory control, accounting, order handling, etc. Thus they are assisting management in its everyday routines but are not giving them the real-time information needed for decision making in mill problems.

The situation is usually such that there is one organization taking care of these (the business) computer systems and still another developing and using the mill computers (process computers). These two usages of digital computers in pulp and paper mills should be combined together so as to form an effective hierarchical total control and information system including process control, coordination between different subprocesses, production planning, production supervising, production evaluation and reporting, energy management, environmental protection and the presentation of plant operating data to management. Figure 2 gives a schematic picture of such a total mill-wide hierarchy.

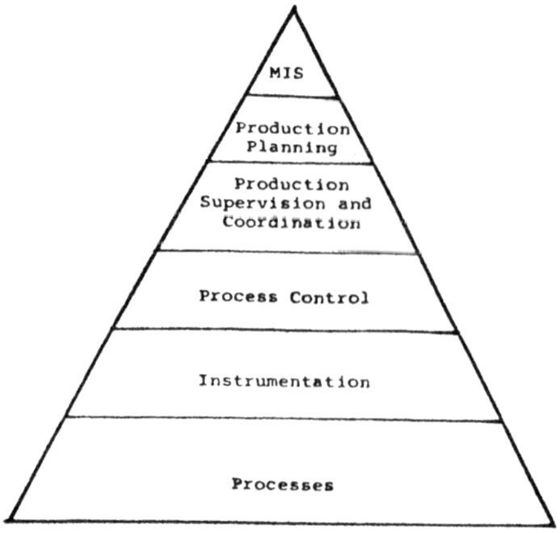

Figure 2. Total mill control hierarchy.

The needs and uses of such computer based hierarchical coordination and mill information systems in the paper industry have been actively studied by several mills and institutes (Uronen, P.: 1981, Integrated computer systems in the Pulp and Paper Industry, IIASA, Austria, WP-81-68, pp. 1-41, Peterson, E. and Rückert, H.: 1978, PPI, April, pp. 77-82, Eriksson, L.: 1978, Survey of multilevel computer control systems in the Pulp and Paper Industry, IFAC World Congress, Proceedings, Helsinki, pp. 213-220) and some applications have already been made. However, the existing systems do not encompass the whole range of tasks and functions at the several levels of a total mill hierarchy. In addition, the rapid development of the concepts of distributed system architecture has made further research and development work essential. It is to believe that this will be the most important future trend in the Pulp and Paper Industry automation, and that the third generation of industrial computer systems in the Pulp and Paper Industry i.e. the era of microcomputer based distributed and hierarchical mill wide systems as depicted in Figure 3 has began.

2. SITUATION TODAY

2.1. Hardware

Figure 4 gives a schematic flow sheet of a typical integrated paper mill, where several paper or paperboard machines or coaters are producing different types of paper products. At the process control level the situation today is that there are digital computer control systems available for each individual subprocess both for sulphate (kraft) and sulphite processes. In addition to this there exist a lot of administrative or EDP-type applications of digital computers. Figure 5 is a visualization of this situation. Further it is to be noted that in a typical mill there exists very little or not at all realtime communica-

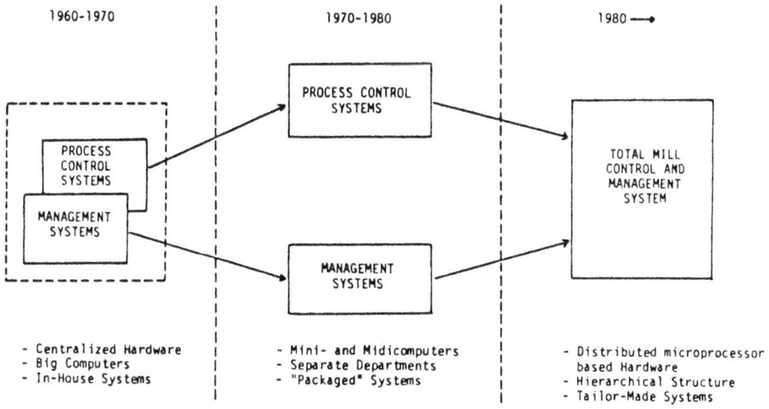

Figure 3. Three generations of control and management systems.

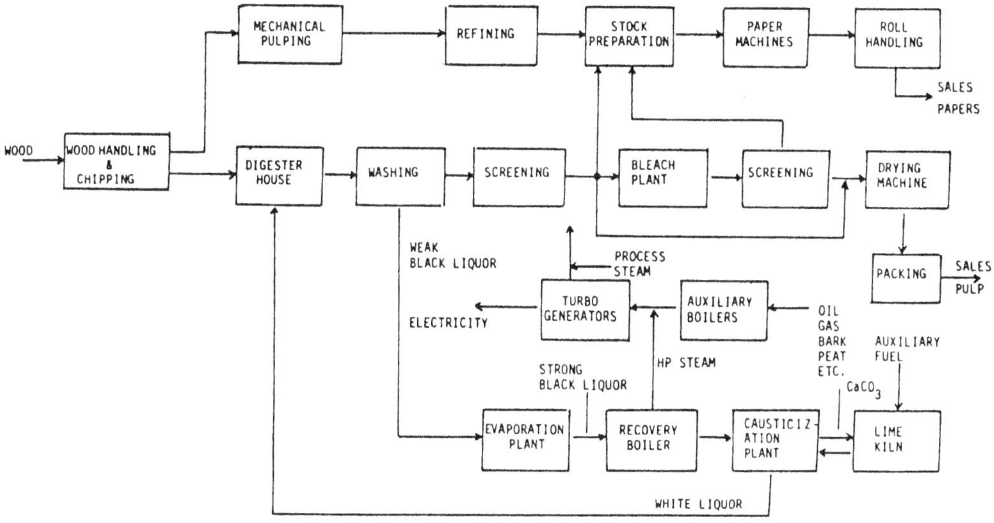

Figure 4. Schematic flow sheet of an integrated paper mill.

tion between different process control computers in the other hand and between process control systems and administrative systems on the other.

There are remarkable differencies between different countries and different companies concerning these applications. Figure 6 gives a comparison of these applications (on average) in some most important paper producer countries. All these process control packages were developed for minicomputers but since 1978 there has been introduced a second generation of these systems based on microprocessors. The rapid development of microelectronics has had also several other major effects on the automation especially in the lowest level of the control hierarchy of Figure 2 e.g. in the instrumentation.

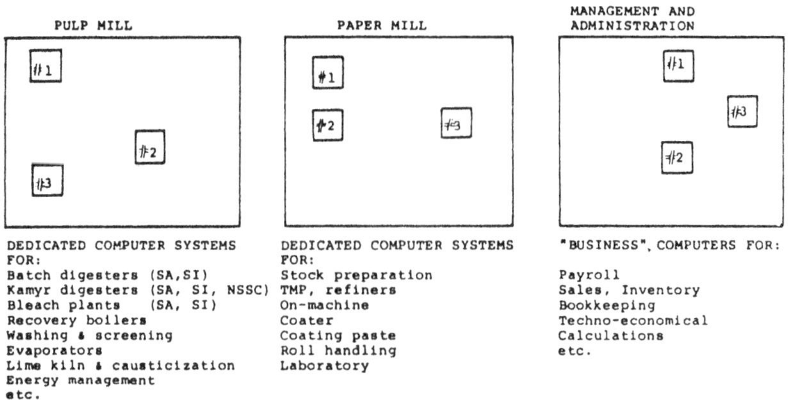

Figure 5. Computer systems in the pulp and paper mill today.

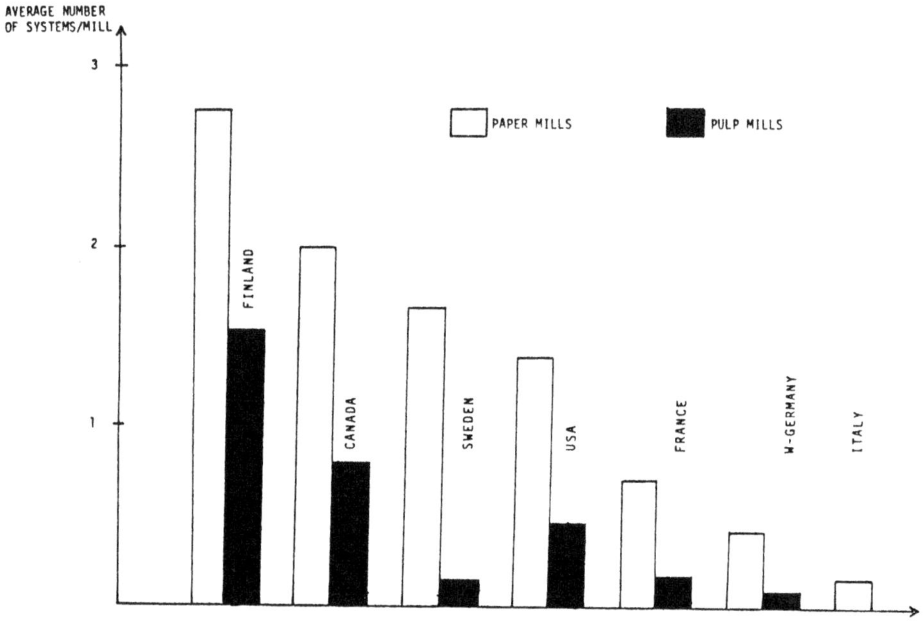

Figure 6. The average number of computer systems in pulp and paper mills in different countries.

We can say today that the control engineering and process automation are under rapid development stages: there is a remarkable change from analog techniques into the distributed digital control with a motto: "distributed control and centralized supervision". This means that the field control stations are distributed and installed near to the sensors and actuators making thus the response of the individual loops very fast. Also the central control room in its earlier form has disappeared and the new control room technique utilizes multicolour videos and other modern display and communication techniques using effectively microprocessors. The main functions in the central control room include higher level controls i.e. plant wide controls, optimization and reporting. This new era started in 1975 when the first digital instrumentation system (Honeywell TDC 2000) was introduced. Since that there are more than 100 different digital instrumentation systems in the market.

An other remarkable feature in the present trend in process automation is the discretizing of the signals and previously continuous functions. For example the conventional PID-controller is now a digital control algorithm on a microprocessor chip. The digital signal transmission has many important advantages like: accuracy, reliability, errorchecking and correction etc. Figure 7 shows a typical arrangement where digital process control is applied to an analog process environment: the border

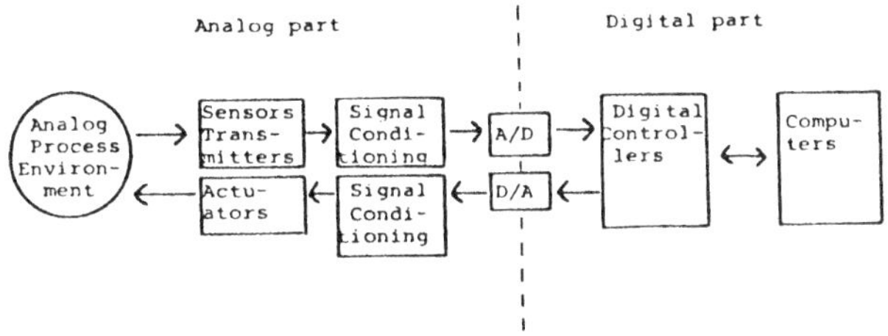

Figure 7. Digital and analog parts of automation.

between the digital and analog parts of the whole system is changing further and further in favour of the digital part: more and more of the equipment and functions will be digital in the future. During this decade a big part of transmitters and actuators will be digital.

The idea of distributed control can be realized in many different ways and using different architectures. Figures 8, 9 and 10 represent the typical analog centralized control and two different distributed control architectures, where the distribution is on one or two levels. The big advantages of the digital distributed control are: speed, accuracy, reliability, self-diagnostics, flexible control algorithms and savings in cabling costs. The vital part of these systems is the high speed data transmission bus (Data highway): this is realized with a coaxial cable and different transmission protocols and microprocessor based traffic controllers are used. The international standardization in this area would be useful and important. The further integration of the control activities in different departments and at different levels of organization having computer systems of different age and from different manufacturers make a local data network by using this kind of transmission buses a necessity. The building of mill-wide local data networks and integrating the process control activities horizontally and vertically into the management and EDP functions of the mill will be one of the key development areas and directions. However, the real implementation of these new technologies varies very much in

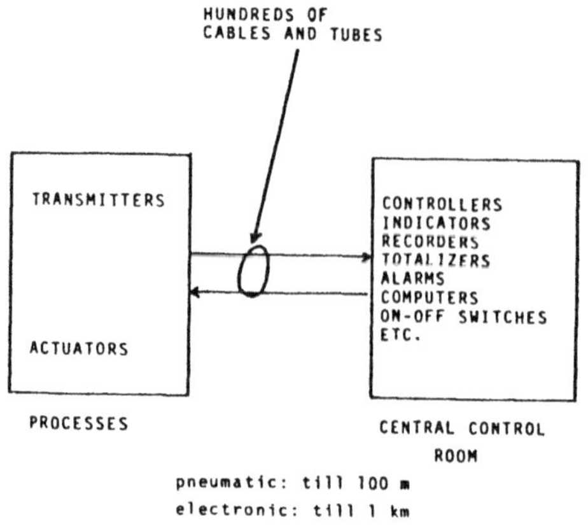

Figure 8. Centralized analog control.

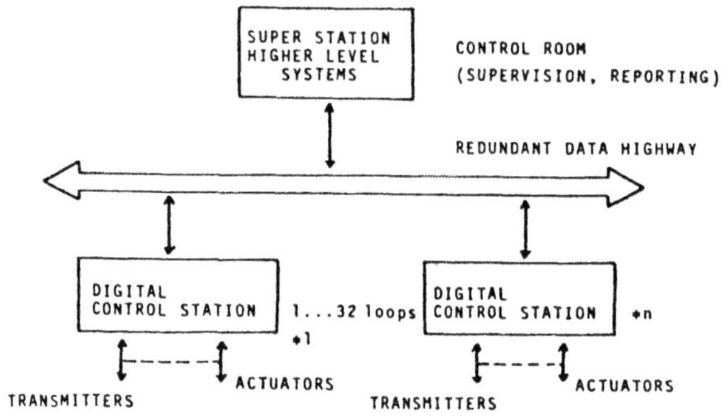

Figure 9. Distributed control system (1-level distribution).

different countries as already stated before. It can also be stated that in connection with big new projects it is natural in most cases today to select the digital instrumentation system; in old plants this can in most cases happen only in connection with a major renewing project.

The use of microprocessors has also lead to the distribution of intelligence, i.e., the use of intelligent sensors and analyzers to collect data and to carry out the necessary filtering, calibration, error checking, linearization, calculation of derived variables, etc., thus liberating the control computer for its main job of controlling the process. Typical examples of this kind of intelligent sensors or instruments in the pulp and paper industry are on-line chemical analyzers, mass flow metering instruments, heat flow transmitters, consistency meters, colour and brightness sensors, basis weight and moisture measurement devices etc.

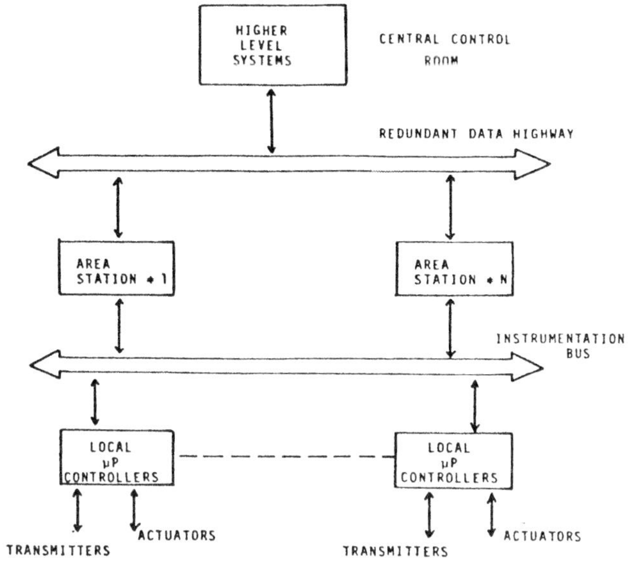

Figure 10. Distributed control system (2-level distribution)

2.2. Software

The cost of software in a typical process computer application tends to be the biggest portion; in some cases 70 % or even more. Therefore the development in software is of vital importance.

There are two main questions in the software area which should be considered. The first one concerns software implementation, i.e., the language problem. Today the situation here is that each vendor has developed his own operating systems and applications programming systems. Assembler languages or real time Fortran have been used for most of these. International standardization concerning the data highway protocols, the operating systems and the application programming would be very valuable and important. However, these standards will not be a reality in the very near future. In last few years there has been certain development in this area but even the new control oriented language ADA is not the final and ideal solution.

The other major software question is that of the methods to be used in controlling pulping and papermaking processes. In general we can state that the methods used at present are rather straightforward and conventional. At the risk of oversimplifying the problem we could say that the basic idea today is to provide for the best operator convenience plus a simple model to be used in a feedforward of feedback compensation manner to control the process.

The control algorithm used most often is the single loop PID algorithm in digital form. An IFAC questionnaire developed in connection with the Hague Conference in 1977 clearly showed this situation for all process industry control applications, not just those in pulp and paper (Duyfjes, G., de Jong, P.J. and Verbruggen, H.D.: 1977, 5th IFAC/IFIP International Conference on Digital Computer Applications to Process Control, The Hague, Netherlands). The PID algorithm has many good features and it has been stated that over 90 % of practical industrial control problems can be successfully solved with the use of a linear single loop theory and the PID algorithm. Again at the risk of overemphasis, the PID algorithm is a powerful algorithm indeed and has the great advantage of being easy to tune. Other algorithms for single input-single output systems have also been developed (Dead-Beat, Kalman, Dahlin, 3P, etc.), but they have had only limited use so far. The practical use of adaptive, multivariable and optimal control strategies is also still quite limited.

Major problems in pulp and paper process control today remain in noisy and unreliable measurements: in a lack of many necessary measurements; and in strongly interacting, poorly understood nonlinear dynamic processes. Therefore, the present situation is fully understandable.

The standard features of pulp and paper mill control applications today are found among the following capabilities.

1. Combined feedforward and feedback control loops.
2. Feedforward control making use of a process model which generally is quite simple and semiempirical.
3. Dead time compensation algorithms.
4. Either supervisory type digital control interfacing a full analog control system or DDC having some degree of backup.
5. Computation of calculated and derived variables which can be controlled. For example, the use of heat and material balances as models for control purposes.
6. Interactions decoupling, for example, moisture and basis weight at paper machines.
7. Adaptive control.
8. Statistical analysis which can be applied to measured data.
9. Trend recording, reporting, etc.

There have also been descriptions of the use of state space methods, self tuning algorithms, on-line identification, Kalman filtering and other state estimation methods, stochastic control applications and optimal control but so far they have had only a very limited number of real applications. The main reason for this limited range of application has been, as already mentioned, problems in instrumentation and process models. Also the modern theories are difficult to apply because they need specially trained personnel and the potential benefits to be achieved are difficult to predict. There is also a question of the attitudes of plant personnel. Thus it is most important that more interest should be placed on the determination of the resulting control benefits. Here, the special needs of the process (safety, start-up, shut-down, grade changes, speed changes, etc.) should be taken into careful consideration in terms of the calculation of their possible monetary benefits.

2.3. Benefits obtained

There are several specific gains in different subprocesses which have been obtained from individual digital computer control systems in terms of increased throughputs, higher yields, lower raw material and energy usage and higher quality. These gains will be discussed in details in next chapter. In addition to these relatively readily recognized gains there is a long list of other, often intangible, gains which are available, particularly from the coordination capabilities of a distributed hierarchy system, which need to be mentioned. Some of these are as follows (Uronen, P., and Williams, T.J.: 1978, PLAIC Report Number 111, Purdue University, West Lafayette, USA):

1. Faster availability and more frequent updating of information flow from the plant to management thus aiding decision making at all such levels.
2. Plant production and operational data (both current and historic) will be much more precise and time coordinated than is possible with current methods. Such should be a major aid to long term planning.
3. Plant data system can detect trends in plant operations. These and other plant data available from the system will readily permit the testing of various plant operating strategies prior to actual implementation.
4. Some literature examples of gains obtained from similar capabilities are the following:
   a) There is about a 90 % average efficiency of utilization of capital equipment in the paper industry today. A 1.0 % increase in this efficiency would mean about 650,000 tons per year of paper for the U.S. Paper Industry for the same capital investment (Hannigan, J.: 1978, Tappi, 61, p. 779).
   b) There is a potential productivity gain of about 5 % through the better coordination of interacting departments and their related buffer storage.
   c) With modern recovery boilers showing efficiencies of the order of 65 % and older systems running as low as 40-45 %, there is a very great potential for energy savings in this area through better control and coordination.
   d) Labor productivity can be improved with a complete measurement and reporting system coordinated with engineering performance standards. Such is readily possible with the proposed hierarchy. According to C.E. Geisel (Geisel, C.E.: 1978, Tappi, 61, pp. 33-36), container corporation has cut labor costs per ton by 19 % during the past eight years

with a productivity measurement system. Another example shows a 2.5 % decrease in material waste on a can line.

5. Full plant optimization studies are possible only with the full availability of plant operating data supplied by a system such as in described here.
6. Fast and effective monitoring of effluents and emissions can often prevent major leaks and the resulting damage or loss.
7. Savings in operational personnel and other staffing.

Some specific benefits of the distributed system architecture are as follows:

1. Flexible system configuration - distributed subsystems may be modified, replaced, or deleted without upsetting the rest of the system.
2. Graceful degradation - failure in one or more components or subsystems will not cause the entire system to fail.
3. High systems reliability due to:
   a) Easy to add parallel redundant units and subsystems which can be incorporated to back up and duplicate the functions of the main components and subsystems.
   b) Transmission of partially processed plant information allowing:
      (1) Decreased data rates since processors are distributed to functional areas and only processed information need be sent between any two subsystems rather than raw data as formerly.
      (2) Use or error detection codes which allow any fault or casualty condition in the system to be detected and identified by the processor in its area of responsibility.
4. Lower cost due to:
   a) Simplified hardware configuration packaging. Since processors need not be large due to the reduced processing requirements of each processor.
   b) Simplified software because functions are carried out by several small, locally responsible processors, not by a large machine that must perform all of the control functions and calculations within the entire control system.
   c) Large scale integration technology.
   d) Multiple use of standard components. Many different subsystems can use identical hardware to perform varied functions.
   e) Ease of incrementally increasing capabilities since units may be added to the system without drastically interfering with the functions of the rest of the system.
   f. Simplified installation since common data channels can be used for processor-to-processor communication. This eliminates the need for individual multiple-wire cables between any two units.

3. TYPICAL APPLICATIONS IN THE PULP AND PAPER INDUSTRY

3.1. Digester control systems

The kraft batch digester control is one of the earliest and most common applications in the pulp mill area and today they will be implemented with digital instrumentation systems and microprocessor technology; only some upper level tasks and reporting will be done with a mini-computer or "super-micro".

The main tasks of these systems are: charging, production scheduling, steam leveling, simulation of the cook cycle by using the H-factor or its modification, and blow control.

The most common criterion is that as low a temperature as possible (without restricting the production) will be used and thus energy savings can be achieved. State models have been developed for pulping kinetics, and some attempts have been made to use them in on-line control systems but so far control is based mainly on an H-factor simulation. Good results have been obtained, i.e., yield increases of 1-1.4 % and steam savings of 10-15 %, etc.

Liquor analysis and chip moisture measurement are the main instrumentation problems remaining. Trial systems for these functions have been applied to both kraft and sulphite production. There are also systems available using one or two samples drawn from the cooking liquor at certain moment(s) for more precise control of endpoint.

The most important continuous pulp digesting process is the so-called Kamyr digester which can be applied to both kraft and sulphite processes. It may be used with or without preimpregnation and a hydraulic or steam-phase. The first computer control systems in the pulp industry were developed for Kamyr digesters in early 1960's.

The existing systems are based mainly on the use of a liquor dosage, temperature profile and residence time (modified H-factor) for predicting the degree of delignification of the pulp at the blow line.

The most important control loops included are alkali to wood ratio, liquor to wood ratio, production rate, chip level, heating zone temperatures, dilution ratio and wash flow rate. The difficulties in Kamyr digester control are caused mainly by long time delays, by measurement problems (chip moisture, chip quality, chip level, delignification rate, chemical analysis of liquors, blow consistency, etc.) and by the possibility of "hanging", i.e., the uneven downwards flow of the chip column in the vessel.

Very important items for the effective control of a Kamyr digester are the special programs developed for either preplanned or emergency changes in production rates or grades. The idea used here is that the "border" between two different grades or production rates will be followed down the digester and all adjustments will be done in such a way that any offspects quality should remain as a minimum. Decreased quality variations, higher yields, steam savings, higher production rates, reduced chemical losses, reduced load on the recovery area and easier grade and rate changes are thus the gains to be made from computer control. These systems today are using microprocessor technology together with digital instrumentation.

3.2. Bleaching

The conventional bleaching sequence, CEHDED (Chlorine, Extraction, Hypochlorite, Chlorine Dioxide, Extraction, Dioxide), has been controlled by computers since 1966 and today there are several such systems in operation. The conventional sequence with towers and washers having long time constants is still the dominant bleaching process even though the oxygen bleaching process will probably displace it in the future in combination with chlorine chemicals. At the same time much research and development work has already been put into the development of the new displacement bleaching process.

It is clear that the displacement method needs a different control strategy from the normal sequence. Another development having a great impact on the existing control strategies is the trend towards a closed operation of bleach plants. The minimization of the usage of water and of the closing of the washing loops in the bleaching sequence will have a very great effect on the material and energy balances of the sequence and they must be considered in the planning of the control system.

The ideas most widely used in today's microprocessor based computer control systems include the use of special intelligent instruments (optical brightness meters, ORP (Oxidation Reduction Potential), residual chlorine analyzers, viscosity meters, consistency measurements, etc.) and empirical predictive models for feedforward control in combination with feedback loops. The control target is to produce a specified amount of pulp having the required properties (brightness, strength, consistency, etc.) with a maximum yield and minimum usage of energy and chemicals. In this connection the total coordination of the digesting and bleaching processes must also be considered. As has already been stated above the existing control systems are based mainly on a set of empirical static correlations and feedforward compensation with the aid of feedback from the on-line special instruments. Here use can also be made of dead time compensation, Kalman filtering techniques, etc. The models used are linear or they can be linearized and thus attempts are also made to use LP-techniques for the overall optimization of the sequence.

The basic problem in a bleach sequence optimization is the optimal distribution of the total bleaching effect taking into account the mechanical properties of the pulp, the yield and the chemical and energy consumption.

3.3. Washing

In connection with continuous Kamyr digesters the first washing stage (hi-heat wash) takes place in the digester. This requires that it is integrated with the cooking process. All changes made here will also have an effect on the control strategy used in the washing and screening processes.

The equipment used in continuous washing includes, in addition to the digester wash already mentioned above, diffuser washing equipment, filter washers, press washing processes and different combinations of all of them. The most important method to be used in the future will probably be the diffuser method.

The major control problem which is encountered in a common drum filter washing plant is that of the optimal usage of washing water. If too high an amount of water is used it means higher fiber losses and higher evaporation costs. If on the other hand too small an amount of water is used it may lead to chemical losses and to a dirty pulp which will mean a higher chemical consumption in bleaching, etc. The most important disturbances in a drum filter washing plant are variations in the kappa number of the incoming pulp and an uneven refining of the pulp. The major control variables include: the vacuum in the filters, the levels in the vat's, the rotating speed of the drums and the flow rates of washing water. The most important output variables are: the consistency of the washed pulp, its basis weight, the thickness of the pulp layer on the filter outlet, the saturation moisture of the pulp, the chemical losses, the flow rate of weak black liquor, the fiber losses and the solids content of the weak black liquor. The control of screening and cleaning is very closely connected with the control of the washing plant and the optimal coupling of different screens, cleaners and refiners whish also may be used here. The measurement problems here include, for example, the need for reliable on-line shive analyzers and mat consistency measurements. All these tasks can easily be handled by digital instrumentation systems.

3.4. Evaporators

While the evaporator plant is the biggest energy consumer in the pulp mill, it also has certain control possibilities affecting the whole mill steam and electrical energy balance. These features must be utilized in the overall energy coordination and control system. The control targets of the evaporator plant itself can be briefly listed as follows:

1. To maintain the solids content of the strong black liquor at an optimum value. This optimum value will be determined mainly by the level of steam generation in the recovery boiler, by the safety factors, and by the scaling of the evaporator effects.

2. To maintain the density of the incoming black liquor at a suitable level as required by the soap skimming operation.
3. To minimize the scaling and to control the washing and boil-outs of the evaporator units.
4. To control the purging of inerts.

The modelling and control of the multiple effect evaporator plant has been studied quite extensively and good mathematical models are available (Williams, T.J. and Rastogi, L.K.: 1975, Modelling and Control of Kraft Production Systems, ISA, Pittsburgh). In these studies also the oscillations which have been observed in real plant operation have been explained. The control systems generally proposed for evaporators are not very extensive and they could possibly be combined with the control of other units such as recovery boilers.

No difficult measurement problems exist since black liquor solids can be reliably measured with, for example, refractometers. Several digital computer systems are already in operation.

3.5. Recovery boiler

The recovery boiler in a kraft mill has a dual function; it is a chemical reactor recovering the digesting chemicals from black liquor into a reusable form and, on the other hand, it is also a steam generator using the organic material of the black liquor as fuel. In these tasks the resulting emissions (particulates and gaseous emissions) must be minimized and the potential risks in its operation (for example, smelt-water explosions) must be eliminated.

The limiting factor in the digital control of recovery boilers has, however, been the lack of some necessary on-line measurements, for example, the bed height, the bed temperature, bed density, some emission rates, etc. Today most of these can be solved in a satisfactory manner. The biggest problem remaining is the measurement of the bed variables (height, temperature, density, etc.), but even this difficulty can be avoided by using mathematical models and state estimation. There are already about 50 computer control systems for recovery boilers in operation today and more will be installed in the near future. The main functions of these systems include:

1. Black liquor spraying controls (pressure and temperature).
2. Air flow and distribution controls.
3. Emission controls.
4. Soot blower controls.
5. Steam generation controls.
6. Dissolving tank controls.
7. Cascade evaporator control.

The reported results are very promising; the biggest savings are from decreased soot blower steam consumption, higher capacity, a higher degree of chemicals reduction, better thermal efficiency and lower emissions. The payback time can be shorter than one year (Uronen, P., Jutila, E., and Pantsar, O.: 1977, 3rd International Symposium on Process Control, CPPA, Vancouver). Also separate microprocessor based sootblower control systems have been developed.

3.6. Power boilers

What has been said above for recovery boilers will be partly valid also for auxiliary power boilers. Here only the fuel is different (bark, gas, oil, coal) and the chemicals to be recovered do not exist. Still the right distribution of air flows to different parts of the boiler, the feedforward compensation for variations in the fuel quality (for example moisture content) and soot blower operations are similar to those for recovery boilers.

The control and operation of power boilers is closely connected to the total energy coordination function and to the operation of the power turbines.

Many energy management computer systems in the pulp and paper industry are already in operation having a considerable variation of combinations of tasks. Promising results and savings have been reported.

3.7. Causticization and lime kiln

The regeneration of white liquor in a kraft pulp mill involves the causticization plant and the lime kiln. The common strategy which has been used in the operation of these plants is that they will be operated separately. Thus their automation has previously been left at a very low level.

During the last few years, more attention has been paid to the modelling and control of these plants because the lime kiln is a very large consumer of energy and also has major effects on environmental pollution. In addition, the present day trend towards a "closed Mill" operation has lead to the collection of various plant waste gases streams for incineration in the lime kiln thus causing severe control problems there.

Achievement of a stable quality of the white liquor on the other hand improves the operation of digesters. Various models for the causticization loop have been developed and presented (Jacobi, E.F., and Williams, T.J.: 1975, A Dynamic Model of White Liquor Preparation, Modelling and Control of Kraft Production systems, ISA, Pittsburgh; Uronen, P., and Aurasmaa, H.: 1977, Modelling and Simulation of a causticization Plant and Lime Kiln, 3rd International Symposium on Process Control, CPPA, Vancouver).

There are some digital control systems with good results already in operation. The main tasks of such system include:

1. Production rate and quality coordination.
2. Control of causticity.
3. Control of green liquor temperature, density and flow distribution.
4. Clarification control.
5. Control of mud washers.
6. Compensation for mud moisture and fuel quality variations.
7. Control of temperature profile in the kiln.
8. Feed rate control.
9. Mud filter control.

3.8. Other pulp mill processes

There are still other departments in a kraft pulp mill area which must also be considered and included in a total mill control concept. These are wood handling machinery, pulp drying machines, baling and storage, raw water and waste water treatment, and preparation of bleaching chemicals.

From among these possibilities those of the bale scales, paper storage operations, water and waste treatment plants and bleaching chemicals preparation have been targets for modelling and automation studies.

The tasks of these systems are highly dependent on the local requirements, and conditions and on the processes equipment used. Because of the demands for environmental protection, it can be predicted that computer systems, at least for the monitoring functions, must be installed in most pulp and paper mills in the near future.

The wood handling department control system should take care of the control of cranes and conveyors, of the measuring and shorting of wood, and of barking and chipping. Most of these operations involve the automation of mechanical operations and the use of sequences controllers or microprocessor based programmable controller will be profitable. The main advantage of this automation is in improving the reliability of the system and in the reduction of labor costs.

Concerning the pulp drying machine and pulp baling operations the following tasks for the control system can be listed: heat economy of the machine, thickness and moisture control of the pulp sheets, control of the bale scale and the automation of different mechanical operations (conveyors, wrapping and labeling machines, etc.). Several digital control systems exist. They are mainly concentrated into the control of bale scales by using pulp moisture meters and drying to keep the dry weight of pulp bales constant. Some remaining instrumentation problems here will have to be solved before more complete control systems will be available.

3.9. Mechanical pulping processes

Preparation of pulp by purely mechanical means includes two main processes; the conventional groundwood process and the disc refiner process producing mechanical pulp from chips. The latter can also be operated after steaming the chips and at elevated pressures and is then called the TMP (or Thermomechanical Pulping)-process. There are digital computer control packages available for all these processes. The target of these systems is to produce a mechanical pulp having the desired quality with a maximum yield and minimum energy usage. It is a well known fact that ground wood and other mechanical pulping processes are big energy consumers and therefore one task of these control systems has been to smooth out and stabilize the energy peaks.

The difficulty here has been the lack of an accurate and reliable online consistency measurement as well as freeness meters and shive analyzers. Therefore the systems used are based on empirically derived and verified correlations between energy consumption, grinding area, grinding pressure and grinder speed. Naturally, the sharpness of the stone (or the refiner burr or disc) also has an effect on the process and thus it must be included in the models.

The yield of a mechanical pulping process is higher than the yield in chemical pulping. Also the environmental load of the mechanical pulping process is much lower than in chemical pulping. The newest process is the TMP process (Thermo-Mechanical Pulping). It has better strength properties than stone ground mechanical pulp or refiner pulp. In the TMP process the mechanical pulping will be carried out at elevated pressures by using steam. Therefore the recovery of this heat energy is an important part of the process and it must be taken into account when designing a control system for a TMP plant. As already mentioned there are systems already working in this application. The basic requirements exist for a higher level system coordination, especially with energy coordination systems.

3.10. Stock preparation

The stock preparation area includes the beating or refining of the pulp in order to treat the raw fibers for better sheet formation and qualities; and for the blending of different pulps and of sizing and dyeing, filling and/or other chemicals. Many different combinations and variations are used depending on machinery, chemicals added and paper qualities to be produced. The target is to achieve a good sheet formation at the wet end (freeness, mechanical properties) with a good resistance to wetting and penetration (sizing) and a suitable colour (dyeing). There are many computer control systems in operation in paper mills under the title "stock preparation" but the tasks and complexity of these systems will vary remarkably.

For the beaters and refiners similar approaches to those presented earlier in this report for mechanical pulping have been developed. The refiners have also been controlled by using the temperature difference between the incoming and outgoing pulp as a measure of the beating efficiency. The lack of reliable, continuous, on-line instruments for freeness for consistency, for mechanical properties, for sheet formation, for colour, etc. are the major difficulties here. One commonly used measure for freeness has been the couch-roll vacuum. The control of dosing and blending operations themselves, although included here, are straightforward operations. The difficulty is to find the right recipes in order to achieve the target quality. The big advantages of digital systems here are in better accuracy, in the ease of changing the recipes and in the possibility of using empirical models and correlations to make any necessary calculations.

There is very intensive research and development work going on in this area for the development of better and "more intelligent" on-line instruments.

3.11. On-machine applications

On-machine applications at different paper and board machines are predominant among the computer control systems installed in the pulp and paper industry; over 60 % of all applications are for on-machine systems.

Typical features of these systems are: use of special intelligent on-line instruments (basis weight, moisture, ash content, caliper, opacity, colour, etc.) and good man-machine interfaces. The software also includes model reference, adaptive, state estimation and decoupling algorithms. These systems have generally proven to be quite successful and the most important benefits achieved include: an increase in machine speed, i.e., capacity; better quality; savings in fiber; savings in chemicals; a decrease in the number of breaks and a decrease of the amount of off-spec quality during production rate or grade changes.

These systems normally have the following main functions: dry weight control, moisture control, head-box control, grade change control, target weight optimization, target moisture optimization and speed optimization. Also crossdirectional basis weight and moisture control and caliper control at the reel have been succesfully implemented. The combination paperboard machines and coaters also have control packages which apply the same principles as the other on-machine applications above. In addition to this, they include measurements of different plies and coat weight, etc. The preparation of coating pastes is also computerized. This process consists of the dosign and blending of different chemicals according to a selected recipe. The number of these recipes in use in a single plant can be over 50. In principle there are no difficulties to computerize this type of batch blending process. Also the finishing and converting machines and corrugators have computer control systems but they are very specialized using special instruments and also taking care of special devices (extruders, etc.). They are not discussed in detail here.

3.12. Roll handling and inventory control systems

These operations include reel weighing, unwinding, rewinding, cutting, control of various conveyors, wrapping, labelling, roll indentification, order entry, inventory control, shipping control and accounting. There are many different digital computer systems already working in this area and the tasks and functions will vary widely. The operations these systems carry out are mainly those of the control and regulation of mechanical operations, of data base functions and of production reporting (document printing). These systems already include functions (order entry, inventory control, shipping control, etc.) which are very closely related to overall production planning and control systems.

3.13. Other systems

The questions concerning energy management, water treatment and environmental monitoring were already mentioned before. One very important department which must be considered here is the quality control laboratory. Most of the quality tests of different papers must be carried out in the laboratory because continuous and reliable on-line instruments are not available. Many tests and analysis require considerable time before the results are ready; therefore there is a need to automate these tests and analyses, as well as the input sample handling and the calculation and monitoring of the results in order to minimize the uncertainties and time lags that are involved since corrective actions to the process should be taken as soon as possible. The role of intelligent analyzers is important here. Many test laboratories have now been automatized having digital computer systems and they will play an important role in the total mill control hierarchy.

4. FUTURE TRENDS AND CONCLUSIONS

In the preceeding chapters it was stated that classical process control will be more and more integrated both horizontally to other operative functions (material handling, maintenance, marketing) and vertically to management and EDP functions. Thus we will have a hierarchical plant or corporate wide control system as depicted in Figure 2. An interesting new development at the top of the hierarchy is DSS (Decision Support Systems) to help top management in unstructured decision making situations. This also necessitates the local data networks as stated already. The use of computers (especially microcomputers) in the automation of pulp and paper industry will continue to increase.

Other typical trends affected by microprocessor technology are briefly:

- The instrumentation will be digital.
- More and more digital transmitters and actuators will be developed and used.
- The use of fiber optics will increase especially in sensors.
- More intelligent sensors will be developed.
- Multivariable transmitters will be developed.
- Application of speech recognition in man-machine communication will be a reality.
- One loop microprocessor controller will be common.
- Modular application oriented HW solutions to specific problems will be developed.
- Modern control methods will be applied on broader basis.
- Plant and corporate wide hierarchical control and information systems will become more and more important.

The progress will, however, be evolutionary in most cases due to the economic and other constraints.

5. REFERENCES

Duyfjes, G., deJong, P.J., and Verbruggen, H.D.: 1977, "Questionnaire on Applications of Modern Control Theory to Computer Control in the Process Industry", Results and Comments, 5th IFAC/IFIP International Conference on Digital Computer Applications to Process Control, The Hague, The Netherlands, June.

Eriksson, L.: 1975, "Datoriseringen inom skogsindustrin med Framåtblickar", Svensk Papperstidning, 78, 81.

Eriksson, L.: 1978, Survey of multilevel computer control systems in the pulp and paper industry. Proceedings of the 7th IFAC World Congress, Helsinki.

Gee, J.W., and Chamberlain, R.E.: 1977, "Digital Computer Applications in the Pulp and Paper Industry", 5th IFAC/IFIP International Conference on the Digital Computer Applications to Process Control, The Hague, The Netherlands, June.

Geisel, C.E.: 1978, "Productivity Measurement, A Prelude to Improvement", TAPPI, 61, pp. 33-36, August.

Hannigan, Judson: 1978, "Productivity as I see It", TAPPI, 61, No 8, pp. 779, August.

Jacobi, E.F., and Williams, T.J.: 1975, "A Dynamic Model of White Liquor Preparation", Modelling and Control of Kraft Production Systems, ISA, Pittsburg .

Keyes, M.A.: 1975, "Computer Control Census", TAPPI, Vol. 58, No 6.

Peterson, E., and Rückert, H.: 1978, PPI, April, pp. 77-82.

Uronen, P.: 1981, WP-81-68, IIASA, Vienna, pp. 1-41.

Uronen, P., and Aurasmaa, H.: 1977, "Modelling and Simulation of a Causticization Plant and Lime Kiln", 3rd International Symposium on Process Control, CPPA, Vancouver.

Uronen, P., Jutila, E., and Pantsar, O.:1977, "Computer Control of a Recovery Boiler", 3rd International Symposium on Process Control, CPPA, Vancouver.

Uronen, P.: 1977, "Översiktliga Synpunkter på Styrsystemtänkandet i Finland", Paper presented at Scandinavian Meeting of Pulp and Paper Engineers, August, Finland.

Uronen, P., and T.J. Williams: 1978, Hierarchical control in the pulp and paper industry. Report No. 111 Purdue Laboratory for Applied Industrial Control, West Lafayette.

Williams, T.J., Rastogi, L.L.: 1975, "Modelling and Control of Multiple Effect, Black Liquor Evaporator Systems", Modelling and Control of Kraft Production Systems, ISA, Pittsburgh.

CHAPTER 19

THE MICROPROCESSOR IN RAILWAY CONTROL SYSTEMS

Satoru Sone
Faculty of Engineering
University of Tokyo
Tokyo, Japan

and

Ikumasa Okumura
Technical Research Inst.
Japanese National
Railways, Japan

Application of microprocessors in railway systems has been as promising as is seen in other fields. After mentioning the sound applications to monitoring on-board equipments, traffic control, passenger information systems, control at marshalling yards, etc., state-of-the-art technologies in this field will be described; on (i) energy saving operation of inter-city trains, (ii) interference-free ac motor drive system and (iii) microcomputerised realisation of fail-safe vital functions of signalling system. Functions of (i) and (ii) would not have been realised without using microprocessors and that of (iii) has been thought realisable only by using non-electronic fale-safe devices such as gravity-operated electromagnetic relays. Sections 1.3 and 2.3 are authored by Dr.Okumura and the rest by Dr.Sone.

1. APPROVED APPLICATIONS OF MICROPROCESSORS IN RAILWAY SYSTEMS

1.1 Monitoring system of on-board equipments

(a) Multiple unit vehicles for advanced trains

Modern railway trains such as Japanese Shinkansen, British High Speed Train (HST) or French TGV (Train a Grande Vitesse) consist of multiple-unit vehicles, which is substantially superior to a locomotive hauled trains from the viewpoint of adhesion and axleload. In case of equipment failure, distributed traction system has both advantage and disadvantage; a single failure does not mostly mean total breakdown if the wrong part can properly be set aside and, if necessary, substituted by the remaining parts, but since the probability of failure increases and fault location is more difficult in a distributed system, it may reduce operating reliability unless prompt treatment against the failure is made. In order to cope with this disadvantage, microcomputerised monitoring systems have been developed.

(b) Functions of train monitoring system

On-board equipment monitoring system is used in two ways; (i) monitoring and treatment in operation, and (ii) checking or instrumentation

S. G. Tzafestas (ed.), Microprocessors in Signal Processing, Measurement and Control, 411–436.

for maintenance purpose in the depot.

The functions needed while in operation are a) indication of operating status of equipments, b) fault location and, if necessary, fault level indication, c) remote cut-out or reconnection control from a driving cab, d) reconfirmation of the treatment done, and e) other operating information to the driver such as train location, temperature or current of equipments, wheel slip/spin, etc.

The following additional functions are added to ease maintenance work; f) assistance to programmable testing facility located in train maintenance depot, and g) data storage in time series of operating conditions including fault and obtained values at testing.

(c) Typical example of the monitoring system

Photograph 1 shows a super express train belonging to Odakyu Electric Railway made in 1980. Because the driving cab locates above observation room and the only access to the cab is by a detachable ladder normally stored inside the cab, all handling by the driver must be done at the front end cab except for a real emergency. Photograph 2 shows the fault location indication board and some of remote control switches such as the pantograph lowering switch as seen at the leftmost.

The system configuration is shown in Fig. 1 where there are as many monitoring devices as the number of tractive units, four in this partic-

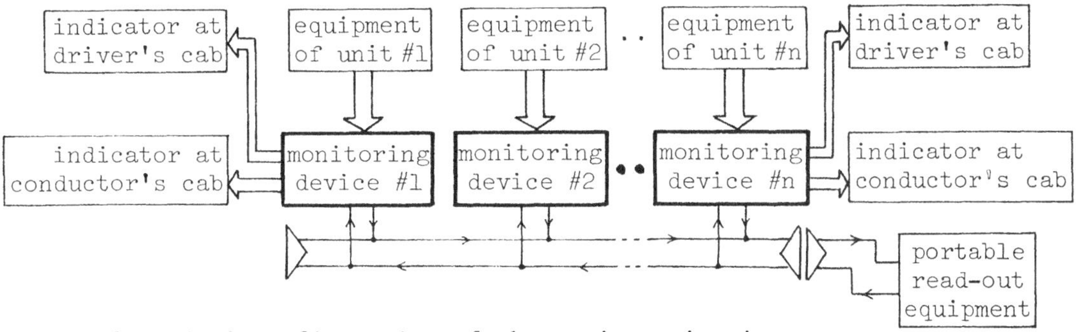

Fig. 1 Typical configuration of the train monitoring system

ular example, connected to the two pairs of communication cables, indicators and remote switches at each driving cab and train conductor's cab connected to the nearest monitoring devices and a connecting coupler at each end of the cable. Monitoring devices are almost identical but the front end device acts as the master which collects and edits the information from the remaining devices and indicates it to the crew. The flow of the control command by the crew is in the reverse direction. A portable read-out equipment of monitored information is also provided which is used either in operation or in the depot by connecting at the cable end.

This particular example uses 8-bit microcomputers, light emitting diodes for indication, 9.6k bits per second frequency shift keying serial transmission, whose sampling period is 50m sec. All these are chosen as proved sound and reliable.

(d) Variety of the monitoring systems

The second generation Shinkansen of Japanese National Railways, which was opened to the public in 1982 between Omiya to Morioka 465km and Omiya to Niigata 275km, uses various types of computers at various purposes. Train monitoring system of Series 200 Shinkansen formation is composed of a master monitoring unit at each driving motor car connected with a character display device using plasma display panels and other driving cab equipments, and slave monitoring terminals as shown in Fig. 2.

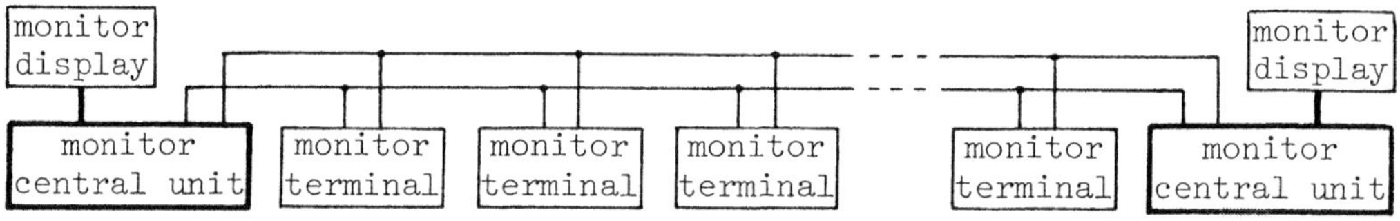

Fig. 2 Monitoring system of Shinkansen train

A slave monitoring terminal is located at each motor car including driving motor car and connected to major equipments and distribution switches. Monitoring units and train radio transmitter are connected with each other by two pairs of cables forming a loop.

Shinkansen's system has several additional functions with those of the typical system shown above such as train location indication, much detailed fault indication using numerals and characters and transmission of the fault information to the central traffic control (CTC) centre via train radio so that fully trained staff at CTC can help the driver in case of emergency.

Some of recent systems experimentally use optic fibres as a transmission line, which is completely free from electromagnetic noise and handling capacity is practically unlimited.

1.2 Total traffic control systems including substation control and information systems for passengers

(a) Computerised traffic control

Automation or computerisation in the field of control of train group movement started at automatic route control device using hard-wired logic early in 1960's. Up to mid 1970's computer application for train traffic control in a broad sense expanded both in complexity and integration of functions and in the area of control which required high speed and big capacity for the central processor as well as data transmission lines. The typical example of this trend is found in the COMTRAC (COMputer-aided TRAffic Control) system for Tokyo-to-Hakata Shinkansen. The major functions of the COMTRAC was realised in 1975, with some improvements afterwards. All necessary informations along the 1069km Shinkansen line are gathered to Tokyo where processing and detailed commands are made and originated. Only one exception of this highly centralised system is the Osaka subcentre, located some 500km west of Tokyo, where non-automated judgments of regulation of trains running west of Osaka are made by local commanders with priority being given at Tokyo centre. The COMTRAC, whose main functions are route control and train regulation, is integrated with passenger information, crew scheduling, maintenance of various subsystems and substation control.

From the cost and security points of view, highly centralised control system, which requires duplex or triplex subsystems at important parts, has been thought disadvantageous once distributed control system using reliable and inexpensive microccmputers was realisable. In recent years function of the central computer has changed to monitoring and inter-subsystem regulation rather than detailed control functions with each subsystem being locally controlled by microcomputer. In order to realise the above mentioned distributed control, the total system should be divided into subsystems whose function and control area is suitable for a microcomputer to control, while the central computer and communication system should be enough to be fail-soft in case of a single subsystem's failure or to be flexible at expansion or alteration of function or area to be controlled.

(b) Major subsystems of total traffic control

(1) Train position indication for monitoring.

(2) Monitoring of important equipments.

(3) Man-machine communication i.e. collaboration of commander and computer at some occasion.

(4) Recording statistical data.

All traffic control systems have had these four functions.

(5) Route control.

Although the major incentive was to automatically execute this control, most system can not control turnout switches inside depots, nor set up a route into or out of depot.

(6) Train regulation.

Most systems have the function of semi-automatic train order control according to the local time-position information of trains that would interact to each other. Some offer a little more complex proposals to the train regulator which will be either accepted or rejected automatically unless otherwise intervened by the regulator in a fixed time after the proposal was given. A few system can regulate order of trains at many stations and for a long period so as to realise the best order and headways at the bottle-neck station.

At any time the train commander can intervene the system manually.

(7) Passenger information system.

Public address, train indication including delay, connection or order at major destinations and departure notice is given automatically or semi-automatically at some systems.

(8) Substation control.

Configuration of power feeding network is sometimes changed by various reasons relevant to time zone, train schedule, maintenance schedule or unforseeable faults. This is done automatically or according to a remotely given manual instruction. Substation control is thought one of the most suitable application fields for microcomputers; local processing needed to execute sequence control at start or stop of equipments is not too heavy for microcomputer and information exchange between adjacent ones as well as between the manned control centre and each substation is also necessary. From system's reliability and security points of view, distributed control system with centre's burden being light but important for optimisation is most appealing.

(9) Train scheduling simulation and training the regulators.

In order to avoid deterioration of centre's staff in making a prompt judgment or decision at abnormal condition, some traffic control systems have such off-line functions as for training, planning, simulation or debugging of a superseding programme.

(10) Automatic transmission of necessary informations.

A few system transmit such informations as train delay, ridership, abnormal conditions or altered status automatically to previously selected points.

(11) Scheduling of the fleet.

(12) Scheduling of the crew.

(13) Route control inside depots.

These three functions are yet to be introduced.

Internal relationship among subsystems of total traffic control is shown in Fig. 3.

1.3 Signalling system and control at marshalling yards

The equipment of the set of sidings for breaking up freight trains

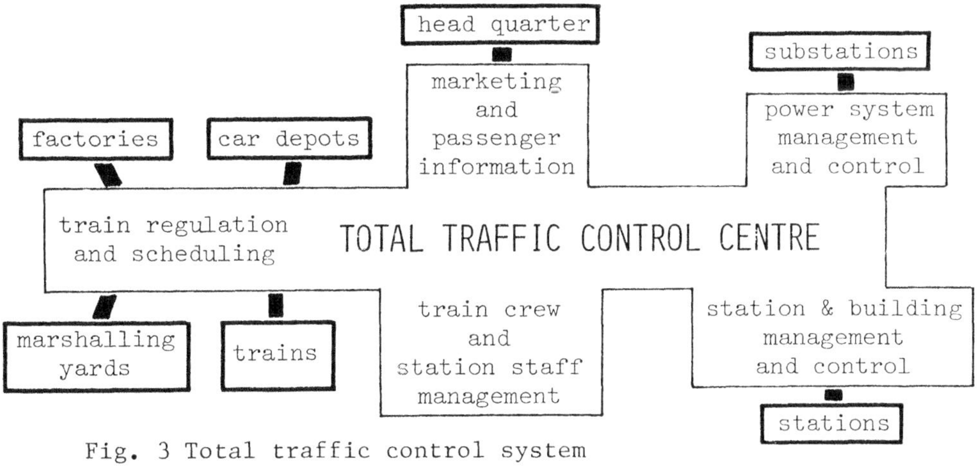

Fig. 3 Total traffic control system

at the marshalling yard was confined to control of the switch points by signal box and manually operated rail brakes (car retarders), the cut of freight cars being stopped on the sorting tracks by means of siding brakes, with a large number of brakesmen. Thier job is hard, not without risk and requires much experience. Therefore, in order to reduce their manpower, the automatic marshalling yard systems using large computers were installed at Conway, U.S.A. in 1955, Koriyama, Japan in 1968, Sotteville, France in 1973 and so on.

A system which is able to control an entire marshalling yard with a centralised method may generally consider to be efficient. Such a system, however, is reliable to develop problems within the signal system between the terminal machines and the central computers, and includes such problems as the system down of the central computers which may influence and extend to the entire system. Nowadays, considering that microcomputers have high reliability and low operating cost, it will be better to equip the microcomputers with a distributed method, assigning appropriate functions to the individual subsystems. The main functions in the automatic marshalling yard systems consist of the following two subsystems.

1) Automatic control of switch points.
2) Automatic control of car speeds.

These subsystems using microcomputers were installed at Inazawa, Japan[1] and at Villeneuve-Saint-Georges, France[2].

(a) Automatic control of switch points

Here is given an example of Inazawa marshalling yard of Japanese National Railways (JNR). An automatic switch point control device using mechanical relays was installed at Inazawa in 1959. In order to reduce the maintenance cost, JNR replaced the relay-based device with the microcomputer-based system in 1978. This microcomputer-based system is described.

The shunting routes are punched on a paper tape by the operator of hump cabin, according to the shunting list for each train. The point operations, punched from the hump cabin and stored in the microcomputer, are activated automatically as the cuts progress through the head of the sorting sidings. This progress is detected by means of track circuits of suitable length. The algorithms of the point operations are as follows.

1) As the cut goes into the track circuit, the next point is switched in the direction of the cut if the next track circuit is free.
2) As the cut goes out of the track circuit, if the following cut exists between itself and the backward point, its point is switched in the direction of the following cut.

The point controls are safely protected by track circuits and supplied by a pneumatic energy, the operation being carried out within 1.5 seconds.

A maximum of seventy cuts for each train and the shunting routes for several trains can be stored. The hump operator is able to cancel and preset an order by the push-buttons on the control desk in the event of an error detected in the destination or breaking of the cuts.

The microcomputer for control of the 33 points consists of Motorola 6800 microprocessor with 8kB ROM's and 2kB RAM's, and these are duplicated to ensure high availability.

(b) Automatic control of car speeds

In this section, automatic marshalling yard at Villeneuve-Saint-Georges of French National Railways (SNCF) is taken as the example. Fig. 4 shows the overall arrangements. It appeared essential to apply the progress achieved with the technique of automation to the yard at Villeneuve, in order to make its operation more productive and more econimic. The control of the points in the shunting sidings is therefore effected by means of a microcomputer. In addition the control of wagon speeds during shunting is effected automatically also with a microcomputer. The latter system is described.

The system of automation employed for the breaking of the wagon while being shunted must provide two types of control as follows;

1) On one hand, in the zone of the points, the cuts must be suitably spaced apart to allow for the operation of the points between successive cuts, taking into account the rapid rate of shunting desired. This result is obtained by giving the cuts sufficient speed in this zone, by means of a high hump.
2) On the other hand, on conventional tracks, the speed of the cuts must be such that they make contact with the raft already at a standstill, at a speed of less than 1.70m/s.

The six primary retarders are high powered so that, in the event of an incident, they can stop a cut engaging the retarder at 8m/s. They

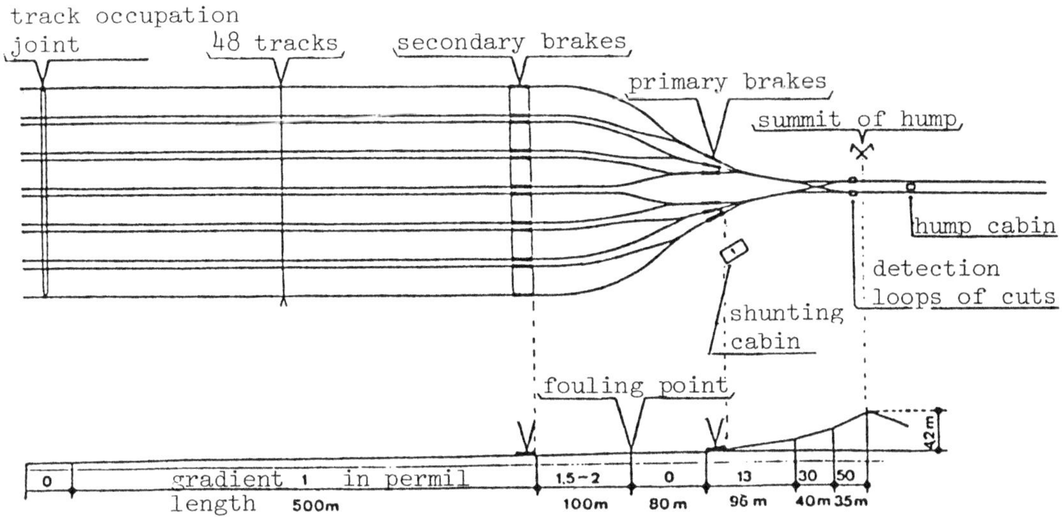

Fig. 4 Sorting sidings of Vileneuve-Saint-Georges marshalling yard

can also brake and release by double-acting hydraulic jack, controlled by electromagnetic valves. The speed of the cut during braking, is measured by a Doppler radar system whose aerial is installed a few metres beyond the retarder and radiates in front of it.

The 48 secondary retarders are similar to the primary ones, and consist of the same mechanical components; however, they are not so powerful, since the cuts engage them at speeds of between 3.5 and 4.5m/s. Therefore, they operate on one rail only. Like the primary retarder, each of the secondary retarders has a Doppler aerial and a controlling electromagnetic valve.

In principle, the cuts approach the retarder when it is in the 'on' position, and it is only released when the speed of the cut has been reduced to a value calculated by the microcomputer, known as the exit speed. The exit speed of the primary retarder is caluculated so that the cuts arrive at the secondary retarder at a speed of 4m/s in principle. This exit speed from the secondary retarders depends on;

- the rolling resistance of the cuts on straight track, and
- the distance they have to run between the secondary retarder and the raft which is being formed.

The microcomputer restricts itself to calculating the exit speeds of the primary and secondary retarders inclusive of calculating the above parameters. The comparison with the actual speeds — measured at the approach to the retarder by means of radar — is effected by the set of electronic logics, the 'servos' independent of the microcomputer; in the event of failure of the microcomputer, this makes it possible to introduce automatic outright braking and, with regard to the secondary retarders, enables the operator in the shunting cabin to preselect manually four exit speeds in relation to the condition at the time. It is thus possible to continue shunting almost normally without affecting the quality of the work too much.

The microcomputer for operating the retarders bases on an Intel 8080 microprocessor. Such are the microcomputerised installations at Villeneuve-Saint-Georges. The progress achieved since the placing in service of the first automated system at Sotteville in 1973 can be assessed by comparing. The prices are obviously cheap, being effected on the SNCF in all intermediate marshalling yards.

Meanwhile, instead of the secondary retarders, the mules powered by trolleys incorporating a linear motor, with the reaction plate fixed to the sleepers, are used to move wagons along classification siding at Shiohama yard, Japan.

The mules run on the inner foot of the rails, and have hinged arms fitted with rollers which engage the wheel flanges of the cars being pushed. In order to control the mules, the microcomputer is also used.

(c) Route control system using a voice data entry device

The fields in which a voice data entry can be utilised for computer systems has expanded rapidly with incorporation of high performance microcomputers with the recent advances in voice discrimination technology. On JNR, the plans are already implemented to utilise voice data entry for route control in shunting operations at a small yard without a hump[3].

The shunting operation at a small yard is a series of operations in which trains arriving at the yard are broken up and then set up to form other trains. This operation is performed by a team of shunters with the operator acting as the supervisor. The individual operations consist mainly of the items listed below.

drawing up the shunting operation plans,
route setting of the wagons,
driving of the shunting locomotive, and
breaking up and setting up of couplers of wagons.

It is the final purpose that the shunting operations of the above items are done by one person (operator) only, eliminating the problem of errors arising from mistakes in communication between the operator, the shunters and the drivers. Therefore, as the first stage, JNR has developed the Radio Freight-car Route Control System (RFRC) where the operator can directly control the routes himself by use of wireless transceivers and voice data entry device, so that the route control can be performed anywhere in the yard without signalmen and switchmen. The light weight transceivers currently in use for the communication in the yard have been modified to serve this purpose.

In the RFRC system, there are two microcomputer systems; for voice data entry and for route control. The voice data entered during the shunting operations is differentiated by the evaluation of the voice data similarity to the voice pattern recorded in the microcomputer. The use of suitable words has been providing satisfactory results. The route control concept is similar to that for switch control shown in item (a). The switch controls are safely protected by track circuit, and the shunting signals indicate safely the route number aspects by means of outputs of the microcomputer and the switch indication relays.

2. NOVEL APPLICATIONS OF MICROPROCESSORS IN RAILWAY SYSTEMS

2.1 Energy saving operation of inter-city trains

Fig. 5 shows typical energy consumption characteristics as a function of running time between stations with given line profile, maximum speed and train performance[4]. At minimum running time, corresponding to no coasting at all, energy consumption at powering is very high, and even if regenerative braking is fully utilised, net energy consumption is also very high. Scheduled running time should therefore be around 110% of minimum running time for a normal scheduling. The 10% margin contributes to stable operation of train group when a disturbance takes place.

In practical scheduling it is difficult to allocate marginal time evenly and even if it is so given, frequent disturbance prevents to run as scheduled; delayed trains run as fast as possible when signal permits but in most occasion fast running ends by the signal's restriction. Once it is forced to stop or run slowly, intention of recovery is no, or sometimes even negative, effect at the expence of increased energy consumption, accelerated deterioration of equipment including tracks and worse riding quality.

Another typical situation subject to optimisation is a long-distance non-stop run where there is plenty of total marginal time given at the driver's disposal. In this case, general principle to attain energy saving operation lies in the two key figures; maximum speed and the speed at which brakes are applied should be set as low as possible inside the limit of a given running time. The reason why human driver can not optimise the train movement is that he can not know exactly at what time each signal turns clear, which depends on the movements of the interactive trains.

Fig. 6 shows a typical running pattern of delayed trains in a single track section. Each train runs as fast as possible to recover delay but train #1, a light freight train, has to wait at station E, heavy freight #2 has to wait at D and C and express passenger train #4 is obliged to follow the slow #2 for some time. If, for some reason, the stations at which one train overtakes or exchange another train can not be changed, an example of clever running pattern is shown in Fig. 7, where the light freight train #1 waits for the slow running heavy train #2 at D

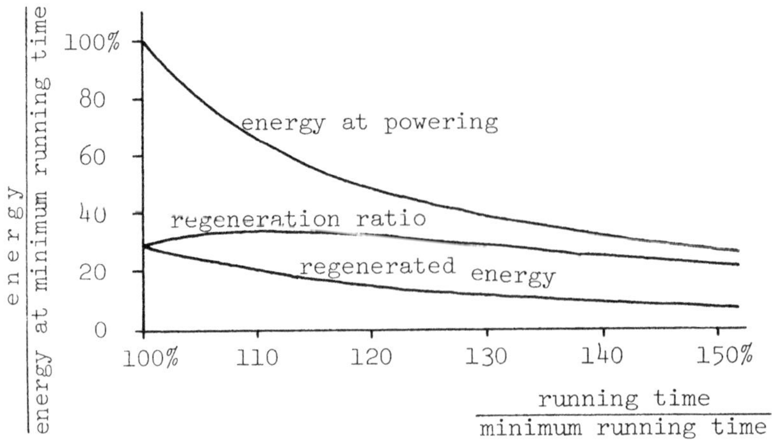

Fig. 5 Typical characteristics of energy consumption vs. running time (armature chopper controlled train)

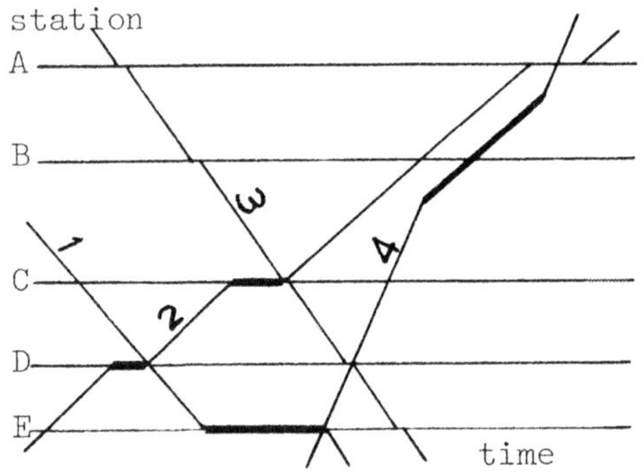

Fig. 6 Train trajectory of conventional delay recovery control

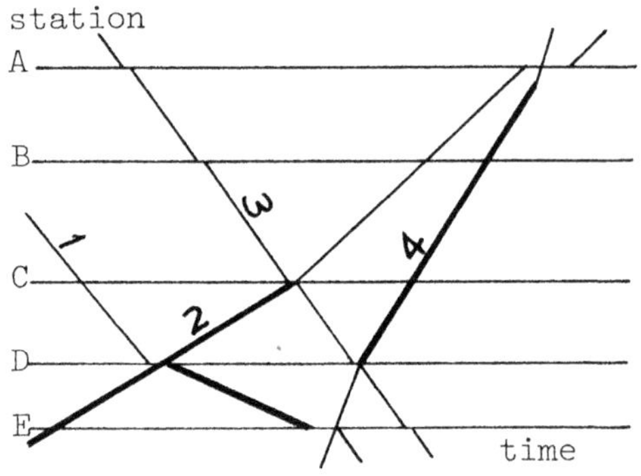

Fig. 7 Train trajectory of energy saving control without additional delay

and all trains run slower at the sections other than critical paths. Thus, unnecessary fast running and long waiting is eliminated without any additional delay. Improved operation as shown in Fig. 7 contributes to energy saving considerably as well as much reduced wear of brake shoes thanks to lower speed and less frequent stops.

In order to improve the situation, introduction of on-board computer was studied in as early days as in 1970 with intention to equip on the next generation Shinkansen trains. The system was named ATOMIC (Automatic Train Operation using Mini-Computer) but practical introduction was postponed by more than 10 years. It is now again to be introduced using microcomputer instead of minicomputer with additional functions of reduced running time and increased stability.

2.2 Interference-free asynchronous motor drive system

(a) Introduction

DC motors or ac commutator motors have long been used for almost all electric traction systems in the world. Several attempts of inverter and ac motor (especially asynchronous motor) drive for traction have been made in Britain and U.S.A. but successful bulky application is seen to date only in West Germany and Switzerland[4]. The aim is to reduce maintenance cost by eliminating commutators and brushes indispensable with dc machines, to improve adhesion performance, to reduce heat or energy loss caused by conventional rheostatic control of dc traction system, and to obtain high power per weight ratio due to high revolution capability of the ac machines.

The biggest technical problem to be solved in the ac motor system working under dc wire is interference problem. Contrary to the chopper and dc motor system, the inverter and ac motor system produces harmonic current on the dc line whose fundamental frequency varies continuously according to train speed. Any signalling or command circuit using audio or lower frequency carrier may be interfered by the harmonic current. Especially, fail-safe system of railway signalling, where no carrier means "danger", may be interfered fatally by the existence of harmonic current.

This interference problem may be solved by the following measures:
(1) development of a quite new signalling system free from interference,
(2) separation of source and load using indirect converter such as chopper and current type inverter or CVCF (constant-voltage constant-frequency) inverter and cycloconverter, or
(3) elimination of harmful noise whose frequency ranges are exclusively used by the signalling systems by adopting a direct type converter (voltage type inverter, etc.).

As an economical countermeasure the author has adopted the idea of (3). Item (1) may only be feasible in an isolated transport system, and (2) does not seem attractive because of its high cost and heavy weight. One of the problems arising from (3) is the need of compatibility for various lines on which the train will operate. Signalling frequencies vary from line to line. The converter should be adjusted so as not to produce harmful noise within the frequency band to be protected of each line or each section of a line. Such an adjustment is hoped to be achieved only by software modification or more easily by memory substitution.

A novel flexible driving system including an advanced harmonic control of (3) is realisable using digital technique with microprocessors.

(b) System configuration[5]

(i) Universal Thyristor Switch (UTS) Fig. 8 shows the attempt of

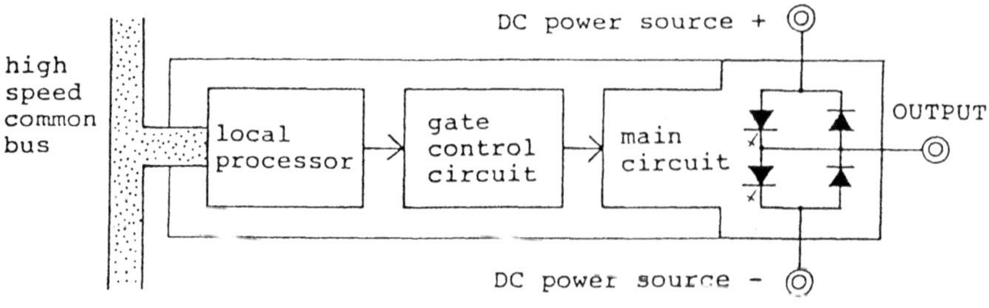

Fig. 8 Universal Thyristor Switch (UTS)

developing a universal use of thyristor circuits as a functional device for power switching.

The three-terminal-switch circuit of UTS can be used commonly for most types of power converters. The local processor generates realtime firing command signal according to the digital information on the high speed common bus and gives it to the gate control circuit. On and off timing is thus controlled by software for the UTS which has now its own intelligence. Moreover UTS provides the system with merit of decentralised operation such as trouble shooting on each unit, separation of disordered units and unbroken but reduced-capacity operation in case of emergency.

By using UTS's as fundamental components of a power circuit, many types of digital controlled power converters are composed easily of the same hardware, which makes it easy to utilise microprocessor's advanced ability or flexibility.

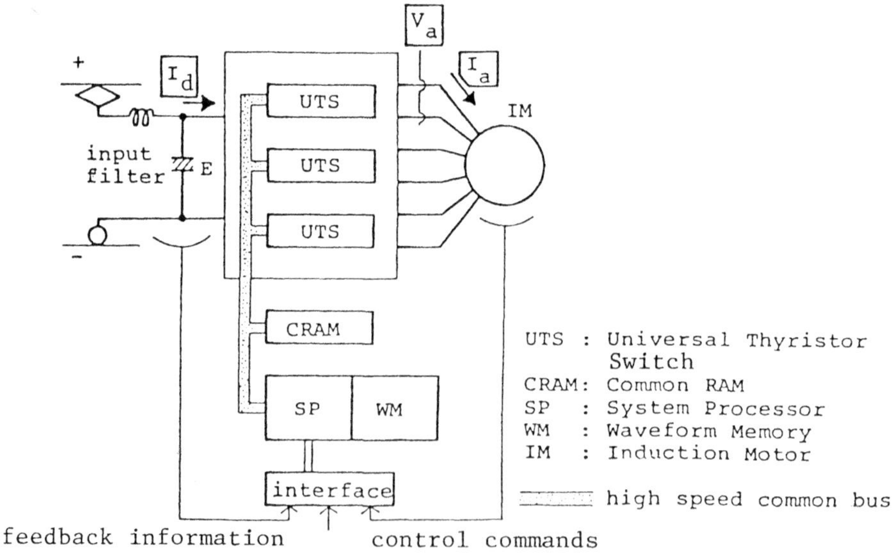

Fig. 9 Schematic diagram of microprocessor-based PWM inverter and induction motor drive system

(ii) PWM inverter and induction motor drive system composed of UTS's
Fig. 9 shows the whole schematic diagram of PWM (pulse width modulated) inverter and induction motor drive system for traction composed of UTS's. Supervisory system processor (SP) interconnects a group of identical UTS's via common memory (CRAM); thus formed a multiprocessor hierarchical structure. System parameters such as source voltage and current, inverter output voltage and current, motor speed, etc. are fed back to SP through the interface circuits.

(iii) Waveform memory Another significant feature than the hierarchical structure is waveform memory (WM) shown in Fig. 10. WM stores many kinds of PWM waveform data $\mathbf{t} = (t_1, \cdots, t_K)$ as 16bit data according to various fundamental frequencies and voltages. The capacity needed for WM is decided by the speed of CPU, required precision of output waveform, word legth of each timing datum and number of data to be prepared. In this system WM needs not more than 20kB capacity with 1μs waveform precision, even if a different waveform mode is provided for each 0.1Hz band over the whole operating frequency range of about 100Hz. Practically lower capacity seems to be sufficient for the same frequency resolution by means of simple interpolation between two proper waveform data. Each PWM waveform stored in WM can be prepared optimally through off-line computation *a priori*. The countermeasure to interference problem or adaptation to a different condition may be achieved by WM arrangement.

SP controls UTS's by selecting most suitable waveform data out of WM according to the control commands and various system conditions. System control, waveform generation and firing control are so separated as to realise sophisticated system function easily and successfully.

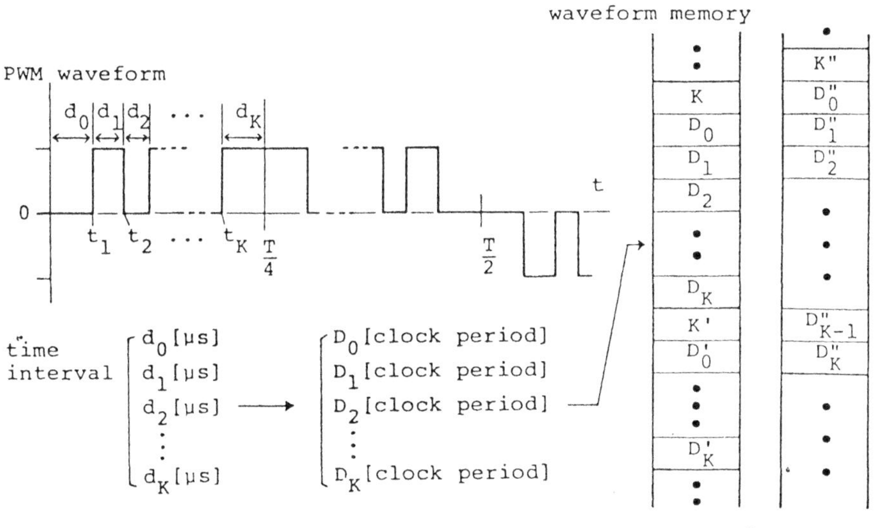

Many PWM waveforms are stored in Waveform Memory as 16bit timing data according to various fundamental frequencies and voltages.

Fig. 10 Waveform memory

(c) Harmonic control method for electric traction[6]

(i) Harmonic characteristics required for the PWM waveforms. The PWM waveform requires the following four characteristics from the viewpoint of interference when applied to electric traction.

(1) Output frequency should be changed continuously or in sufficiently fine step.
(2) The fundamental component of load voltage should be adjusted to a desired value corresponding to powering or braking torque required.
(3) Amplitude of low order harmonics in the load voltage or current should be sufficiently small for the motor to run smoothly without much additional heat or energy loss, and
(4) The frequency band used for railway signalling system should be protected against harmful harmonics in the source current, which means that the source current harmonics in the frequency band be almost completely eliminated.

It is easy to obtain the PWM waveforms to satisfy (1), (2) and (3). Difficulty lies in synthesising ones to satisfy the requirements (1) to (3) and (4) simultaneously.

(ii) Harmonic calculation for the PWM waveform synthesis Fig. 11 shows typical waveforms of PWM load voltage Va(t), load current Ia(t) and input current Id(t). A set of the timing data t_1, t_2, $\cdots$, t_k decide the PWM waveform. $V_a(t)$ is expressed by a Fourier series:

$$V_a(t) = E \cdot v_a(t) = E \cdot \sum_{n=1}^{\infty} a_n \cdot \sin(2n-1)\omega_0 t \qquad (1)$$

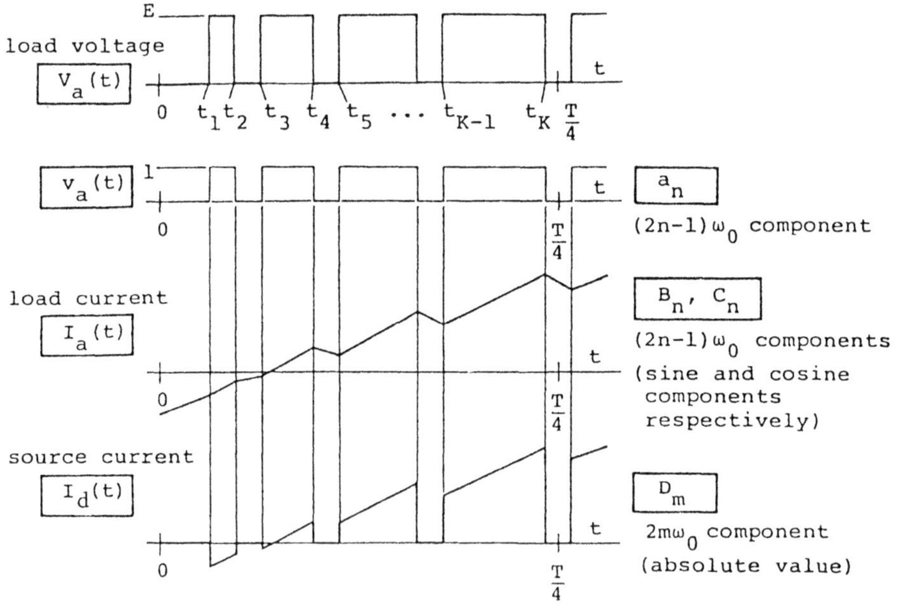

Fig. 11 Symbol definition for harmonic calculation

where $$a_n = \frac{4}{(2n-1)\pi} \sum_{h=1}^{k} (-)^{h+1} \cos(2n-1)\omega_0 t_h \qquad (2)$$

and $$\omega_0 = 2\pi / T \qquad (3)$$

T is the period of the fundamental wave and k is the number of pulses in half a period. a_n represents the (2n-1)th order harmonic of the normalised load voltage. Similarly, load current $I_a(t)$ can be expressed as (4).

$$I_a(t) = \sum_{n=1}^{\infty} [\, B_n \cdot \sin(2n-1)\omega_0 t + C_n \cdot \cos(2n-1)\omega_0 t \,] \qquad (4)$$

B_n and C_n, which represent (2n-1)th order sine and cosine components of $I_a(t)$ respectively, can be calculated from impedances of the induction machine for each symmetric component.

The input current $I_d(t)$ of an inverter per phase is given as a product of $v_a(t)$ and $I_a(t)$.

$$I_d(t) = v_a(t) \cdot I_a(t) \qquad (5)$$

In the period when $v_a(t) = 0$ the load current circulates in the main circuit via the feedback diode and the input current is zero. As the cycle time of $I_d(t)$ is T / 2, $I_d(t)$ contains a dc component and harmonic components of $2\omega_0$, $4\omega_0$, $6\omega_0$, $\cdots$, but two phase inverter produces no harmonics other than $4\omega_0$, $8\omega_0$, $12\omega_0$, $\cdots$ and three phase structure only $6\omega_0$, $12\omega_0$, $18\omega_0$, $\cdots$ components exist.

D_m, $2m\omega_0$ harmonic component of $I_d(t)$, is obtained as follows by the expansion of eq. (5). The product of (2i-1)th component of $v_a(t)$ and (2j-1)th component of $I_a(t)$ generates 2(i+j-1)th, 2(i-j)th and 2(j-i)th components in $I_d(t)$. By putting each term 2m and picking up the pair of (i,j) to satisfy this relation, the expansion of eq.(5) is successfully achieved.

$$\begin{aligned} I_d(t) = \frac{1}{2} \sum_{i=1}^{\infty} a_i \cdot B_i + \frac{1}{2} \sum_{m=1}^{\infty} [\, (& \sum_{i=1}^{m} a_i C_{m+1-i} + \sum_{i=m+1}^{\infty} a_i C_{i-m} \\ & - \sum_{i=1}^{\infty} a_i C_{m+i}) \sin 2m\omega_0 t + (- \sum_{i=1}^{m} a_i B_{m+1-i} \\ & + \sum_{i=m+1}^{\infty} a_i B_{i-m} + \sum_{i=1}^{\infty} a_i B_{m+i}) \cos 2m\omega_0 t \,] \end{aligned} \qquad (6)$$

The dc component of $I_d(t)$ is

$$D_0 = \frac{1}{2} \sum_{i=1}^{\infty} a_i B_i \qquad (7)$$

and the $2m\omega_0$ components are given by (8).

$$\begin{aligned} D_m = \frac{1}{2} [\, (& \sum_{i=1}^{m} a_i C_{m+1-i} + \sum_{i=m+1}^{\infty} a_i C_{i-m} - \sum_{i=1}^{\infty} a_i C_{m+i})^2 \\ & + (- \sum_{i=1}^{m} a_i B_{m+1-i} + \sum_{i=m+1}^{\infty} a_i B_{i-m} + \sum_{i=1}^{\infty} a_i B_{m+i})^2 \,]^{\frac{1}{2}} \end{aligned} \qquad (8)$$

(iii) PWM waveform synthesis method Simultaneous equations (9) should be solved in order to synthesise the PWM waveform that satisfy the four requirements mentioned in (i).

$$V_{a1} = U, \quad D_{m1} = 0, \quad D_{m2} = 0, \quad \cdots, \quad D_{mM} = 0 \qquad (9)$$

The first equation decides the fundamental voltage (V_{a1}) to be a desired value (U: see Fig. 12 for example), and the following equations eliminate M kinds of harmful harmonic components in input current selectively.

Equation (9) can only be solved numerically, because they are non-linear transcendental simultaneous equations. Moreover, as a nature of periodical functions, the solution of them is not unique.

In order to obtain a nearly optimum solution from the viewpoint of total harmonic reduction,

(1) the triangular method or its modification is adopted to get initial values of iterative numerical calculus, known as the excellent method to reduce low order harmonics effectively (see Figs. 13 & 14), and
(2) maximum gradient method is used to solve eq. (9). The method gives the final solution closest to the initial value.

The objective function $g(\mathbf{t})$ is defined as eq.(10), and the iterative calculus to minimise it to zero is put forward by maximum gradient method.

$$g(\mathbf{t}) = (V_{a1} - U)^2 + D_{m1}^2 + D_{m2}^2 + \cdots + D_{mM}^2 \qquad (10)$$

where $\mathbf{t} = (t_1, t_2, \cdots, t_k)$ (11)

(k-1) kinds of harmonics can be eliminated in general. One degree of freedom of timing data is used for determination of the fundamental voltage. Therefore, eq. (9) has the solution only when

$$M \leq k-1 \qquad (12)$$

The remaining degree of freedom, k-(M+1), may be used for reduction of lower order harmonics.

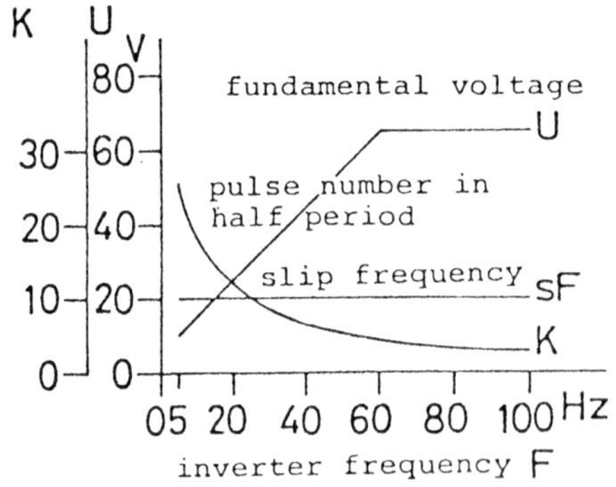

Fig. 12 Typical characteristics of PWM inverter

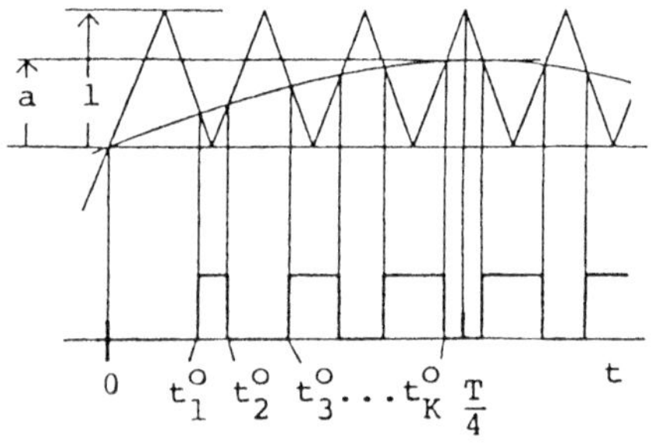

Fig. 13 The triangular method to find initial values

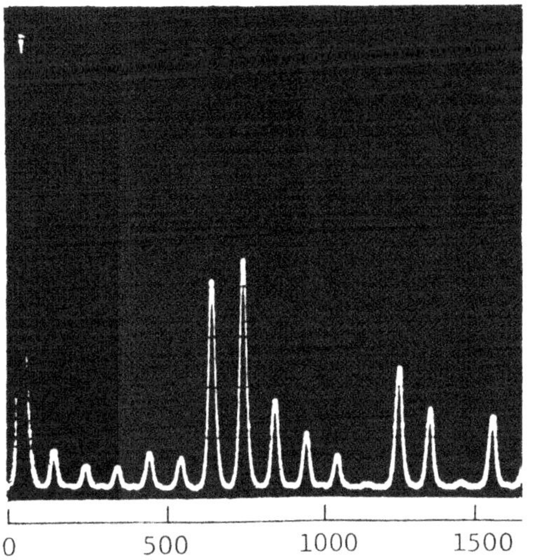

Fig. 14 Example of harmonic distribution of load voltage V_a

This harmonic distribution with effectively reduced low order harmonics is that of the final solution.

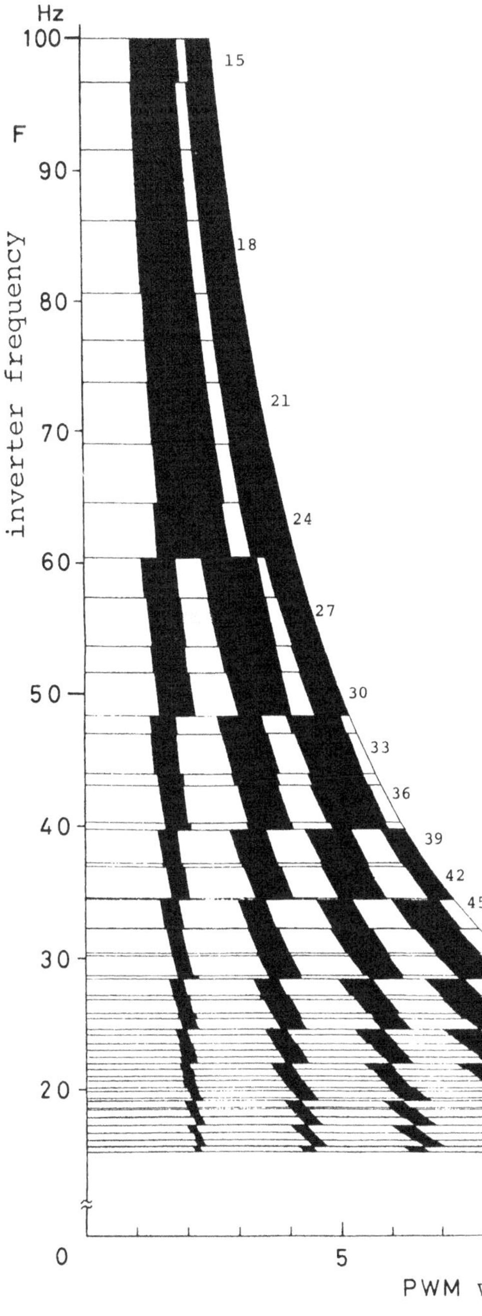

Fig. 15 Synthesised PWM waveforms to be stored in the waveform memory

This figure shows only a quarter cycle of each PWM waveform.
White parts of them indicate zero level output of the PWM waveform.

Fig. 15 shows the synthesised PWM waveforms within an operating frequency range from 15Hz to 100Hz. Fundamental voltage and pulse number in half a period are decided by Fig. 12. The figures 15, 18, 21, ••• indicate the order of eliminated input current harmonics which drop in the protected frequency band (between 2800Hz and 3000Hz in this example).

Only the waveform data at the modal change points of inverter frequency are stored into the waveform memory. PWM waveform with a desired fundamental frequency inside an identical mode can be generated by means of simple interpolation between two proper waveform data.

(iv) Experimental results Experiments whose results are shown in Fig. 16 confirm the effectiveness of the harmonic control method mentioned above. In this case, three successive components (2800Hz, 2900Hz and 3000Hz) are to be eliminated.

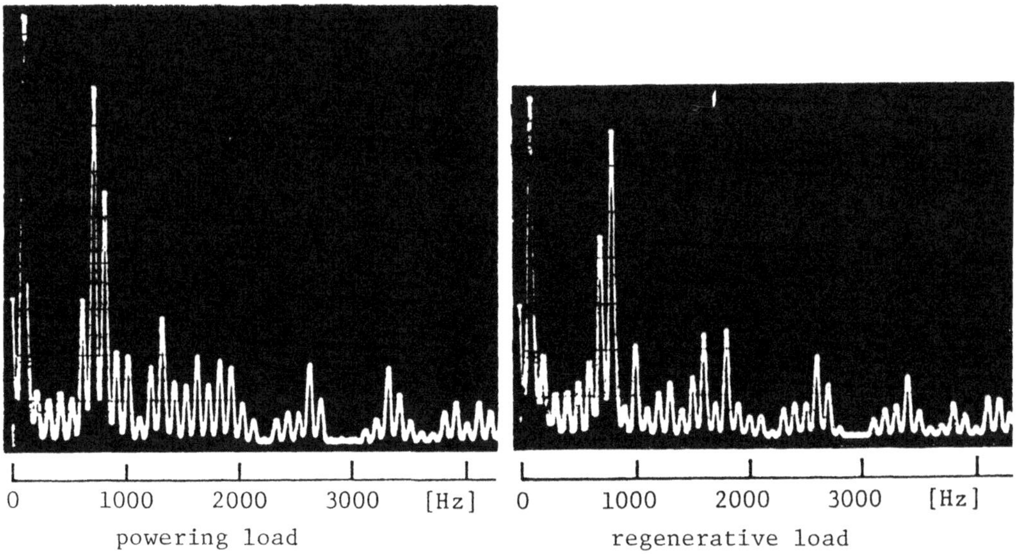

Fig. 16 Harmonic characteristics obtained by this method

2.3 Microcomputerised realisation of fail-safe vital functions of railway signalling system

There are two basic levels of control found in railway signalling. Those functions concerned with regulating the flow of traffic, such as train description or programmed route setting, have already been computerised quite successfully. Those mentioned in 1.2 are some of good examples.

The situation is very different when it comes to the safety control loop which includes interlockings and block working. This must be fail-safe in the sense that no equipment mulfunction should give rise to an unsafe situation. Only recently have computers been introduced in this area, and as far as the autor is aware there are only two or three systems actually in use[7,8].

Table 1 shows computerised systems of fail-safe vital functions of railway signalling system inclusive of those under development.

JNR has made several attempts to apply computers to vital signalling functions; the first being in 1966 when a fail-safe computer using parametrons was constructed[9]. This functioned as expected, but it was too expensive for practical use. Microcomputer technology has developed rapidly since then, so an interlocking system using microcomputers was installed for trial at Ishiuchi on JNR's Joetsu line in March 1981[10]. The following description will be restricted to Japanese system.

(a) System organisation

When analising the function of an interlocking system, the system can be divided into two major parts; one which mainly manages man-machine communications through a control desk, and the other which controls the track-side equipments through processing lock and unlock operations. Whereas the former is required to have high performance and high flexibility, the latter is severely required to have a high standard of safety because it affects the safety of train operations. There is a conflict, though, between having many programs to attain these objectives and minimising their number so as to achieve reliable software. This contradiction has been resolved by the use of hierarchical architecture whereby a general processing minicomputer relieves the safety interlocking of non-vital tasks (Fig. 17). There are two subsystems; an interlocking subsystem and a communication subsystem. In the interlocking subsystem, JNR's superior fail-safe computer using multiple i-8085 microprocessors and programs limited only to vital processes ensure high levels of safety and reliability.

Conventional signalling systems generally have 'single-thread' architecture using fail-safe components. A single-thread approach to

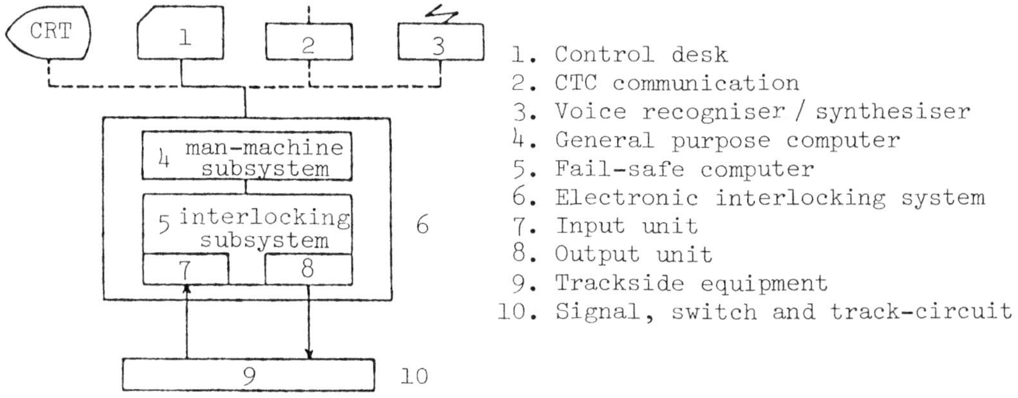

Fig. 17 Hierarchical architecture of an electronic interlocking system

Table 1 A list of microcomputerised interlockings

Country	Station		Interlocking system			Reference
	Name	Size	Microcomputer	Redundancy	Software	
Sweden	Göteborg (1978) Malmö (1980)	20 stations 800 devices	UAC 1610 (16 bits) minicomputer	Single	Dual diversity programs	7
Denmark	Jatland (1980) (in parallel with relay interlocking)	14 stations 20 - 170 devices	APN 163 (16 bits)	Single	Dual diversity programs	8
Japan	Ishiuchi (1981) (in field trials)	106 routes	Intel 8085 (8 bits)	2 out of 3	Same programs	10
Britain	——	378 devices	Motorola 6800 (8 bits)	2 out of 3	Same programs	11
France	——	50 routes	? (16 bits)	Dual	Diversity programs	12
West Germany	——	25 devices	SAB 8080 (8 bits)	Dual	Same programs	13

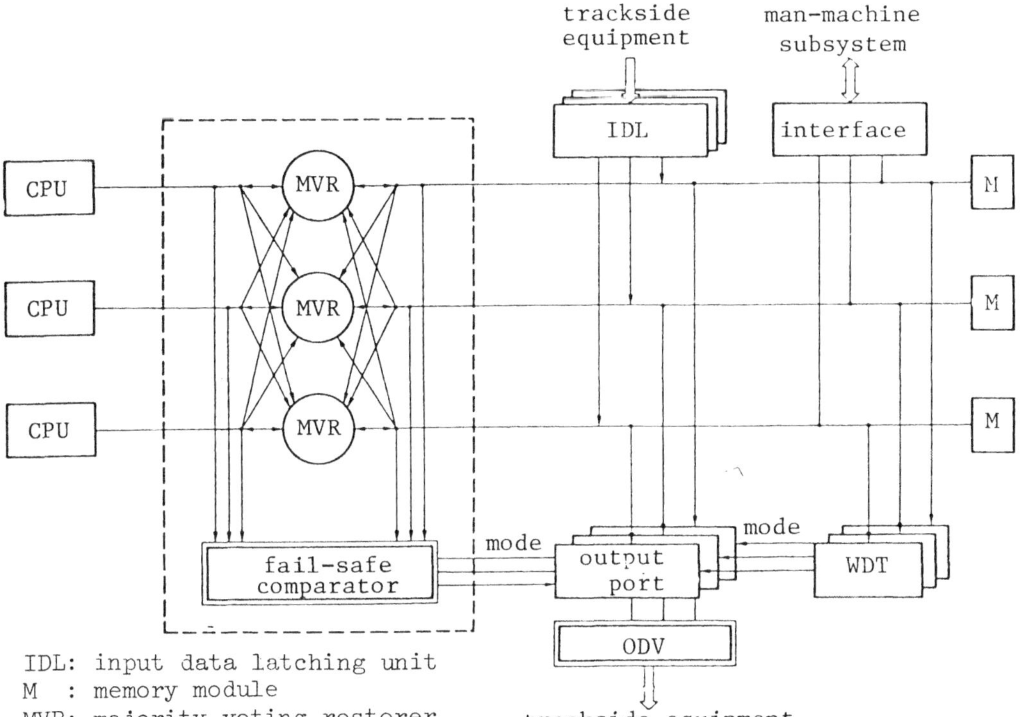

IDL: input data latching unit
M : memory module
MVR: majority voting restorer
ODV: output data voting unit
WDT: watchdog timer

Fig. 18 Overall configuration of a fail-safe computer

electronic implementation of vital function would be hazardous because component failure modes are unpredictable. Fortunately, multiple-thread configuration can be implemented without difficulty in a way which uses 'fault-tolerant' and 'fail-safe' technologies. Internal redundancy, bus-levelled majority voting and self-checking technology are all incorporated in the system (Fig. 18).

(b) The architecture of a fail-safe computer

There are three i-8085 and three memory modules. The triplicated majority voting restorers (MVR) are inserted in internal buses.

Data exchange with the communication subsystem is performed through an interface unit (IFU). The data from external devices such as signal aspects, switch point positions and track circuit occupancy are fed into the computer through an input data latching unit (IDL), and are distributed to triplicated input ports. The output data voting unit (ODV) produces external outputs by two-out-of-three voting on data from the three output ports. Voting is necessary because of the need to achieve fault-tolerance.

Several fault detecting functions are included. The fail-safe comparator (FSC) watches continuously to see whether or not failures occur in components. The fault location and mode control unit (FLU) decides which module is in a faulty condition, and the nature of the fault. FSC and FLU are themselves made fail-safe.

To detect certain kinds of software failures, a memory illegal access watching unit and fail-safe watchdog timer (WDT) are provided. but the space does not allow to describe these in detail.

(c) Circuit concept

The triplex processors are synchronised by a master clock, and in the period of data exchange through buses, triplex data are voted on bit by bit in order to achieve fault-tolerance. A fault in any of the data components is corrected immediately without allowing it to propagate through data exchange.

Incidentally, in this system faults in almost all components appear as disagreements of data among the three buses. FSC supervises MVR's at every period of data accessing through the bus, and any disagreement is detected infallibly.

According to the results of a comparison, FLU identifies the fault location and decides whether the fault is parmanent or transient. If the disagreement is judged to be caused by a parmanent failure, FLU cuts off the power supply to the corresponding parts in the ODV; i.e. the external output of a faulty module will be prevented from voting, even though it takes part in internal bus-level voting. Then, the system operates as a dual system.

If another disagreement occurs between the remaining two buses, FLU judges this system to have multiple failures and kills the output. Of course, in such a case external outputs from this system are all forced to the safe condition.

Using this function, the system operating mode is controlled automatically. Consequently, system safety and reliability can be maintained securely.

Most failures can be detected by these functions, but one unsafe circumstance is conceivable. That is a fault in output circuits such as output ports or the output data voting unit. To conserve system safety here, fail-safe logic elements and a fail-safe circuit are used.

In the output data voting unit, a C-type fail-safe logic element is introduced, the basic conception of which was proposed by Tsuchiya[14]. This element is composed of a fail-safe threshold oscillator, an amplifier and a rectifier, and is formed as a thin-film integrated circuit.

External outputs in the output data voting unit drive the safety relays for the signals and switches. Almost all kinds of failures in this unit also discontinue the external outputs, so the output relays are not picked up.

(d) Reliable software

When considering the safety of a computerised system, the reliability of the software is every bit as important as that of the hardware. In constructing the software, the two main themes of a data structure and a programming structure have been mentioned, the objective being to develop universally available programs and to achieve high reliability.

Interlocking conditions which have to be met uniquely at each station are filed as geographical data, linked to each other in the same way as the track layout of the station. These interlocking tables are held in read-only memories (ROM's); they are known as route files. The interlocking program, which is generally applicable, is also stored in ROM's. To construct a high security program, most of the algorithms for performing interlocking functions are specified according to traditional methods of tabulation, as far as this is possible.

Furthermore, in designing the programming structure, a composite structured method (otherwise known as a 'moduling method') is applied. The interlocking program is divided into a number of simple functional blocks. All modules are arranged in a hierarchical structure, and all upper-level modules are connected with each other in a single-thread routine. This thread is scanned continuously at 200ms intervals.

This programming structure makes it possible to assure very reliable processing without a complex operating system. Under this software structure, any change to an interlocking condition can be achieved easily merely by exchanging or adding the route files.

(e) Control desk

The control desk consists of two universal input-output microprocessors; a control panel with push buttons for route setting; and an illuminated track layout diagram showing the route set, track-circuit occupancies, and signal aspect lamps. Light-emitting diodes are used for all these displays so as to ensure high reliability.

(f) Field trials

The microcomputerised interlocking system is now in field trials at Ishiuchi station on Joetsu line. The existing relay-based interlocking there covers 106 routes, 25 switches and 36 track circuits. The system is placed alongside the relay-based interlocking, receiving the same inputs from track circuits, signal aspects displayed, etc. It gives output instructions, but these are simply recorded for comparison with the parallel relay-based interlocking. Thus the performance of the microcomputerised interlocking system can be monitored over a substantial period without in any way affecting the safety of trains. Furthermore, a lot of information on both the existing and the microcomputerised interlocking systems are automatically sent to the Railway Technical Research Institute located in Tokyo over telephone line by a remote monitoring system.

The microcomputerised interlocking system is performing satisfactorily in the field trials. It is expected that it will prove safe and reliable during the field trials at Ishiuchi, and the microcomputerised interlocking system will be in practical use in JNR in the near future.

REFERENCES

1. Ohmura, I: Automatic switchpoint control system for wagons, *Signalling and Safety* p.552-557, in Japanese, 1978

2. Huet, J: The marshalling yard at Villeneuve-Saint-Georges modernisation of shunting operation, *Rail International* p.920-924, 1979

3. Ohishi, M: The application of voice data entry devices for operation of the railways, *Japanese Railway Engineering* p.12-14, 1981

4. Sone, S & Hori, Y: Ac drive electric trains in the world, *Symposium in 1982 Annual Meeting Records, I.E.E.,Japan* S9-2, in Japanese, 1982

5. Sone, S & Hori, Y: Microprocessor-based universal thyristor switch and its application to a PWM inverter for traction, *IEEE Trans. IECI IECI 28-2* p.162-167, 1981

6. Sone, S & Hori, Y: Seventeen years' experience of power electronics traction systems of Japanese railways with special emphasis on interference problems, *IEE Conf. Publ. 203* p.111-116, 1981

7. Mathson, R: Computers replacing safety relays in railway signalling, *Digital Computer Application to Process Control* p.321-328, 1977

8. Jonasen, A.A. et al: Microcomputers take over the interlocking function, *Railway Gazette International* p.1028-1030, 1981

9. Okumura, I: Studies on railway signalling interlocking device using fail-safe computer, *Railway Technical Research Report #803*, in Japanese 1972

10. Okumura, I: Electronic interlocking to be tried in Japan, *Railway Gazette International* p.1043-1046, 1980

11. Editor: Joint venture to develop solid-state interlocking, *Railway Gazette International* p.660-661, 1981

12. Aulair, J.P. et al: Information processing and signalling, *O R E coloquim on safe electronic systems* 1980

13. Stelow, H. et al: Das sichere microcomputersystem SIMIS, *Signal und Draht* p.82-86, 1978

14. Tsuchiya, S: Studies on fail-safe logical system, *Research of Electrotechnical Laboratory, Japan #695* in Japanese, 1969

SUBJECT INDEX

GPSR Compliance

The European Union's (EU) General Product Safety Regulation (GPSR) is a set of rules that requires consumer products to be safe and our obligations to ensure this.

If you have any concerns about our products, you can contact us on ProductSafety@springernature.com

In case Publisher is established outside the EU, the EU authorized representative is:

Springer Nature Customer Service Center GmbH
Europaplatz 3
69115 Heidelberg, Germany

Batch number: 10816188

Printed by Printforce, the Netherlands